SCHUMAKER—Spline Functions: Basic Theory
SHAPIRO—Introduction to the Theory of Numbers
SIEGEL—Topics in Complex Function Theory
Volume 1—Elliptic Functions and Uniformization Theory
Volume 2—Automorphic Functions and Abelian Integrals
Volume 3—Abelian Functions and Modular Functions of Several Variables
STAKGOLD—Green's Functions and Boundary Value Problems
STOKER—Differential Geometry
STOKER—Nonlinear Vibrations in Mechanical and Electrical Systems
TURÁN—On A New Method of Analysis and Its Applications
WHITHAM—Linear and Nonlinear Waves
ZAUDERER—Partial Differential Equations of Applied Mathematics

TEICHMÜLLER THEORY AND QUADRATIC DIFFERENTIALS

TEICHMÜLLER THEORY AND QUADRATIC DIFFERENTIALS

FREDERICK P. GARDINER
Department of Mathematics
Brooklyn College, City University of New York
Brooklyn, New York

A Wiley-Interscience Publication

JOHN WILEY & SONS

New York Chichester Brisbane Toronto Singapore

Library of Congress Cataloging in Publication Data:

Gardiner, Frederick P.
Teichmüller theory and quadratic differentials.

(Pure and applied mathematics)
"A Wiley-Interscience publication."
Bibliography: p.
Includes index.
1. Teichmüller spaces. 2. Quasiconformal mappings.
3. Riemann surfaces. I. Title. II. Series: Pure and applied mathematics (John Wiley & Sons)
QA331.G328 1987 515'.223 87-8144
ISBN 0-471-84539-6

To Joanie

PREFACE

A Riemann surface is an oriented topological surface together with a system of local parameters whose domains of definition cover the surface. The local parameters have the property that if any two of them have overlapping domains of definition, the transition mapping expressing the first parameter in terms of the second is holomorphic. Another system of local parameters is compatible with the first system if the union of the local parameters for the first system together with the local parameters for the second still satisfies the condition that in overlapping domains of definition transition mappings are holomorphic. The union of all systems compatible with a given system is a Riemann surface structure. Two such Riemann surface structures are equivalent if there is a biholomorphic homeomorphism taking one surface structure onto the other.

In general, there are many inequivalent Riemann surface structures on a given topological surface. The set of such structures is called the moduli space. The moduli space can be very complicated, although if you begin with a compact surface it will be a finite dimensional complex variety but not in general a manifold.

Teichmüller's approach to the problem of moduli was to change the equivalence relation to a relation on the space of orientation preserving topological mappings from a fixed base surface onto a variable Riemann surface. Two such topological mappings f_1 and f_2 are equivalent if their images are conformal by a mapping which makes f_1 and f_2 homotopic. Teichmüller was able to show that this space of equivalence classes is homeomorphic to a cell, of real dimension $6g - 6$ when the genus g of the original surface is more than 1 and of real dimension 2 when the genus is 1. The space of these equivalence classes is called Teichmüller space. The group of homotopy classes of orientation preserving homeomorphisms of the base surface is called the modular group. It has a natural action on Teichmüller space which identifies points corresponding to conformally equivalent Riemann surfaces. Factoring Teichmüller space by the modular group yields moduli space.

The most essential tool in Teichmüller's approach to the problem of moduli was quasiconformal mapping. Although it is not necessary for quasiconformal mappings to be differentiable in the classical sense, it is easiest to describe the quasiconformality condition when they are. Any C^1-mapping from a plane domain into a plane domain maps, on the infinitesimal level, a small circle into a small ellipse. The local dilatation K_x at a point x is the ratio of the length of the major axis of this ellipse to the length of its minor axis. The global dilatation, or simply the dilatation, of the mapping is the supremum K of the values K_x for all x in the domain. An orientation-preserving mapping for which the dilatation K is finite is called quasiconformal. In order to make this definition apply to mappings which are not of class C^1, one needs a considerable amount of analysis and the theory of derivatives in the sense of distributions.

Although this theory was in its early stages during Teichmüller's life, he developed it sufficiently to apply it to the problem of moduli. He showed that any homotopy equivalence class of quasiconformal mappings from a compact Riemann surface R to a compact Riemann surface R' contains a unique mapping whose maximal dilatation K is minimum. Moreover, this unique mapping can be described geometrically in terms of a holomorphic quadratic differential on R. Such a differential gives a way of cutting up the surface R into Euclidean rectangles, and the unique extremal quasiconformal mapping corresponds to stretching by an amount $K^{1/2}$ along the horizontal lines in these rectangles and shrinking by an amount $K^{-1/2}$ along the vertical lines in these rectangles. Teichmüller's metric is defined by letting the distance from R to R' be $\frac{1}{2}\log K$. The motion determined by stretching along the horizontal trajectories by the amount $(tK)^{1/2}$ and shrinking along the vertical trajectories by the amount $(tK)^{-1/2}$, $(0 \leqslant t \leqslant 1)$ determines a geodesic in the Teichmüller metric.

Later, Ahlfors and Bers showed that the Teichmüller space has a natural complex structure and that it can be realized as a bounded domain in $\mathbb{C}^n$. Moreover, the modular group acts as a group of biholomorphic mappings on this domain, and the usual complex invariants associated with a Riemann surface, such as entries in the period matrix, are complex holomorphic functions with respect to this complex structure. Then Earle, Eells, and O'Bryne showed that Teichmüller's metric is equal to the integral of its infinitesimal form, for finite as well as infinite dimensional Teichmüller spaces. Finally, Royden showed that Teichmüller's metric is determined by the complex structure on Teichmüller space. In fact, he showed that for finite dimensional Teichmüller spaces Teichmüller's metric coincides with the Kobayashi metric, which is defined in a way which depends only on the noneuclidean metric for the unit disk and the family of holomorphic mappings from the disk into Teichmüller space.

Royden also showed that when the genus is more than 2 the modular group is the full group of biholomorphic self-mappings of Teichmüller space.

While these results were developing, Reich and Strebel found an ingenious way to express the extremal property for extremal quasiconformal mappings relative to quadratic differentials. This is through what we call the Reich–Strebel inequality. Among their many applications of this inequality, they were able to show that the Hamilton–Krushkal necessary condition for a quasiconformal mapping to be extremal is also sufficient. Using the length–area principle, Strebel added to the work of Jenkins by proving many theorems about the existence of quadratic differentials with closed trajectories.

Independently, Thurston developed a deformation theory for Riemann surfaces from a topological point of view. Essential for his theory was a theory of measured foliations on topological surfaces. Briefly, a measured foliation is a foliation of a surface with singularities of a certain type and with a vertical measure which measures the amount which an arbitrary curve crosses the foliation. The types of singularities permitted are those which have the same topological structure as the singularities in the horizontal trajectory structure of a holomorphic quadratic differential. The vertical measure is nonnegative; it measures the total amount a curve crosses the leaves of the foliation without regard to the direction of the crossing. An example of a measured foliation is obtained by letting the leaves of the foliation be the horizontal trajectories of a holomorphic quadratic differential $\varphi(z)\,dz^2$ which has at most simple poles and letting the vertical measure be $|\mathrm{Im}(\sqrt{\varphi(z)}\,dz)|$.

A measured foliation gives a height function on the space of homotopy classes of simple closed curves on the surface. The height of a closed curve is the line integral over the curve with respect to the vertical measure for the foliation. The height of a homotopy class is the minimum of the heights of every closed curve in the homotopy class. Two measured foliations are said to be in the same measure class if they induce the same height function on homotopy classes of simple closed curves. Hubbard and Masur showed that every measure class of measured foliations on a compact Riemann surface is represented uniquely by a holomorphic quadratic differential.

The purpose of this book is to give an exposition of all of these results, including, where possible, the part of the theory which remains true for infinite dimensional spaces and for surfaces with elliptic and parabolic punctures. There are two currents which run through the whole subject. The first consists of the various uniqueness theorems which follow, in general, from the length–area principle of Grötzsch. A powerful version of this principle was given by Marden and Strebel. They call it the minimum norm principle for holomorphic quadratic differentials. Marden and Strebel stated the principle by way of comparison with harmonic quadratic differentials. We find it useful to give two improvements of this principle. In the first version, one takes a minimum over all L_1-measurable quadratic differentials. These differentials satisfy an inequality of line integrals taken over arcs which are segments of regular vertical trajectories of a given holomorphic quadratic differential. In the second version, the minimum is taken over all continuous quadratic differentials satisfying an inequality of line integrals over all homotopy classes of simple closed curves. These principles lead to the following results: the inequality of Reich and Strebel, the uniqueness part of Teichmüller's theorem, the sufficiency of the Hamilton–Krushkal condition for extremality, and the uniqueness of a quadratic differential with given heights.

The second current is the basis for the various existence theorems. Many of these theorems can be seen as issuing from the theorem on the existence of curves of trivial Beltrami coefficients tangent to a given infinitesimally trivial Beltrami differential. Among these are the existence part of Teichmüller's theorem and the necessity of the Hamilton–Krushkal condition. The existence of a quadratic differential with heights equal to the heights of a given measured foliation can be seen as coming from the existence theorem for Jenkins–Strebel differentials (differentials with closed trajectories). That theorem comes more directly from normal families arguments and Weyl's lemma.

Along the way, we present two additional subjects. The first is the surjectivity of the Poincaré theta series operator for quadratic differentials, and the second is the Ahlfors–Bers density theorem for quadratic differentials with simple poles. These results are used to generalize the Reich–Strebel inequality to Riemann surfaces of infinite type. They also provide background for generalizing to arbitrary surfaces the theorem on the necessity and sufficiency of the Hamilton–Krushkal condition for extremality and the theorem on the equality of Teichmüller's and Kobayashi's metrics.

Another subject we cover is the discontinuous action of the modular group on Teichmüller space. This is one of several components of Royden's proof that when the genus is bigger than or equal to 3, the modular group is identical to the full group of biholomorphic self-mappings of Teichmüller space.

A great deal of helpful introductory material can be found in the books by Abikoff [Ab], Ahlfors [Ah4, Ah6], Farkas and Kra [FarK], Kra [Kr5], and Lehto and Virtanen [LehtV]. The most important background material is presented in the first chapter. In particular, the following results are presented without proof.

1. *The uniformization theorem.* For much of the book this theorem is unnecessary if one takes the viewpoint that the surfaces under consideration can all be obtained by factoring the complex plane or the hyperbolic plane by a discontinuous group. The fact that all Riemann surfaces with complex structure except the complex sphere can be so obtained merely shows that the theory is as general as possible.

2. *The Ahlfors–Bers theorem giving the solution of the Beltrami equation as a holomorphic function of the Beltrami coefficient.* It is hoped that the summary of this material given in Chapter 1 will give the reader sufficient background to proceed without going into a detailed study of quasiconformal mappings.

3. *The formula for the dimension of the vector space of holomorphic quadratic differentials on a Riemann surface of finite type.* This dimension can be calculated from topological considerations, as will be seen in the last chapter on measured foliations. It also can be deduced easily from the Riemann–Roch theorem. The calculation can be found in Farkas and Kra [FarK].

Apart from these results and the results normally contained in a first-year graduate course in complex analysis, the development of Teichmüller theory presented here is for the most part self-contained. Two exceptions to this rule stand out. Chapters 8 and 9 contain several references to the Riemann–Roch theorem. Also, in Chapter 11 we assume some topological results about foliations with closed leaves and transverse foliations. We postpone discussion of these topics until later. Complete details of the topological theorems are contained in the book *Travaux de Thurston sur les Surfaces*, by A. Fathi, F. Laudenbach and V. Poénaru [FatLP].

In a number of places material is included which already appears in other books. It is included for the sake of completeness of exposition. This remark applies to nearly all of what appears in Chapter 1. In Chapter 4 the part which concerns Bers's density theorem, Bergman kernel functions, and the surjectivity of Poincaré theta series is a condensation of more general results from the book by I. Kra, *Automorphic Forms and Kleinian Groups* [Kr5]. The material on trajectory structure of quadratic differentials in Chapter 2 and on Jenkins–Strebel differentials in Chapter 10 appears in much more detail in the book by K. Strebel, *Quadratic Differentials* [St4].

Much of that part of Chapter 5 which concerns Teichmüller space of a Riemann surface appears in Ahlfors's book, *Lectures on Quasiconformal Mappings* [Ah4].

Material on Teichmüller's theorem and Fenchel–Nielsen coordinates for Fuchsian groups is given in Abikoff's book, *The Real Analytic Theory of Teichmüller Space* [Ab]. He does not consider infinite dimensional spaces. His approach has fairly little intersection with the approach given in this book.

There is also a book by S. L. Krushkal on Teichmüller theory [Kru2]. It also does not cover much of the same material as is covered here.

At the end of each chapter there are bibliographical notes that give references to books and articles used as background. It is regrettable and unavoidable that these references are incomplete. The bibliographical notes also indicate some of the directions of new research in the field. These directions do not make a complete list and are selected merely because of their appeal to this author.

Also, at the end of each chapter (except Chapter 3) there are illustrative exercises. Exercises for Chapter 3 were omitted because, except for unsolved problems, nothing I could devise seemed appropriate. In many cases, an understanding of the exercises is essential. This will become apparent as the reader proceeds through the book and encounters references to the exercises in the subsequent text.

Brooklyn, New York
January 1987

FREDERICK P. GARDINER

ACKNOWLEDGMENTS

I owe much to many colleagues and friends with whom I have studied over the years and I owe the most to my teacher, Lipman Bers. A number of people have read preliminary versions of the manuscript. I particularly want to thank Professors William Abikoff, Daniel M. Gallo, Fred W. Gehring, Andrew Haas, Linda Keen, Ravi Kulkarni, Irwin Kra, and Patricia Sipe. They have made helpful suggestions and corrected many errors. The errors that remain are mine.

I have also benefited from lectures of Lars V. Ahlfors, Lipman Bers, Clifford J. Earle, and Irwin Kra. Of particular help has been a set of handwritten notes on Royden's theorems by C. J. Earle.

Outside the world of mathematics, I am most indebted to my wife and to my mother. They have supported me in many essential ways. Also, the Research Foundation of the City University of New York has paid for typing and duplicating preliminary versions of the manuscript.

CONTENTS

1

RESULTS FROM RIEMANN SURFACE THEORY AND QUASICONFORMAL MAPPING

Most of the material of this chapter is summarized without proof. Proofs are included when they are elementary or when they can be presented in a simplified form. The topics are selected to provide the necessary background for Teichmüller theory.

After the definition of a Riemann surface, the first topic considered is the uniformization theorem. It says that, except for the case of the Riemann sphere, the study of Riemann surfaces can be regarded as part of the study of discrete groups acting on the upper half plane or on the whole complex plane. The most important case is that of discrete groups acting on the upper half plane, the so-called Fuchsian groups. Whereas the uniformization theorem is not proved, those parts of the chapter which concern elementary properties of Fuchsian groups are treated in detail with proofs included.

We first present an account of the elementary properties of the Poincaré metric. Then a few general theorems from metric space theory are used to show how the upper half plane factored by a Fuchsian group (a discrete group of isometries in the Poincaré metric) yields a metric space which is a Riemann surface. These metric space theorems also are of use in a later chapter which concerns the action of the modular group on Teichmüller space.

A section on the classification of Möbius transformations into elliptic, parabolic, and hyperbolic elements is included because it provides terminology used repeatedly throughout the book. There is also a section on fundamental domains included for the same reason.

Probably the deepest topics discussed are from quasiconformal mapping. In particular, we need the holomorphic dependence of a normalized solution of the Beltrami equation on the Beltrami coefficient. This result provides a way to introduce a natural complex structure on Teichmüller space. It is also important to know that the Beltrami equation can be solved for L_∞-Beltrami coefficients. In this generality, the existence theorem for the Beltrami equation is sometimes called the measurable Riemann mapping theorem. Other essential parts of the theory of quasiconformal mapping which are summarized without proof are various convergence criteria for sequences of quasiconformal mappings.

There are also sections on extremal length and Grötzsch's problem for an annulus. These sections provide a way of viewing extremal length as a functional on the Teichmüller space of an annulus and show how the length–area principle yields a formula for the infinitesimal variation of extremal length. Moreover, the section on Grötzsch's problem contains the proof of Teichmüller's theorem for an annulus. The method of proof is analogous to the method used in Chapter 6 to prove that theorem in a general setting.

The final section of the chapter gives the formula for the dimension of the vector space of holomorphic quadratic differentials on a Riemann surface of finite type.

1.1. DEFINITION OF A RIEMANN SURFACE

Definition. *A Riemann surface R is a connected one dimensional complex analytic manifold. This means R is a connected Hausdorff topological space and there is a covering of R by open sets U_α and there are homeomorphisms z_α from U_α into $\mathbb{C}$ such that the transition mappings $f_{\alpha\beta} = z_\alpha \circ z_\beta^{-1}$ from $z_\beta(U_\alpha \cap U_\beta)$ to $z_\alpha(U_\alpha \cap U_\beta)$ are holomorphic.*

Remark. We use the words *holomorphic* and *analytic* synonymously.

The pairs (U_α, z_α) are called charts. Two systems of charts (U_α, z_α) and (V_β, w_β) are called compatible if whenever $U_\alpha \cap V_\beta$ is nonempty, $w_\beta \circ z_\alpha^{-1}$ is an analytic mapping from $z_\alpha(U_\alpha \cap V_\beta)$ to $w_\beta(U_\alpha \cap V_\beta)$. It is obvious that compatibility is an equivalence relation between systems of charts because the composition of analytic mappings is analytic.

The equivalence class of compatible systems of charts on R is the complex analytic structure of R. We also call it the Riemann surface structure of R.

Any system of charts for R determines an orientation of R. The mapping $z_\alpha: U_\alpha \to \mathbb{C}$ puts an orientation on U_α by taking the preimage of the usual orientation for $\mathbb{C}$. (A counterclockwise rotation is considered positive.) On $U_\alpha \cap U_\beta$ the orientation is consistently determined because the Jacobian of $f_{\alpha\beta}$ is $|f'_{\alpha\beta}|^2 > 0$.

In the classical terminology, a compact Riemann surface is called closed while a noncompact Riemann surface is called open.

A continuous mapping f from a Riemann surface R_1 to a Riemann surface R_2 is called analytic if for any chart z_α on R_1 and any chart w_β on R_2 the mapping $w_\beta \circ f \circ z_\alpha^{-1}$ is analytic.

R_1 and R_2 are analytically equivalent if there is an analytic homeomorphism from R_1 onto R_2.

1.2. THE UNIFORMIZATION THEOREM

The uniformization theorem can be stated in many different forms. The deepest part of it is called Koebe's planarity theorem, which says that a noncompact planar Riemann surface is analytically equivalent to a domain in the plane. A surface R is planar if any simple closed curve on R divides R into two components.

This result can be combined with the topological theory of covering surfaces. In particular, R is analytically equivalent to its universal covering $\tilde{R}$ factored by a group of deck transformations Γ which is isomorphic to the fundamental group of R. One is led to the following result, which also is sometimes called the uniformization theorem.

Theorem 1. *Let R be a Riemann surface. Then R is analytically equivalent to one of the following:*

(1) *The Riemann sphere* $\hat{\mathbb{C}} = \mathbb{C} \cup \{\infty\} = \mathbb{C}P^1$.
(2) *The complex plane* $\mathbb{C}$.
(3) *The punctured plane* $\mathbb{C} - \{0\} = \mathbb{C}/(z \to z + 1)$.
(4) *The plane modulo a lattice* $\mathbb{C}/L$, *where* L *is isomorphic to* $\mathbb{Z} \times \mathbb{Z}$ *and is spanned by two vectors in* $\mathbb{C}$, *linearly independent over* $\mathbb{R}$.
(5) *The upper half plane* $\mathbb{H} = \{z\colon \operatorname{Im} z > 0\}$ *modulo a properly discontinuous, torsion free group* Γ *of holomorphic homeomorphisms of* $\mathbb{H}$.

In the last case we mean to include the possibility that Γ is the identity, in which case the quotient space is $\mathbb{H}$ itself. Proofs of this theorem can be found in Ahlfors [Ah6], Ahlfors and Sario [AhS], Farkas and Kra [FarK], Springer [Sp], as well as numerous other books.

None of the five types of surfaces listed in this theorem can be analytically equivalent. It is left to the reader to verify this, but we point out that in most instances the inequivalence is topological. For example, $\hat{\mathbb{C}}$ is inequivalent to $\mathbb{C}$ because $\hat{\mathbb{C}}$ is compact and $\mathbb{C}$ is not. $\mathbb{C}/L$, which is a torus and has Euler characteristic 0, is inequivalent to $\hat{\mathbb{C}}$, which has Euler characteristic 2. Although $\mathbb{C}$ is topologically equivalent to $\mathbb{H}$, it cannot be analytically equivalent since any analytic mapping $f\colon \mathbb{C} \to \mathbb{H}$ composed with the mapping from $\mathbb{H}$ to the unit disk given by $z \mapsto (z - i)/(z + i)$ would be a bounded entire function. From Liouville's theorem it would therefore have to be constant.

1.3. METRICS OF CONSTANT CURVATURE

A consequence of Theorem 1 is that every Riemann surface carries a complete Riemannian metric which induces the conformal structure and which has constant curvature. Moreover, the metric is conformal in the sense that its fundamental form is given by the simple expression

$$ds^2 = \rho^2(dx^2 + dy^2)$$

or $ds = \rho|dz|$, where $\rho > 0$.

For a metric $\rho|dz|$ of class C^2, the quantity

$$K(\rho) = -\rho^{-2}\,\Delta \log \rho, \tag{1}$$

where Δ is the Laplacian ($\Delta u = \partial^2 u/\partial x^2 + \partial^2 u/\partial y^2$), is known as the Gaussian curvature. We will not give a geometric definition of curvature, but point out that curvature is an analytic invariant. To see this, let $w = f(z)$ be an analytic homeomorphism from a domain in the z-plane to a domain in the w-plane and let $\tilde{\rho}$ be the induced metric in the w-plane. That is,

$$\tilde{\rho}(w)|dw| = \rho(z)|dz|. \tag{2}$$

Then $K(\tilde{\rho}) = K(\rho)$, where the curvature for $\tilde{\rho}$ is calculated with respect to the Laplacian in the w-plane. Because $\rho(z) = \tilde{\rho}(f(z))|f'(z)|$ and because $\log|f'(z)|$ is harmonic, it follows that $\Delta \log \rho(z) = \Delta \log \tilde{\rho}(w)$, where both Laplacians are computed with respect to z. Since the Laplacian obeys the change of variable rule

$$\Delta_z \log \tilde{\rho} = |f'(z)|^2\,\Delta_w \log \tilde{\rho},$$

we find that

$$K(\tilde{\rho}) = K(\rho).$$

A metric of constant curvature for the sphere is

$$ds = \frac{|dz|}{1 + |z|^2}. \tag{3}$$

In order to calculate the curvature of this metric, it is convenient to use the complex notations:

$$\frac{\partial}{\partial z} = \frac{1}{2}\left(\frac{\partial}{\partial x} - i\frac{\partial}{\partial y}\right),$$

$$\frac{\partial}{\partial \bar{z}} = \frac{1}{2}\left(\frac{\partial}{\partial x} + i\frac{\partial}{\partial y}\right),$$

$$\Delta u = 4\frac{\partial^2 u}{\partial z \partial \bar{z}}.$$

Applying these notations to (1) and (3), one sees easily that the curvature of the spherical metric is $+4$. Although it will not be needed in the sequel, we make note of the fact that the global chordal metric for the sphere is

$$d(z, w) = \frac{|z - w|}{\sqrt{1 + |z|^2}\sqrt{1 + |w|^2}}$$

for z and w in $\hat{\mathbb{C}} = \mathbb{C} \cup \{\infty\}$. The formula is assumed to take its limiting value if z or w is ∞. It is an easy exercise to show that this metric makes the extended complex plane $\hat{\mathbb{C}}$ a complete metric space. Moreover, substituting $w = z + dz$ into this formula yields the infinitesimal form (3). However, integration of the infinitesimal metric (3) yields the spherical metric, which is larger than the chordal metric.

In the case of the complex plane, the Euclidean metric $|dz|$ is complete and has curvature 0. For the punctured plane, $\mathbb{C} - \{0\}$, $|z|^{-1}|dz|$ is complete and has zero curvature. For the plane modulo a lattice, the Euclidean metric $|dz|$ gives an infinitesimal form on $\mathbb{C}/L$ which obviously has curvature zero and is complete because the quotient torus is compact.

We turn now to the most interesting case, the case of the hyperbolic plane $\mathbb{H} = \{z\colon \operatorname{Im} z > 0\}$. Here, the infinitesimal metric is

$$\rho(z)|dz| = \frac{|dz|}{|z - \bar{z}|}. \tag{4}$$

Of course, formula (4) gives an infinitesimal metric on any of the orbit spaces $\mathbb{H}/\Gamma$ as long as Γ acts as a group of isometries. The metric (4) is often called the Poincaré metric. Equally often it is called the hyperbolic or the noneuclidean metric.

It is useful to be able to pass back and forth between the upper half plane and the unit disk. In general, if Ω is an arbitrary simply connected domain and f a biholomorphic mapping from Ω onto $\mathbb{H}$, we define

$$\rho_\Omega(z) = \rho(f(z))|f'(z)|, \tag{5}$$

where ρ is given in formula (4). If f_1 is another biholomorphic mapping from Ω onto $\mathbb{H}$, formula (5) yields the same metric ρ_Ω. The reader should postpone verification of this until reading formula (10), below. By letting Ω be the unit disk and $f(z) = w$, where $z = (w - i)/(w + i)$, we find from (4) and (5) that the

Poincaré metric for the unit disk Δ is

$$\rho_\Delta(z)|dz| = \frac{|dz|}{1 - |z|^2}. \tag{6}$$

An application of formula (1) shows that the Poincaré metrics (4) and (6) have curvature equal to -4.

The standard procedure for obtaining a global metric from an infinitesimal form is to first define the length of a path. If $\alpha(t)$ is a smooth path lying in the unit disk Δ parameterized by $t_0 \leqslant t \leqslant t_1$, the length of α is defined by

$$\ell(\alpha) = \int_{t_0}^{t_1} \rho_\Delta(\alpha(t))|\alpha'(t)|\, dt.$$

The length of a piecewise smooth path is the sum of the lengths of each of its smooth parts. Then the distance function is defined by

$$d(a, b) = \inf\{\ell(\alpha): \alpha \text{ is a piecewise smooth path joining } a \text{ to } b\}. \tag{7}$$

This function is obviously symmetric and satisfies the triangle inequality. For the unit disk, let α be a path which joins $a = 0$ to $b = x$, $0 < x < 1$. We can estimate the length of α in the following way: For $\alpha(t) = \alpha_1(t) + i\alpha_2(t)$,

$$\ell(\alpha) = \int_{t_0}^{t_1} \frac{|\alpha'(t)|\, dt}{1 - |\alpha(t)|^2} \geqslant \int_{t_0}^{t_1} \frac{\alpha_1'(t)}{1 - \alpha_1(t)^2}\, dt = \int_0^x \frac{d\alpha_1}{1 - \alpha_1^2} = \frac{1}{2}\log\frac{1 + x}{1 - x}. \tag{8}$$

On letting $\alpha(t) = tx, 0 \leqslant t \leqslant 1$, we get equality in (8) and we see that the infimum in (7) is achieved by this curve and

$$d(0, x) = \frac{1}{2}\log\frac{1 + x}{1 - x}. \tag{9}$$

Moreover, for equality to hold in (8), one must have $\alpha(t)$ real and $\alpha'(t) \geqslant 0$. Thus the segment $[0, x]$ is the unique geodesic joining 0 to x. In order to find the formula for the distance between an arbitrary pair of points, first observe that bianalytic self-mappings of the unit disk leave the infinitesimal form (6) invariant. This statement hinges on the identity

$$\frac{|A'(z)|}{1 - |A(z)|^2} = \frac{1}{1 - |z|^2}, \tag{10}$$

where A is a general transformation of the unit disk of the form

$$A(z) = e^{i\theta} \cdot \frac{z - w}{1 - \bar{w}z}, \tag{11}$$

with $|w| < 1$ and θ real. Therefore, the lengths of piecewise smooth paths are left invariant by the transformations A and so is the distance function (7). Since, for an arbitrary pair of points z and w in the unit disk, there is a transformation A of the form (11) which takes w onto 0 and z onto a point on the line segment between 0 and 1, from (9) one can deduce

$$d(z, w) = \frac{1}{2} \log \frac{1 + \left|\dfrac{z - w}{1 - \bar{w}z}\right|}{1 - \left|\dfrac{z - w}{1 - \bar{w}z}\right|}. \tag{12}$$

The foregoing considerations lead to the following proposition.

Proposition 1. *The Poincaré metric on the unit disk has infinitesimal form given in formula* (6) *and global form given in* (12). *It is a Riemannian metric with curvature constantly equal to* -4 *and it is complete for the unit disk and induces the usual topology there. Any two points in the unit disk are joined by a unique geodesic which lies on a circle orthogonal to the unit circle* $\{z: |z| = 1\}$.

Any other simply connected domain conformally equivalent to the unit disk has a Poincaré metric with all of the same properties except that geodesics may not lie on circles.

Proof. We have already derived the global form of the metric from its infinitesimal form and we have calculated the curvature. The fact that the topology coming from the metric d is the usual Euclidean topology follows from (12). Completeness of the metric follows from the formula

$$d(0, z) = \frac{1}{2} \log \frac{1 + |z|}{1 - |z|}. \tag{13}$$

A sequence z_n which is Cauchy for the metric d must be bounded in that metric. Thus from (13) there is a number $k < 1$, with $|z_n| \leqslant k$. One obtains a subsequence which converges to a point in the interior of the unit disk and therefore the original Cauchy sequence z_n converges in the unit disk.

Since transformations A in (11) preserve circles which are orthogonal to the unit circle and since they are isometries in the metric (12), the geodesic segment $[0, x]$, lying on the x-axis, is taken by A onto a segment of a circle orthogonal to the unit circle. Since the segment $[0, x]$ is a unique geodesic, so are all of its images under the isometries A.

The rest of the proposition is self-evident.

1.4. FUCHSIAN GROUPS

A group G is said to act on a topological space X if every element A of G can be viewed as a homeomorphism of X and if the group operation in G is compatible with composition of functions in the sense that $AB(x) = A(B(x))$ for every A and

B in G and every x in X. The group is called torsion free if the identity I in G is not equal to a power of any element of G except for the identity itself.

Definition. *A group G of homeomorphisms of a topological space X is said to act properly discontinuously on X if, for every compact set K in X, the set of A in G for which $A(K)$ intersects K is finite.*

Because of the uniformization theorem, groups of holomorphic homeomorphisms of the upper half plane $\mathbb{H}$ which are torsion free and act properly discontinuously on $\mathbb{H}$ play a central role in Riemann surface theory. It is useful to drop the condition that these groups be torsion free. Accordingly we make the following definition.

Definition. *A group of holomorphic homeomorphisms of the upper half plane $\mathbb{H}$ which acts properly discontinuously on $\mathbb{H}$ is called a Fuchsian group.*

Lemma 1. *Suppose $A\colon \mathbb{H} \to \mathbb{H}$ is a bijective, holomorphic mapping. Then there exists real numbers a, b, c, d and $ad - bc = 1$ such that*

$$A(z) = \frac{az + b}{cz + d}. \tag{14}$$

The numbers a, b, c and d are determined uniquely up to simultaneous multiplication of all of them by $+1$ or -1. Thus a Fuchsian group is actually a subgroup of $\mathrm{PSL}(2, \mathbb{R}) = \mathrm{SL}(2, \mathbb{R})/\{\pm 1\}$.

Proof. We consider the transformation $T(z) = (z - i)/(z + i)$. It is a bijective holomorphic mapping from $\mathbb{H}$ to the interior of the unit disk. The hypotheses on A imply that $T \circ A \circ T^{-1}$ is a bijective holomorphic self-mapping of the unit disk.

If the mapping $T \circ A \circ T^{-1}$ takes 0 into α, we can postcompose it with the mapping $C(z) = (z - \alpha)/(1 - \bar{\alpha}z)$, which preserves the unit disk and takes α to 0. Thus $C \circ T \circ A \circ T^{-1}$ is a holomorphic bijection of the unit disk which keeps the origin fixed. By the uniqueness part of Schwarz's lemma, it must be a rotation $z \to e^{i\theta}z$.

On solving the matrix equation

$$C \circ T \circ A \circ T^{-1}(z) = e^{i\theta}z,$$

it is clear that A has a matrix representation (14) for some complex numbers a, b, c, and d with $ad - bc \neq 0$, which, after suitable scaling, we may assume to satisfy $ad - bc = 1$. Since A extends to a mapping which preserves $\mathbb{R} \cup \{\infty\}$, we see that $A(\infty) = a/c$ and $A(0) = b/d$ must be real. Since $A'(x) = (cx + d)^{-2}$ must be real for all real x, this shows that a, b, c, and d are either all real or all imaginary. If they were all imaginary, then A would map the upper half plane to the lower half plane. However, we assume $A(\mathbb{H}) = \mathbb{H}$, so the case where a, b, c, and d are imaginary is ruled out.

Assuming $ad - bc = 1$, the transformation completely determines the numbers a, b, c, d except for one possible ambiguity: a, b, c, d will give the same transformation as $-a$, $-b$, $-c$, and $-d$. Thus a Fuchsian group is a subgroup of PSL(2, $\mathbb{R}$), the projective special linear group of 2×2 matrices with entries in $\mathbb{R}$ (see Exercise 4).

Since a Fuchsian group is a properly discontinuous group of isometries of a metric space, the following lemma from metric space theory is useful. For a group G acting on a set X, define the isotropy group of a point to be the subgroup of G which fixes the point. Define the orbit of a point p to be the set of points $A(p)$ for A in G.

Lemma 2. *Let G be a group of isometries of a metric space X with metric d. Then G acts properly discontinuously on X if the isotropy groups are finite and the orbits are discrete. Conversely, if G acts properly discontinuously and X is locally compact, then the isotropy groups are finite and the orbits are discrete.*

Proof. Suppose K is a compact subset of X and that $A_n(K)$ intersects K for an infinite sequence of distinct elements A_n. Without changing notation for subsequences, we obtain an infinite sequence of distinct elements A_n in G and points p_n in K with $r_n = A_n(p_n)$ in K such that p_n converges to p and r_n converges to r.

Using the fact that elements of G are isometries and the triangle inequality, it is elementary to show that (see the next paragraph)

$$d(A_m \circ A_n^{-1}(r_k), r_k) \leqslant 2d(r_k, r_n) + d(p_n, p_m) + d(r_m, r_n). \tag{15}$$

Now select n_0 so that for n and $m \geqslant n_0$ we have $d(r_m, r_n) \leqslant \varepsilon/4$ and $d(p_n, p_m) < \varepsilon/4$. Then for n, m, and $k \geqslant n_0$ one sees from (15) that $d(A_m \circ A_n^{-1}(r_k), r_k)) < \varepsilon$. Taking the limit as $k \to \infty$, we obtain

$$d(A_m \circ A_n^{-1}(r), r) \leqslant \varepsilon$$

for infinitely many elements $A_m \circ A_n^{-1}$. But if orbits are discrete, we can select ε so that no points in the orbit of r which are distinct from r have distance less than ε from r. The only possibility is that $A_m \circ A_n^{-1}(r) = r$. But that is also impossible since $A_m \neq A_n$ and isotropy groups are finite.

To prove (15) observe that

$$\begin{aligned} d(A_m \circ A_n^{-1}(r_k), r_k) \leqslant\ & d(A_m \circ A_n^{-1}(r_k), A_m \circ A_n^{-1}(r_n)) \\ & + d(A_m \circ A_n^{-1}(r_n), r_n) + d(r_n, r_k). \end{aligned}$$

Since $A_m \circ A_n^{-1}$ is an isometry, we get

$$d(A_m \circ A_n^{-1}(r_k), r_k) \leqslant 2d(r_n, r_k) + d(A_m(p_n), r_n).$$

But

$$d(A_m(p_n), r_n) \leqslant d(A_m(p_n), A_m(p_m)) + d(A_m(p_m), r_n)$$
$$= d(p_n, p_m) + d(r_m, r_n).$$

Putting these inequalities together, we obtain (15)

To obtain the converse part of the lemma, notice that since a single point set $\{p\}$ is compact the proper discontinuity implies that its isotropy group is finite. Secondly, for an arbitrary point p, the local compactness of X implies there exists $\varepsilon > 0$ such that $K = \{q \mid d(p, q) \leqslant \varepsilon\}$ is compact. Let $A_1, \ldots, A_n$ be the finite set of elements of G for which $A_j(K)$ intersects K. Let ε_1 be the minimum value of $d(p, A_j(p))$ for all the A_j with $1 \leqslant j \leqslant n$ for which $A_j(p) \neq p$. Then every point which is in the orbit of p and not equal to p has distance greater than or equal to ε_1 from p. Thus the orbits are discrete.

From this lemma, we can now prove the following result which characterizes Fuchsian groups.

Proposition 2. *A subgroup* Γ *of* PSL(2, $\mathbb{R}$) *is discrete if, and only if, it acts properly discontinuously on* $\mathbb{H}$.

Proof. Let S^1 be the set of complex numbers with absolute value 1 and let Δ be the interior of the unit disk. The Möbius transformation $T(z) = (z - i)/(z + i)$ transforms the upper half plane into the open unit disk and the conjugation mapping $A \to T \circ A \circ T^{-1}$ is an isomorphism of PSL(2, $\mathbb{R}$) into the holomorphic bijections of the unit disk, which are uniquely expressible in the form

$$C(z) = \lambda \frac{z - \alpha}{1 - \bar{\alpha} z}, \qquad \text{where } \lambda \text{ is in } S^1 \text{ and } \alpha \text{ is in } \Delta.$$

It is easy to check that the mapping

$$\mathrm{PSL}(2, \mathbb{R}) \ni \begin{pmatrix} a & b \\ c & d \end{pmatrix} \mapsto (\lambda, \alpha) \in S^1 \times \Delta \tag{16}$$

is a homeomorphism and the orbit of i in $\mathbb{H}$ under a group $\Gamma \subset$ PSL(2, $\mathbb{R}$) is transformed by T to the orbit of 0 under the group $T \circ \Gamma \circ T^{-1}$.

Suppose Γ is discrete and A_n is an enumeration of the elements of Γ. Then clearly the set of corresponding numbers (λ_n, α_n) under the mapping (16) is discrete. Therefore, for any $r < 1$, the set of (λ_n, α_n) for which $|\alpha_n| \leqslant r$ is a finite set. We conclude that the set of numbers $-\lambda_n \alpha_n$, which sweep out the orbit of zero, is a discrete set. Moreover, the isotropy group of 0 cannot be infinite because, if it were, then α_n would be zero for infinitely many n and there would be an infinite discrete set of numbers λ_n in the compact set S^1.

On replacing T by the transformation $(z - z_0)/(z - \bar{z}_0)$ this same argument shows that orbits under Γ are discrete and isotropy groups are finite for all points z_0 in $\mathbb{H}$.

Conversely, suppose Γ acts discontinuously on $\mathbb{H}$. Then the image of Γ under the mapping (16) must give a discrete set of values $-\lambda_n\alpha_n$ since these values are the orbit of 0. Thus there can be only finitely many values (λ_n, α_n) lying inside the compact set $S^1 \times \bar{\Delta}_r$, where $\bar{\Delta}_r = \{z:|z| \leqslant r\}$ and $r < 1$. This shows that Γ is a discrete subset of PSL(2, $\mathbb{R}$).

Remark 1. This proposition shows that any discrete subgroup Γ of PSL(2, $\mathbb{R}$) is a Fuchsian group. If B is any element of PSL(2, $\mathbb{C}$), we also call the group $B \circ \Gamma \circ B^{-1}$ a Fuchsian group. Instead of having $\mathbb{H}$ as invariant domain, the group $B \circ \Gamma \circ B^{-1}$ has $B(\mathbb{H})$ as invariant domain. $B(\mathbb{H})$ is either the interior or the exterior of a circle or a half plane in the extended complex plane.

Remark 2. The same method used to prove the previous proposition can be used to show that a subgroup of PSL(2, $\mathbb{C}$) is discrete if, and only if, it acts discontinuously on $\mathbb{H}^3$, three dimensional hyperbolic space. However, discrete subgroups of PSL(2, $\mathbb{C}$) may fail to act discontinuously on any subdomain of $\hat{\mathbb{C}}$.

Let $\bar{p}$ denote the orbit of the point p. Thus

$$\bar{p} = \{A(p)\colon A \in G\}.$$

Let X/G denote the pace of such orbits. When X is a metric space and G is a group of isometries, the following lemma explains how X/G becomes a metric space.

Lemma 3. *Suppose G is a group of isometries of a metric space X with metric d. Suppose further that G acts properly discontinuously on X. Moreover, assume X is locally compact. Then X/G is metrized by the metric*

$$\bar{d}(\bar{p}. \bar{r}) = \inf_{A \text{ in } G} d(A(p), r) \tag{17}$$

and the natural mapping $\pi\colon X \to X/G$ is open and continuous and is a local homeomorphism except at points for which the isotropy group is nontrivial.

Proof. It is clear that formula (17) for $\bar{d}$ gives a nonnegative symmetric function satisfying the triangle inequality. If $d(\bar{p}, \bar{r}) = 0$, we must show the orbit of p equals the orbit of r. But if $d(\bar{p}, \bar{r}) = 0$, then there exist elements A_n in G such that $d(A_n(p), r) \to 0$. We may assume infinitely many of the A_n are distinct, because if they are not, then the set $\{A_n(p)\colon n \text{ an integer}\}$ is a finite set and, therefore, for some value n_0, $A_{n0}(p) = r$. This would show p is in the orbit of r.

Let $r_n = A_n(p)$. Using the triangle inequality and the fact that elements of G are isometries, in the same way that we arrived at (15), one can show that

$$d(A_m \circ A_n^{-1}(r_k), r_k) \leqslant 2d(r_n, r_k) + d(r_m, r_n). \tag{18}$$

Since X is assumed to be locally compact, we may choose $\varepsilon > 0$ so that every point in the orbit of r which is distinct from r has distance larger than ε from r. Then choose n_0 so that $n \geqslant n_0$ and $m \geqslant n_0$ imply $d(r_m, r_n) < \varepsilon/3$. This can be done because we know that r_n converges to r. From (18) we obtain

$$d(A_m \circ A_n^{-1}(r_k), r_k) < \varepsilon \tag{19}$$

for m, n, and $k \geqslant n_0$. Taking the limit in (19) as $k \to \infty$, we get $d(A_m \circ A_n^{-1}(r), r) \leqslant \varepsilon$. By choice of ε, this implies $d(A_m \circ A_n^{-1}(r), r) = 0$. The assumption that isotropy groups are finite is therefore contradicted unless $\{A_m \circ A_{n_0}^{-1}\}$ is a finite set and thus $\bar{r} = \bar{p}$.

From the obvious inequality $\bar{d}(\bar{p}, \bar{r}) \leqslant d(p, r)$, it follows that the natural mapping $\pi\colon X \to X/G$ is continuous. Since the ball of radius ε about $\bar{p}$ is contained in the image under π of the ball of radius ε about p, π is open.

Suppose p is a point whose isotropy group is trivial. Since the orbit of p is discrete, $\varepsilon > 0$ can be chosen so that every point in the orbit of p and not equal to p has distance more than 2ε from p. If $d(p, r) < \varepsilon$ and an element A of G fixes r, then obviously $d(p, A(p)) \leqslant d(p, r) + d(A(r), A(p)) < 2\varepsilon$. Hence, A must be the identity and so r has trivial isotropy group.

Let $N = \{r\colon d(p, r) < \varepsilon\}$. We have just shown that every point of N has trivial isotropy group. It is also true that no two points of N can be identified by G. For suppose r_1 and r_2 are distinct points of N and $A(r_1) = r_2$ for some A in G. Then $d(p, A(p)) \leqslant d(p, r_2) + d(A(r_1), A(p)) < 2\varepsilon$. But this contradicts our assumption on p unless A is the identity. Thus, the projection $\pi\colon X \to X/G$ is one-to-one when restricted to N and it is open and continuous. The lemma is proved.

Definition. *A group G acting on a metric space X has the identity property if whenever an element A in G fixes a set of points in X with an accumulation point in X, then A is the identity.*

Notice that $\mathrm{PSL}(2, \mathbb{R})$ acting on $\mathbb{H}$ has the identity property.

Definition. *An elliptic point p for G acting on X is a point in X whose isotropy group is nontrivial.*

Remark. In the case where G is a Fuchsian group acting on $\mathbb{H}$, the isotropy group of any point in $\mathbb{H}$ is automatically finite. This is because such an isotropy group is a discrete subgroup of the compact group of noneuclidean rotations about the point. Of course, many points on the real axis have infinite isotropy groups. However, the real axis is not part of the space where G acts as a group of isometries.

Lemma 4. *Let G be a group of isometries acting on a metric space X and suppose orbits are discrete and isotropy groups are finite. Suppose further that G has the identity property. Then every point p in X has an open neighborhood N such that $N - \{p\}$ consists of nonelliptic points.*

Proof. Where p is nonelliptic we already proved the existence of the neighborhood N in the second to last paragraph of the proof of the preceding lemma.

Suppose r is an elliptic point and p_n is a sequence of distinct elliptic points converging to r. It is obvious that there exists n_0 such that for $n \geqslant n_0$ the isotropy group of p_n is contained in the isotropy group of r. Since the isotropy group of r is finite and the isotropy group of each p_n is nontrivial, we can extract a subsequence p_{n_k} and a nontrivial element A which fixes each point p_{n_k}. This contradicts the identity property.

At the moment we are interested in applying Lemmas 2, 3, and 4 to case where the metric space X is the upper half plane $\mathbb{H}$, the metric d is the Poincaré metric, and the group G is a discrete subgroup of PSL(2, $\mathbb{R}$). Later, we will apply them to the case where X is Teichmüller space, the metric d is Teichmüller's metric, and G is the Teichmüller modular group.

Theorem 2. *Let Γ be a Fuchsian group acting on $\mathbb{H}$. Then $\mathbb{H}/\Gamma$ is a complete metric space with the quotient metric given by $\bar{d}$ in (17). The set of elliptic points for Γ in $\mathbb{H}$ is a discrete set. Moreover, there is a unique complex structure on $\mathbb{H}/\Gamma$ making $\mathbb{H}/\Gamma$ into a Riemann surface and the mapping $\pi:\mathbb{H} \to \mathbb{H}/\Gamma$ into a holomorphic mapping. At nonelliptic points π is a local homeomorphism, and at elliptic points π is locally an n-to-1 mapping where n is the order of the isotropy group of the elliptic point.*

Proof. In view of Lemmas 3 and 4 we only need to exhibit the complex structure for $\mathbb{H}/\Gamma$ and to show why the metric $\bar{d}$ is complete. Suppose $\bar{p}_n$ is a Cauchy sequence in $\mathbb{H}/\Gamma$. Then for every $\varepsilon > 0$ there exists n_0 such that, for n and m larger than or equal to n_0, $\bar{d}(\bar{p}_m, \bar{p}_n) < \varepsilon$. Because of the definition of $\bar{d}$ in Lemma 3, if we take a representative p_{n_0} of $\bar{p}_{n_0}$, then we may find a representative p_n of each $\bar{p}_n$ such that $d(p_{n_0}, p_n) < \varepsilon$. This means the sequence p_n is a bounded sequence in hyperbolic space $\mathbb{H}$ and therefore it has a convergent subsequence p_{n_k}. Because π is continuous, $\bar{p}_{n_k}$ is then a convergent subsequence of the Cauchy sequence $\bar{p}_n$, which therefore converges.

To show $\mathbb{H}/\Gamma$ has a Riemann surface structure, first assume $\bar{p}$ in $\mathbb{H}/\Gamma$ is the image of a nonelliptic point. From Lemma 3, we may choose an open set U containing p in $\mathbb{H}$ such that π restricted to U is a homeomorphism. Let V be the open set $\pi(U)$ and let the local coordinate z for V be $(\pi|U)^{-1}$. Here, the notation $\pi|U$ means the restriction of π to U. If p_1 is another point in $\bar{p}$, and A is the element of Γ taking p to p_1, then $U_1 = A(U)$ will be a neighborhood of p_1 such that π restricted to U_1 is a local homeomorphism. Let $z_1 = (\pi|U_1)^{-1}$ be a local coordinate for $V_1 = \pi(U_1)$. Clearly, $z_1 \circ z^{-1} = A$, which is an analytic mapping [it is even in PSL(2, $\mathbb{R}$)]. These local coordinates determine a complex structure on a system of neighborhoods which cover all nonelliptic points of $\mathbb{H}/\Gamma$.

Now let p be an elliptic point (which we know by Lemma 4 must be isolated from other elliptic points). By conjugating Γ with a Möbius transformation that takes p to the origin and $\mathbb{H}$ to the unit disk, we may assume the isotropy group is a cyclic group of order n generated by a rotation $\zeta \to \lambda\zeta$, where $\lambda = \exp(2\pi i/n)$. (See Exercise 5.) In a sufficiently small neighborhood U of 0, the only identifications made by Γ arise from this rotation. On letting $z = \zeta^n$, z becomes a local coordinate for the neighborhood $\pi(U)$ in $\mathbb{H}/\Gamma$. It is easy to check that the transition functions to nearby nonelliptic points are holomorphic.

It is obvious that these local coordinates make π into a holomorphic mapping and that they determine the Riemann surface structure.

1.5. CLASSIFICATION OF ELEMENTS OF PSL(2, $\mathbb{R}$)

A transformation

$$A(z) = \frac{az + b}{cz + d}, \qquad ad - bc = 1, \qquad a, b, c, d \text{ in } \mathbb{R},$$

which is not the identity, obviously has at most two fixed points on $\hat{\mathbb{C}}$, since a fixed point satisfies the quadratic equation $A(z) = z$. The trace of A is not well defined since the matrix for A is only determined up to plus or minus sign. But $(\operatorname{tr} A)^2$ is well defined and we can classify any transformation A which is not the identity according to its value:

A is *hyperbolic* if $(\operatorname{tr} A)^2 > 4$.
A is *parabolic* if $(\operatorname{tr} A)^2 = 4$.
A is *elliptic* if $(\operatorname{tr} A)^2 < 4$.

Since the roots of the equation $A(z) = z$ are

$$\frac{a - d + \sqrt{(a + d)^2 - 4}}{2c} \quad \text{when } c \neq 0,$$

a hyperbolic element has two fixed points z_1 and z_2 on the real axis. When $c = 0$ one of the fixed points is at ∞. When $c \neq 0$, on letting

$$B(z) = \frac{z - z_1}{z - z_2},$$

the conjugate $B \circ A \circ B^{-1}$ has fixed points at 0 and ∞ and therefore $B \circ A \circ B^{-1}(z) = \lambda z$ for some positive real number λ. When $c = 0$, the hyperbolic element A has a fixed point at z_1 and ∞. If we let $B(z) = z - z_1$, once gain we obtain $B \circ A \circ B^{-1}(z) = \lambda z$ for some $\lambda > 0$. The number λ cannot be equal to 1 unless A is the identity. If $0 < \lambda < 1$ by interchanging the two points z_1 and z_2, we can make $\lambda > 1$. The number $\lambda > 1$ is called the multiplier of the transformation A. Because of the equation $(\lambda^{1/2} + \lambda^{-1/2})^2 = \operatorname{tr}(A)^2$, the multiplier depends only on the conjugacy class of A.

From the formula for the fixed points, a parabolic transformation has just one fixed point, z_1. If z_1 is not already equal to ∞, then by letting $B(z) = (z - z_1)^{-1}$, we see that BAB^{-1} fixes ∞ and thus is an affine mapping of the plane with no fixed points except at ∞. Thus $BAB^{-1}(z) = z + b$ in the parabolic case.

In the elliptic case, there are two fixed points z_1 and z_2 which are complex conjugates of each other. In this case, the transformation

$$B(z) = \frac{z - z_1}{z - z_2}$$

takes the real axis onto the unit circle, z_1 into 0, and z_2 into ∞. Thus BAB^{-1} is a rotation about the origin, that is, $BAB^{-1}(z) = e^{i\theta}z$.

1.6. FUNDAMENTAL DOMAINS FOR FUCHSIAN GROUPS

A fundamental domain for a Fuchsian group Γ acting on $\mathbb{H}$ is an open subset ω of $\mathbb{H}$ such that:

(i) Every point of $\mathbb{H}$ is Γ-equivalent to at least one point of the closure of ω.

(ii) No two points of ω are identified by an element A of Γ.

(iii) The boundary $\partial\omega$ of ω in $\mathbb{H}$ can be written as a countable union of piecewise analytic arcs γ_j in such a way that:

(iv) For every arc γ_j there is an arc γ_k and an element A of Γ such that $A(\gamma_j) = \gamma_k$.

In the sequel we will not need to use all of the properties of a fundamental domain. For our purposes it will suffice to know the existence of a measurable fundamental set.

It is obvious that if B is a Möbius transformation and ω is a fundamental domain for Γ acting on $\mathbb{H}$, then $B(\omega)$ is a fundamental domain for $B \circ \Gamma \circ B^{-1}$ acting on $B(\mathbb{H})$.

Here we give without proof the methods for constructing the Dirichlet and Ford fundamental domains. We also show that after conjugation by a suitable Möbius transformation and restriction to the unit disk, the Ford fundamental domain coincides with the Dirichlet domain. This observation together with an explanation of its geometric meaning is given in Beardon's book [Bea, pp. 176 and 234]. For the proof that these domains satisfy the properties listed in the definition the reader is referred to Beardon [Bea], Bers [Ber1], Ford [For], or Lehner [Leh].

Method I. The Dirichlet Fundamental Domain of a Fuchsian Group Γ

Select a point z_0 in $\mathbb{H}$ which is not an elliptic fixed point. For each A in $\Gamma - \{id\}$ construct the noneuclidean half plane

$$h(A) = \{z \in \mathbb{H} : d(z, z_0) < d(z, A(z_0))\}$$

where d is the noneuclidean metric. Finally, the Dirichlet fundamental domain is

$$\omega_D(z_0) = \bigcap h(A) \tag{20}$$

where the intersection is over all elements A in $\Gamma - \{id\}$. The boundary of $\omega_D(z_0)$ in $\mathbb{H}$ consists of segments of hyperbolic lines (or half lines or entire lines) which are perpendicular bisectors of hyperbolic segments joining z_0 to $A(z_0)$ for certain elements A of Γ. These segments of perpendicular bisectors bound the set $\omega_D(z_0)$ which is convex in the sense of noneuclidean geometry and, hence, connected and simply connected. There may be segments of the real axis which bound $\omega_D(z_0)$. Such segments are called *free sides*. Of course, a free side is not part of the boundary of $\omega_D(z_0)$ in $\mathbb{H}$.

Method II. The Ford Fundamental Domain of a Fuchsian Group Γ

Corresponding to any Möbius transformation $A(z) = (az + b)/(cz + d)$ with $c \neq 0$ and $ad - bc = 1$ there is an associated isometric circle $I(A) = \{z : |cz + d| = 1\}$. Since $A'(z) = (cz + d)^{-2}$, the isometric circle is the set of z for which $|A'(z)| = 1$. A group Γ may contain an element A which is not the identity and which fixes the point at infinity. This happens precisely if $c = 0$. Such an element does not have an isometric circle and prevents the construction of the Ford fundamental region. Therefore, if there is an element of $\Gamma - \{id\}$ fixing the point at infinity, we select a transformation B such that $B \circ \Gamma \circ B^{-1} - \{id\}$ does not have any element fixing infinity. Obviously, B^{-1} applied to a fundamental domain for $B \circ \Gamma \circ B^{-1}$ is a fundamental domain for Γ.

Thus, to construct a fundamental domain, we may and do assume that every element of $\Gamma - \{id\}$ has an isometric circle. For each element A of $\Gamma - \{id\}$, let $E(A)$ be the region in the extended complex plane $\hat{\mathbb{C}}$ which is exterior to the isometric circle of A. The Ford fundamental domain for Γ is

$$\omega_F = \left(\bigcap E(A)\right) \cap \mathbb{H}, \tag{21}$$

where the first intersection is over all A in $\Gamma - \{id\}$.

To see the relationship between the Ford and Dirichlet fundamental domains, let z_0 be a point in $\mathbb{H}$ which is nonelliptic for Γ and let $B(z) = (z - z_0)/(z - \bar{z}_0)$. Note that B transforms $\mathbb{H}$ into the interior of the unit disk Δ and z_0 into the origin and the group $B \circ \Gamma \circ B^{-1}$ has no elements except the identity fixing infinity or the origin. The Dirichlet fundamental domain $\omega_D(0)$ centered at the origin is the intersection of all half planes determined by the inequalities

$$d_\Delta(z, 0) < d_\Delta(z, A(0)), \tag{22}$$

where z is in Δ, where d_Δ is the noneuclidean metric for Δ, and where A is an arbitrary element of $B \circ \Gamma \circ B^{-1} - \{id\}$. The main point is that inequality (22) determines precisely the exterior of the isometric circle for A^{-1} intersected with Δ. To see this note that any Möbius transformation A which preserves the unit disk is represented by a matrix

$$\begin{pmatrix} \alpha & \beta \\ \bar{\beta} & \bar{\alpha} \end{pmatrix},$$

where α and β are complex numbers satisfying $|\alpha|^2 - |\beta|^2 = 1$ (see Exercise 6). Inequality (22) with A replaced by A^{-1} can be rewritten as $d_\Delta(z, 0) < d_\Delta(A(z), 0)$, which is equivalent to $|\bar{\beta}z + \bar{\alpha}z|\,|z| < |\alpha z + \beta|$. After some elementary algebra, this inequality combined with the condition $|\alpha|^2 - |\beta|^2 = 1$ reduces to $|\bar{\beta}z - \bar{\alpha}| > 1$.

In summary, let Γ be a Fuchsian group acting on the unit disk and let 0 be a nonelliptic point. Then the Dirichlet fundamental domain $\omega_D(0)$ centered at the origin coincides with the intersection of the unit disk with the Ford fundamental domain.

There is an important problem which is converse to finding fundamental domains. A domain with sides bounded by circular arcs and a finite set $A_1, \ldots, A_n$ of Möbius transformations identifying these arcs pairwise are given. The problem is to give conditions under which the group generated by $A_1, \ldots, A_n$ is Fuchsian and has the given domain as fundamental domain. A great deal of work has been done on this problem and on analogous problems for Kleinian groups. See, for example, Keen [Ke1], Maskit [Mask], and Macbeath [Mac1].

We now return to Fuchsian groups Γ acting on the upper half plane $\mathbb{H}$. The area of $\mathbb{H}/\Gamma$ is defined to be

$$\text{area}(\omega) = 4 \iint_\omega \rho^2(z)\, dx\, dy,$$

where $\rho(z) = |z - \bar{z}|^{-1}$ is the infinitesimal noneuclidean metric and ω is any fundamental domain for Γ. This integral is independent of the choice of fundamental domain ω. First, note that for any measurable subset S of $\mathbb{H}$, $\text{area}(S) = \text{area}(A(S))$ for A in $\text{PSL}(2, \mathbb{R})$. Second, since the boundary of ω is a countable union of analytic arcs, $\mathbb{H} - \cup_{A \in \Gamma} A(\omega))$ is a set of measure zero. Thus, for any two fundamental domains ω_1 and ω_2 we have

$$\text{area}(\omega_1) = \sum \text{area}(\omega_1 \cap A(\omega_2)) = \sum \text{area}(A^{-1}(\omega_1) \cap \omega_2) = \text{area}(\omega_2),$$

where the two summations are over all A in Γ.

In general, when it is a finite number, the area of $\mathbb{H}/\Gamma$ depends on the genus of the surface $\mathbb{H}/\Gamma$ as well as the number of parabolic conjugacy classes in Γ and the number of elliptic conjugacy classes as well as their orders. The Gauss–Bonnet theorem gives a formula for the area of $\mathbb{H}/\Gamma$ in terms of these topological invariants [see formula (2) of Section 9.1].

The limit set Λ of a Fuchsian group Γ is the set of accumulation points of the orbit of any point z_0 in $\mathbb{H}$. It is a closed Γ-invariant set contained in the real axis. Γ is said to be of the first kind if $\Lambda = \hat{\mathbb{R}}$ and of the second kind if Λ is not all of $\hat{\mathbb{R}}$.

Although the following theorem will not be needed in this book, we state it as a matter of general information. The proof is long and may be found in Kra [Kr5] or Lehner [Leh].

Theorem 3. *Let Γ be a Fuchsian group acting on H and ω a fundamental domain defined by* (20). *Then the following conditions are equivalent:*

(a) *Area* $(\mathbb{H}/\Gamma)$ *is finite.*
(b) *ω has a finite number of sides and no free sides.*
(c) *Γ is finitely generated and of the first kind.*

When Γ is a group of the second kind, the Riemann surface $R = \mathbb{H}/\Gamma$ has a border. The border of $\mathbb{H}/\Gamma$ is obtained by adjoining to $\mathbb{H}/\Gamma$ the set $(\hat{\mathbb{R}} - \Lambda)/\Gamma$ with the natural topology induced by the covering $\pi\colon \mathbb{H} \cup (\hat{\mathbb{R}} - \Lambda) \to (\mathbb{H} \cup (\hat{\mathbb{R}} - \Lambda))/\Gamma$. If $\tilde{\Gamma} = B \circ \Gamma \circ B^{-1}$ for some Möbius transformation B, the transformation B induces a homeomorphism of the border of the surface $\mathbb{H}/\Gamma$ onto the border of the surface $\mathbb{H}/\tilde{\Gamma}$.

1.7. QUASICONFORMAL MAPPINGS: GEOMETRIC AND ANALYTIC DEFINITIONS

A Jordan region on a Riemann surface R is a connected and simply connected open subset of R whose boundary is a simple closed curve contained in R. By a generalized quadrilateral Q on R we mean a Jordan region on R together with two disjoint closed arcs β_1 and β_2 on the boundary of Q. The module of Q, $m(Q)$, is determined by the conformal mapping of Q onto a rectangle which takes the disjoint closed arcs onto the vertical sides of the rectangle. If this rectangle has width a and height b, then $m(Q) = a/b$. If Q^* is the same rectangle as Q but with two disjoint closed arcs α_1 and α_2 complementary in the boundary of Q except for common endpoints, then $m(Q^*) = m(Q)^{-1}$.

Definition (Geometric Form). *Let f be a sense preserving homeomorphism from a region Ω to a region Ω'. Then f is K-quasiconformal if for every quadrilateral Q in Ω, $m(f(Q)) \leqslant Km(Q)$. The smallest possible value of K for which this inequality is satisfied for all quadrilaterals Q is called the dilatation of f.*

Clearly, $m(f(Q^*)) \leqslant Km(Q^*)$ and $f(Q^*) = f(Q)^*$, so the definition implies

$$K^{-1}m(Q) \leqslant m(f(Q)) \leqslant Km(Q) \tag{23}$$

for every quadrilaterial Q. Using this geometric definition, it is obvious that if f_1 and f_2 are K_1- and K_2-quasiconformal, then $f_1 \circ f_2$ is K_1K_2-quasiconformal.

The analytic definition of a quasiconformal mapping depends on the notion of absolute continuity on lines, which we abbreviate by ACL. The function $f(z) = u(x, y) + iv(x, y)$ is ACL if for every rectangle in Ω with sides parallel to the x- and y-axes, both $u(x, y)$ and $v(x, y)$ are absolutely continuous on almost every horizontal and almost every vertical line in R. The functions u and v will then have partial derivatives u_x, u_y, v_x, v_y almost everywhere in Ω. The complex partial derivatives are, by definition,

$$f_z = \tfrac{1}{2}(f_x - if_y) \quad \text{and} \quad f_{\bar{z}} = \tfrac{1}{2}(f_x + if_y). \tag{24}$$

Definition (Analytic Form). *Let f be a homeomorphism from a domain Ω to a domain Ω'. Then f is K-quasiconformal if*

(i) *f is* ACL *in Ω; and*
(ii) *$|f_{\bar{z}}| \leqslant k|f_z|$ almost everywhere, where $k = (K - 1)/(K + 1) < 1$.*

The minimal possible value of K for which (ii) *is satisfied is called the dilatation of f.*

Theorem 4. *The geometric and analytic definitions of K-quasiconformality are equivalent. Moreover, the partial derivatives $f_{\bar{z}}$ and f_z of a quasiconformal mapping f are locally square integrable.*

For the proof of this theorem see Ahlfors [Ah4] or Lehto and Virtanen [LehtV].

1.8. THE BELTRAMI EQUATION

We have seen that if f is a topological (homeomorphic) ACL mapping and if $|f_{\bar{z}}/f_z| \leqslant k < 1$ almost everywhere, then f is quasiconformal. Let $\mu(z)$ be a measurable complex valued function defined in a domain Ω for which $\|\mu\|_\infty = k < 1$. The Beltrami equation is

$$f_{\bar{z}}(z) = \mu(z)f_z(z), \tag{25}$$

where the partial derivatives are assumed to be locally square integrable and taken in the sense of distributions. The function μ is called the Beltrami coefficient of the mapping f.

Distributional derivatives are defined by means of the appropriate integration-by-parts formula. In particular, h is the $(\partial/\partial x)$-derivative of f if

$$\iint f g_x \, dx \, dy = -\iint h g \, dx \, dy$$

for every C^∞-function g with compact support in the domain of definition of f. The $(\partial/\partial y)$-distributional derivative is defined analogously and thus $\partial/\partial z$ and $\partial/d\bar{z}$ are determined by (24).

In the theory of the Beltrami equation ([Ah4], [AhBer], [LehtV]) a solution f can be expressed as a power series in μ, where the power series is made up by taking compositions of singular integral operators. The following theorem asserts the existence of normalized global solutions to (25) on $\hat{\mathbb{C}} = \mathbb{C} \cup \{\infty\}$. It also expresses in a very specific way the analytic dependence of the solution f on the Beltrami coefficient μ. The analyticity of this dependence is essential in determining a complex structure for Teichmüller space.

Theorem 5. *The equation* (25) *gives a one-to-one correspondence between the set of quasiconformal homeomorphisms of* $\hat{\mathbb{C}}$ *which fix the points* 0, 1, *and* ∞ *and the set of measurable complex valued functions* μ *on* $\hat{\mathbb{C}}$ *for which* $\|\mu\|_\infty < 1$. *Furthermore, the normalized solution* f^μ *to* (25) *depends holomorphically on* μ *and for any* $R > 0$ *there exists* $\delta > 0$ *and* $C(R) > 0$ *such that*

$$|f^{t\mu}(z) - z - tF(z)| \leqslant C(R)t^2 \qquad \text{for } |z| < R \quad \text{and} \quad |t| < \delta, \tag{26}$$

where

$$F(z) = -\frac{z(z-1)}{\pi} \iint_{\mathbb{C}} \frac{\mu(\zeta)\, d\xi\, d\eta}{\zeta(\zeta-1)(\zeta-z)},$$

and $\zeta = \xi + i\eta$.

Frequently, we need solutions f of (25) which map the upper half plane to itself and preserve the real axis and have arbitrary Beltrami coefficient μ with support in $\mathbb{H}$. Let $M(\mathbb{H})$ be the space of complex valued L_∞-functions μ with support in $\mathbb{H}$ and with $\|\mu\|_\infty < 1$. For μ in $M(\mathbb{H})$ let $\hat{\mu}$ be identically equal to μ on $\mathbb{H}$ and equal to $\overline{\mu(\bar{z})}$ on $\mathbb{H}^*$, the lower half plane. Then solve the equation (25) with μ replaced by $\hat{\mu}$. (The values of $\hat{\mu}$ on the real axis are unimportant because it is a set of measure zero.)

Let $f = f^\mu$ be the solution to (25) which has Beltrami coefficient $\hat{\mu}$ and which is normalized to fix the points 0, 1, and ∞. (Any three points on the real axis would do.) Then $\overline{f(\bar{z})}$ has the same Beltrami coefficient as f and it fixes the same three points on the real axis. Therefore, by uniqueness of solutions to (25), $\overline{f(\bar{z})} = f(z)$ and f is a quasiconformal mapping which preserves the real axis in an orientation-preserving manner. Moreover, f preserves the upper and lower half plane.

We obtain a corollary to the preceding theorem.

Corollary. *To every μ in $M(\mathbb{H})$, there exists a unique quasiconformal self-mapping f of $\mathbb{H}$ satisfying $f_{\bar{z}}(z) = \mu(z) f_z(z)$ on $\mathbb{H}$ which extends continuously to the closure of $\mathbb{H}$ and is normalized to fix the points 0, 1, and ∞. Moreover, to every $R > 0$ there exist $C(R) > 0$ and $\delta > 0$ such that*

$$|f^{t\mu}(z) - z - tF(z)| \leqslant C(R)t^2$$

for $|z| < R$ and t real and $|t| < \delta$, where

$$F(z) = -\frac{1}{\pi} \iint_{\mathbb{H}} [\mu(\zeta)R(\zeta, z) + \bar{\mu}(\zeta)R(\bar{\zeta}, z)]\, d\xi\, d\eta,$$

$$R(\zeta, z) = \frac{z(z-1)}{\zeta(\zeta - 1)(\zeta - z)},$$

and $\zeta = \xi + i\eta$.

Proof. We have already shown how the first part of the corollary follows from the theorem. To obtain the infinitesimal formula for $f^{t\mu}$ we apply the corresponding formula in the theorem to the Beltrami coefficient $\hat{\mu}$.

We frequently need the following closely related convergence principles for quasiconformal mappings.

Lemma 5. *Suppose μ_n is a sequence of Beltrami coefficients and $\|\mu_n\| \leqslant k < 1$ and $\mu_n(z)$ converges to $\mu(z)$ pointwise almost everywhere. Let $f_n(z)$ and $f(z)$ be the unique quasiconformal homeomorphisms of the extended complex plane $\hat{\mathbb{C}}$ which have Beltrami coefficients μ_n and μ, respectively, and which are normalized to fix 0, 1, and ∞. Then $f_n(z)$ converges to $f(z)$ uniformly on compact subsets of $\mathbb{C}$.*

Lemma 6. *Let f_n be a sequence of quasiconformal homeomorphisms of $\hat{\mathbb{C}}$, normalized to fix 0, 1, and ∞. Let $K(f_n)$ be the dilatation of f_n and assume $K(f_n) \leqslant K_0$ for every n. Then there is a subsequence of f_n converging uniformly on compact subsets of $\mathbb{C}$ to a normalized quasiconformal mapping f. Moreover, $K(f) \leqslant K_0$.*

1.9. EXTREMAL LENGTH

Let F be a family of curves on a Riemann surface. Every γ in F is assumed to be a countable union of open arcs or closed curves. The extremal length of F, $\Lambda(F)$, is a sort of average minimum length of the curves in F. It is an important quantity because it is invariant under conformal mappings and quasiinvariant under quasiconformal mappings, in the sense to be described precisely in the next proposition. First, we define the set of allowable metrics. A metric $\rho(z)|dz|$ is allowable if

(i) it is invariantly defined for different local parameters z, i.e., $\rho_1(z_1)|dz_1| = \rho_2(z_2)|dz_2|$, where ρ_1 and ρ_2 are representatives for ρ in terms of the parameters z_1 and z_2;

(ii) ρ is measurable and $\geqslant 0$ everywhere; and

(iii) $A(\rho) = \iint \rho^2\, dx\, dy \neq 0$ or ∞ (the integral is taken over the whole Riemann surface).

For such an allowable ρ, define

$$L_\gamma(\rho) = \int_\gamma \rho|dz|$$

if ρ is measurable along γ; otherwise define $L_\gamma(\rho) = +\infty$. Let $L(\rho) = \inf L_\gamma(\rho)$, where the infimum is over all curves γ in F. The extremal length of the curve family F is

$$\Lambda(F) = \sup_\rho \frac{L(\rho)^2}{A(\rho)}, \tag{27}$$

where the supremum is taken over all allowable metrics. Notice that the ratio in this supremum is invariant if ρ is multiplied by a positive scalar. Thus, in attempting to evaluate $\Lambda(F)$ we may normalize so that $L(\rho) = 1$ and try to make $A(\rho)$ as small as possible.

Example. Suppose R is the interior of a rectangle $\{z\colon 0 \leqslant x \leqslant a, 0 \leqslant y \leqslant b\}$ and F is the family of arcs in R which join the left vertical side of R to the right vertical side. Then $\Lambda(F) = a/b$.

To see this let $\rho \equiv 1$ in R. Then $L(\rho) = a$ and $A(\rho) = ab$ and we obtain $\Lambda(F) \geqslant a/b$. On the other hand, if ρ is any allowable metric on R, by multiplying ρ by a suitable scalar, we can make $L(\rho) = a$. Thus,

$$a \leqslant \int_0^a \rho(x + iy)\, dx$$

for every y with $0 \leqslant y \leqslant b$. Integration over y and an application of Schwarz's inequality yield

$$ab \leqslant \iint_R \rho\, dx\, dy, \qquad (ab)^2 \leqslant ab \iint_R \rho^2\, dx\, dy.$$

$\langle \rho, g\rangle = \iint \rho g\, dx dy$

$(ab)^2 \leq \|\rho\|^2 = |\langle \rho, \rho\rangle| \leq \iint \rho\cdot 1\, dx dy = \iint 1\, dx dy \cdot \iint \rho^2\, dx dy$

Thus,

$$\frac{L(\rho)^2}{A(\rho)} \leqslant \frac{a^2}{ab} = \frac{a}{b}.$$

Notice that if F^t is the family of arcs in R which join the bottom of the rectangle R to its top, then $\Lambda(F^t) = b/a$ and $\Lambda(F^t)\Lambda(F) = 1$.

Example. Let $R = \{z\colon r_1 < |z| < r_2\}$ and let F be the family of closed curves in R which are homotopic to the core curve γ_1, where $\gamma_1(\theta) = e^{i\theta}(r_1 + r_2)/2$, $0 \leqslant \theta \leqslant 2\pi$. Let $\rho_0(z) = (2\pi|z|)^{-1}$ and $\gamma_r(\theta) = re^{i\theta}$. For any curve γ homotopic to γ_1, we must have

$$1 = (2\pi i)^{-1} \int_\gamma \frac{dz}{z}.$$

Thus $1 \leqslant \int_\gamma \rho_0(z)|dz|$ and $1 \leqslant L(\rho_0)$. Moreover,

$$A(\rho_0) = \iint_R \frac{r\,dr\,d\theta}{(2\pi)^2 r^2} = \frac{1}{2\pi}\log(r_2/r_1)$$

and, therefore, $\Lambda(F) \geqslant 2\pi(\log(r_2/r_1))^{-1}$.

On the other hand, for any allowable metric $\rho|dz|$,

$$L(\rho) \leqslant \int_0^{2\pi} \rho(re^{i\theta})r\,d\theta,$$

$$\frac{L(\rho)}{r} \leqslant \int_0^{2\pi} \rho(re^{i\theta})\,d\theta,$$

$$L(\rho)\log(r_2/r_1) \leqslant \iint_R \rho\,dr\,d\theta,$$

$$(L(\rho))^2(\log(r_2/r_1))^2 \leqslant \iint \frac{1}{r}\,dr\,d\theta \iint \rho^2 r\,dr\,d\theta,$$

$$\frac{L(\rho)^2}{A(\rho)} \leqslant 2\pi(\log(r_2/r_1))^{-1}.$$

By similar methods one can show that if F^t is the curve family whose elements join the two boundary components of R, then $\Lambda(F^t) = (2\pi)^{-1}\log(r_2/r_1)$.

Remark. Notice that in both examples there is an extremal metric for which the supremum in (1) is achieved. One can say more. In both examples the extremal metric ρ_0 is of the form $\rho_0(z)\,|dz| = |\varphi_0(z)|^{1/2}\,|dz|$, where $\varphi_0(z)$ is a holomorphic function. Later we will see that it is natural to consider the expression $\varphi_0(z)\,dz^2$, which we call a holomorphic quadratic differential. For the rectangle $\varphi_0(z)\,dz^2 = dz^2$ and for the annulus $\varphi_0(z)\,dz^2 = z^{-2}\,dz^2$. In Chapter 11 it is shown that for a general extremal problem associated with a measured foliation, the infinitesimal variation of the extremal length is realized by the same quadratic differential which realizes the foliation. This phenomenon is illustrated for the case of an annulus in Proposition 4 below. First we need

Proposition 3. *Suppose f is a quasiconformal mapping with dilatation K taking a Riemann surface R onto a Riemann surface R′ and a curve family F onto the curve family F′. Then* $K^{-1}\Lambda(F) \leqslant \Lambda(F') \leqslant K\Lambda(F)$.

Proof. Let w be a local parameter on R' and z a local parameter on R and assume the mapping f takes z into w. For a given allowable metric ρ on R let the metric $\tilde{\rho}$ on R' be defined by

$$\tilde{\rho}(w) = \left|\frac{\rho(z)}{|w_z| - |w_{\bar{z}}|}\right| \circ f^{-1}(w).$$

We leave it to the reader to verify that $\tilde{\rho}$ is a metric on R'. Then for $\gamma' = w(\gamma)$, and $\zeta = \xi + i\eta = w(z)$, we have

$$\int_{\gamma'} \tilde{\rho}(\zeta)\,|d\zeta| \geqslant \int_{\gamma} \rho(z)\,|dz|$$

$$\iint_{R'} \tilde{\rho}(\zeta)^2\,d\xi\,d\eta = \iint \rho(z)^2\,\frac{|w_z| + |w_{\bar{z}}|}{|w_z| - |w_{\bar{z}}|}\,dx\,dy \leqslant KA(\rho).$$

This proves $\Lambda(F) \leqslant K\Lambda(F')$. The other inequality follows by applying the same argument to w^{-1}.

To extend this result to arbitrary quasiconformal mappings, one can use the analytic definition of quasiconformality together with existence of locally L_2-derivatives w_z and $w_{\bar{z}}$, in the distributional sense. For details we refer either to Lehto and Virtanen [LehtV] or Ahlfors [Ah4].

There is another way in which the extremal metrics for extremal length problems appear. To illustrate, consider the curve family F of arcs which join the two boundary components of the annulus $\{z\colon 1 < |z| < r\}$. Let μ be an L_∞-function on the annulus with $\|\mu\|_\infty < 1$ and w^μ a quasiconformal mapping from R onto a doubly connected domain R^μ. The previous proposition tells us that for $F^\mu = w^\mu(F)$ the extremal lengths satisfy $K^{-1}\Lambda(F) \leqslant \Lambda(F^\mu) \leqslant K\Lambda(F)$.

Proposition 4. *The extremal length of the annulus* $\Lambda(F^{\mu})$ *is a differentiable function of* μ *and, for small real numbers* t,

$$\log \Lambda(F^{t\mu}) = \log \Lambda(F) + 2t \operatorname{Re} \iint \mu(z)\varphi(z)\, dx\, dy + o(t), \tag{28}$$

where $\varphi(z) = z^{-2}(2\pi \log r)^{-1}$.

Remark. The constant factor $(2\pi \log r)^{-1}$ in $\varphi(z)$ is selected so that

$$\iint_{1<|z|<r} |\varphi(z)|\, dx\, dy = 1.$$

Proof of the proposition. The correspondence $z = \exp(\zeta)$ determines a mapping from the rectangle $\{\zeta : 0 \leqslant \xi \leqslant \log r, 0 \leqslant \eta \leqslant 2\pi\}$ onto the annulus $\{z: 1 < |z| < r\}$. Without changing the Beltrami coefficient of the mapping $w = w^{\mu}$, we can assume by uniformization that w maps onto an annulus $\{w: 1 < |w| < r_1\}$. The correspondence $w = \exp(f)$ determines a mapping from the rectangle $\{f : 0 \leqslant \operatorname{Re} f \leqslant \log r_1,\ \ 0 \leqslant \operatorname{Im} f \leqslant 2\pi\}$, onto the annulus $\{w: 1 < |w| < r_1\}$.

Let γ be a horizontal segment connecting the two vertical sides of the ζ-rectangle. Then $f(\gamma)$ is a path connecting $\operatorname{Re} f = 0$ to $\operatorname{Re} f = \log r_1$ and, therefore,

$$\log r_1 \leqslant \int_{f(\gamma)} |df| = \int_0^{\log r} |f_\zeta|\,|1 + \tilde{\mu}|\, d\xi,$$

where $\tilde{\mu} = f_{\bar{\zeta}}/f_\zeta$ and $\zeta = \xi + i\eta$. Integrating from $\eta = 0$ to $\eta = 2\pi$, one obtains

$$2\pi \log r_1 \leqslant \iint |f_\zeta|\,|1 + \tilde{\mu}|\, d\xi\, d\eta,$$

where the integral is over the rectangle in the ζ-plane. Introducing a factor of $\sqrt{1 - |\mu|^2}$ in both the numerator and the denominator and applying Schwarz's inequality yields

$$(2\pi \log r_1)^2 \leqslant \iint |f_\zeta|^2 (1 - |\tilde{\mu}|^2)\, d\xi\, d\eta \iint \frac{|1 + \tilde{\mu}|^2}{1 - |\tilde{\mu}|^2}\, d\xi\, d\eta.$$

Since the first integral on the right-hand side is the area of the rectangle in the f-plane, we get

$$2\pi \log r_1 \leqslant \iint \frac{|1 + \tilde{\mu}|^2}{1 - |\tilde{\mu}|^2}\, d\xi\, d\eta.$$

Since $z = e^{\zeta}$, we have $\tilde{\mu} = \mu z/\bar{z}$, where $\mu = f_{\bar{z}}/f_z$. If we set $\varphi(z) = z^{-2}(2\pi \log r)^{-1}$, the last inequality can be rewritten as

$$\frac{\Lambda(F^{\mu})}{\Lambda(F)} \leqslant \iint\limits_{1<|z|<r} \frac{|1 + \mu\varphi/|\varphi||^2}{1 - |\mu|^2} |\varphi| \, dx \, dy. \tag{29}$$

By applying the same argument to a vertical segment connecting the two horizontal sides of the ζ-rectangle, one obtains

$$\frac{\Lambda(F)}{\Lambda(F^{\mu})} \leqslant \iint\limits_{1<|z|<r} \frac{|1 - \mu\varphi/|\varphi||^2}{1 - |\mu|^2} |\varphi| \, dx \, dy. \tag{30}$$

The infinitesimal formula in Proposition 4 is a straightforward consequence of (29) and (30).

1.10. GRÖTZSCH'S PROBLEM FOR AN ANNULUS

If A_1 is the annulus $\{z: 1 < |z| < r_1\}$ and A_2 is the annulus $\{z: 1 < |z| < r_2\}$, then there is no conformal map from A_1 onto A_2 unless $r_1 = r_2$. This can be proved using the conformal invariance of the extremal length of the curve family which joins the inner contour to the outer contour; it is $(\log r_1)/2\pi$ for A_1 and $(\log r_2)/2\pi$ for A_2.

Grötzsch's problem is to find the most nearly conformal mapping from A_1 to A_2 [Gr1, Gr2]. There is some question about what is the best way to define "most nearly conformal." One way is to ask for a mapping with the smallest possible maximal dilatation. A second possibility is to ask for a mapping with the minimum average dilatation, although this approach will not be important to us. In either case, the problem has a solution which is unique up to postcomposition by conformal self-mappings of A_2.

To prove this we let K be the ratio of the extremal lengths of the annuli A_1 and A_2, that is, $K = (\log r_2)/(\log r_1)$ and we let f_K be the mapping

$$f_K(re^{i\theta}) = r^K e^{i\theta}.$$

To calculate the Beltrami coefficient of f_K, use the notation $re^{i\theta} = z = \exp(\zeta)$ and $w = \exp(f)$. Then $w = K\xi + i\eta$ and $w_{\bar{\zeta}}/w_{\zeta} = k = (K-1)/(K+1)$. Thus, $(f_K)_{\bar{z}}/(f_K)_z = kz/\bar{z}$.

Suppose f is an arbitrary quasiconformal mapping taking the annulus A_1 onto the annulus $A_2 = f_K(A_1)$. First assume that $K \geqslant 1$ or, equivalently, that $k \geqslant 0$. From inequality (29) of the previous section we obtain for $\varphi(z) = z^{-2}(2\pi \log r_1)^{-1}$,

$$K \leqslant \iint\limits_{1<|z|<r_1} \frac{|1 + \mu\varphi/|\varphi||^2}{1 - |\mu|^2} |\varphi| \, dx \, dy \leqslant \frac{1 + \|\mu\|_\infty}{1 - \|\mu\|_\infty}. \tag{31}$$

Note $\iint |\varphi| \, dx dy = 1$ with change of variable.

Hence, $\|\mu\|_\infty \geqslant k$ with equality only if $\mu = k|\varphi|/\varphi = kz/\bar{z}$ almost everywhere.

Next assume $K < 1$ or, equivalently, $k < 0$. From inequality (30), we obtain

$$\frac{1+|k|}{1-|k|} = \frac{1}{K} \leqslant \iint\limits_{1<|z|<r} \frac{|1 - \mu\varphi/|\varphi||^2}{1 - |\mu|^2} |\varphi| \, dx \, dy = \frac{1 + \|\mu\|_\infty}{1 - \|\mu\|_\infty}. \tag{32}$$

Once again, it follows that $\|\mu\|_\infty \geqslant |k|$ with equality only if $\mu = k|\varphi|/\varphi$ almost everywhere.

This shows that the mapping $f_K: A_1 \to A_2$ is extremal in the sense that it has the smallest possible maximal dilation of any quasiconformal mapping from A_1 to A_2. Moreover, if another quasiconformal mapping from A_1 to A_2 has the same maximal dilatation as f_K, then it must have the same Beltrami coefficient. In other words, f_K is uniquely extremal up to postcomposition by conformal self-mappings of A_2.

Remark. Our proof of Teichmüller's uniqueness theorem for a compact Riemann surface will follow exactly the same outline and depends on establishing an inequality of the same type as (31). There are two differences. The first is that the integration is over a surface R instead of the annulus. The second is that the quadratic differential φ can be any holomorphic quadratic differential with

$$\iint\limits_R |\varphi(z)| \, dx \, dy = 1$$

1.11. THE DIMENSION OF THE SPACE OF QUADRATIC DIFFERENTIALS

We need to know the dimension of the space of holomorphic quadratic differentials on a Riemann surface R. A holomorphic quadratic differential φ is an assignment of a holomorphic function $\varphi_1(z_1)$ to each local coordinate z_1 such that if z_2 is another local coordinate, then $\varphi_1(z_1) = \varphi_2(z_2)(dz_2/dz_1)^2$. Along a border arc α in the border of R we require that $\varphi_1(\alpha)$ be real if z_1 is a local coordinate taking real values along α. If there is a point p in the boundary of R isolated from other boundary points of R (such that by adding the point p to R, p becomes an interior point of disk), then we require that φ have at most a simple pole at p. We do not permit poles of φ to occur on the border curves of R. Denote the vector space of holomorphic quadratic differentials φ satisfying these properties by $A(R)$.

Suppose the genus of R is g and R is obtained from a compact surface by deleting m disjoint closed disks and n isolated points not lying on these disks.

Theorem 6.

$$\dim_{\mathbb{R}} A(R) = 6g - 6 + 3m + 2n$$

except in cases shown in the following table:

g	m	n	$\dim_{\mathbb{R}} A(R)$
0	0	0, 1, 2	0
0	1	0, 1	0
1	0	0	2

In the case where $m = 0$, $A(R)$ is a complex vector space and the same formula yields

$$\dim_{\mathbb{C}} A(R) = 3g - 3 + n$$

for any $g \geqslant 2$, for $g = 1$ and $n \geqslant 1$, and for $g = 0$ and $n \geqslant 3$.

This theorem is most easily deduced from the Riemann–Roch theorem, although it can be viewed as a consequence of Teichmüller's theorem and any of the various techniques which give the dimension of the space of deformations of a Riemann surface. We omit the proof. It can be found in Farkas and Kra [FarK].

Notes

It is too difficult even to begin to give complete bibliographical notes for the material presented in this chapter. Partial references are given for most of the theorems within the text. For more complete references we refer to the books listed in the bibliography [Ah4, Ah6], [AhS], [Bea], [FarK], [Kr5], [Leht], [Ts], [We]. I would like also to mention the unpublished notes by Bers [Ber1] and Macbeath [Mac1] which give more complete introductions to much of the basic material. The question of how to construct a Fuchsian group by prescribing transformations identifying the sides of a fundamental domain is treated by many authors. We point out the papers of Keen [Ke1] and Maskit [Mask].

In the case where Γ is a Fuchsian group with elliptic elements, the Riemann surface $R = \mathbb{H}/\Gamma$ has additional structure at the points of which are images of elliptic fixed points in $\mathbb{H}$. This structure is called an orbifold structure by Thurston [Th1] and generalizes to Kleinian groups.

Early results on the existence of the Poincaré metric are found in the work of Poincaré [Po1] and [Po2].

EXERCISES

SECTION 1.3

1. Find the global form of the infinitesimal metric $|z|^{-1}|dz|$ on the domain $\mathbb{C} - \{0\}$. Show that this metric is complete. *Hint:* $\mathbb{C} - \{0\}$ is isomorphic to $\mathbb{C}/(z \to z + 2\pi i n)$ and the covering mapping is $z = e^w$.

2. Show that the mapping (11) preserves the unit disk. Then, using Schwarz's lemma, show that it is the most general form of bianalytic self-mapping of the unit disk.
3. Verify the identity (10).

SECTION 1.4

4. For $A(z) = (az + b)/(cz + d)$ with $ad - bc \neq 0$, we say that A is the linear fractional transformation corresponding to the matrix

$$\begin{pmatrix} a & b \\ c & d \end{pmatrix}.$$

Let $B(z) = (\alpha z + \beta)/(\gamma z + \delta)$ with $\alpha\delta - \beta\gamma \neq 0$. Show that the composed mapping $A(B(z))$ is the linear fractional transformation corresponding to the matrix product of

$$\begin{pmatrix} a & b \\ c & d \end{pmatrix} \quad \text{and} \quad \begin{pmatrix} \alpha & \beta \\ \gamma & \delta \end{pmatrix}.$$

5. Show that for a Fuchsian group the isotropy subgroup fixing any point in $\mathbb{H}$ must be cyclic and finite.
6. Show that an arbitrary Möbius transformation which preserves the unit disk can be written in the form $A(z) = (\alpha z + \beta)/(\bar{\beta} z + \bar{\alpha})$, where α and β are complex numbers satisfying $|\alpha|^2 - |\beta|^2 = 1$.

SECTION 1.5

7. If A is hyperbolic with fixed points at a and b, the semicircle orthogonal to $\mathbb{R}$ with endpoints at a and b is called the axis of A. Show that A leaves the axis of A invariant. Show that the hyperbolic distance from a point p to $A(p)$ is $\frac{1}{2}\log\lambda$, where λ is the multiplier for A. Show that if q is not on the axis of A, then $d(q, A(q)) > \frac{1}{2}\log\lambda$.
8. If an elliptic element belongs to a Fuchsian group, show that at a fixed point it rotates through an angle which is a rational multiple of π.
9. Show that if A is elliptic, then $(\operatorname{tr} A)^2 = 2(1 + \cos\theta)$, where θ is the angle of rotation. Show that $A^2 = I$ and $A \neq I$ if and only if $(\operatorname{tr} A)^2 = 0$.
10. Consider the action of a Fuchsian group Γ acting on the upper half plane $\mathbb{H}$. Show that a point p in $\mathbb{H}$ is elliptic for Γ in the sense of the definition given in Section 1.4 if and only if there is an elliptic element A in $\Gamma - \{I\}$ with fixed point at p.

SECTION 1.6

11. Let A be a hyperbolic Möbius transformation of the unit disk. Show that the isometric circles $I(A)$ and $I(A^{-1})$ contain the attracting and repelling fixed points of A, respectively. Show that $I(A)$ and $I(A^{-1})$ do not intersect if and only if A is hyperbolic. Show that $I(A)$ and $I(A^{-1})$ intersect the axis of A at right angles.

12. Show that a point p is in the half plane $h(A^{-1})$ defined in Section 1.6 if, and only if, $A(p)$ is not in the closure of $h(A)$.

13. Show that no two points of the Dirichlet domain defined in (20) can be equivalent under the action of Γ. Show that the Dirichlet domain is open and that every point in $\mathbb{H}$ is Γ-equivalent to at least one point of the closure of the Dirichlet domain.

14. Show that the limit set of a Fuchsian group Γ must be a subset of the real axis. Show that the limit set is a closed set invariant under the action of elements of Γ.

15. Find the noneuclidean area of the noneuclidean triangle bounded by the y-axis ($y \geqslant 1$), the unit circle $|z| = 1$, and the vertical line whose x-coordinate is x_0 with $0 \leqslant x_0 \leqslant 1$.

SECTION 1.8

16. Let f and g be C^1 orientation-preserving homeomorphisms of a domain onto itself which have Beltrami coefficients μ and ν, respectively. Show that the Beltrami coefficient of $g \circ f$ is

$$\frac{\mu(z) + \nu(f(z))\theta_\mu(z)}{1 + \bar{\mu}(z)\nu(f(z))\theta_\mu(z)},$$

where $\theta_\mu(z) = \bar{p}/p$ and $p = f_z$. *Hint:* Use the chain rule in complex form: for $w = f(z)$

$$(g \circ f)_z = g_w f_z + g_{\bar{w}} \bar{f}_z \quad \text{and} \quad (g \circ f)_{\bar{z}} = g_w f_{\bar{z}} + g_{\bar{w}} \bar{f}_{\bar{z}}.$$

Remark. This exercise reveals the similarity between transformations of the unit ball of Beltrami coefficients and hyperbolic isometries of the unit disk of the form $z \to (\alpha + z)/(1 + \bar{\alpha}z)$, where $|\alpha| < 1$. Note that the Beltrami coefficient of $g \circ f$ depends holomorphically on the Beltrami coefficient of g.

17. Let $\mu \in M(\mathbb{H})$ and f be a quasiconformal self-mapping of $\mathbb{H}$ with Beltrami coefficient μ. Consider the mapping R_μ: $M(\mathbb{H}) \to M(\mathbb{H})$ defined by

$$R_\mu(\nu) = \frac{\mu(z) + \nu(f(z))\theta(z)}{1 + \overline{\mu(z)}\nu(f(z))\theta(z)},$$

where $\theta = \theta_\mu$ is defined in the previous exercise. Show that R_μ is a one-to-one mapping of $M(\mathbb{H})$ onto $M(\mathbb{H})$. Calculate the complex derivative

$$\lim_{t \to 0} t^{-1}[R_\mu(tv) - R_\mu(0)]$$

for small complex numbers t.

Remark. The complex differentiability of the mappings R_μ provides the approach for introducing the natural complex structure for Teichmüller space.

18. If g has Beltrami coefficient $v(z)$, use the result of Exercise 16 to find the Beltrami coefficient of g^{-1}.

SECTION 1.10

19. Let $f_K(z) = r^K e^{i\theta}$ map the annulus $A_1 = \{z\colon 1 < |z| < r_2\}$. Assume $K \geqslant 1$ and let $\varphi(z) = z^{-2}(2\pi \log r_1)^{-1}$. For any other quasiconformal mapping f from A_1 onto A_2 with Beltrami coefficient μ, define

$$\operatorname{AvgDil}(f, \varphi) = \iint_{1<|z|<r_1} \frac{|1 + \mu\varphi/|\varphi||^2}{1 - |\mu|^2} |\varphi| \, dx\, dy.$$

Prove that $\operatorname{AvgDil}(f_K, \varphi) \leqslant \operatorname{AvgDil}(f, \varphi)$ with equality only if $f_K = c \circ f$ for some conformal self-mapping c of A_2.

SECTION 1.11

20. Let D be a plane domain. Given any holomorphic function $g(z)$ defined on D, the expression $g(z)\, dz^2$ is a holomorphic quadratic differential. Conversely, show that any holomorphic quadratic differential on D is of the form $g(z)\, dz^2$, where the $g(z)$ is a globally defined holomorphic function on D.

2

MINIMAL NORM PROPERTIES FOR HOLOMORPHIC QUADRATIC DIFFERENTIALS

A nonzero holomorphic quadratic differential imposes on its underlying Riemann surface a geometric structure of a very special type. Such a differential has horizontal and vertical trajectories which naturally give a way of cutting up the surface into Euclidean rectangles. The coordinate function on each rectangle maps horizontal trajectories into horizontal lines in the complex plane, and the transition functions are compositions of rotations by 180° and translations. At the zeroes of the quadratic differential there are singular points where more than two rectangles are sewn together along sides which meet at a vertex with three or more prongs.

In this chapter we shall be concerned with holomorphic quadratic differentials of finite norm on a Riemann surface which is compact except for a finite number of punctures. Such a surface is called a surface of finite analytic type. A surface which has holes is not considered to be of finite analytic type.

For such surfaces we prove two minimal norm properties for holomorphic quadratic differentials φ of finite norm. The first minimal norm property says that φ has minimal norm compared to all measurable quadratic differentials satisfying side conditions which depend on the comparing heights along all vertical trajectories of φ. The second minimal norm property says that φ has minimal norm compared to all continuous quadratic differentials satisfying side conditions which depend on comparing heights along homotopy classes of simple closed curves on the surface.

The proofs of both minimal norm properties are based on two ideas. The first is to analyze the local Euclidean structure induced by the holomorphic quadratic differential on the Riemann surface away from the singular points. The second is to apply the length–area method together with the fact that a curve which is transversal to the horizontal trajectories has minimum height among homotopic curves with fixed endpoints.

We will show that these minimal norm properties lead to a general extremal length inequality discovered by Reich and Strebel. In later chapters we will see that the minimal norm properties explain the appearance of integrable holomorphic quadratic differentials in the following:

(1) Teichmüller's theorem.
(2) The Hamilton–Krushkal condition for a quasiconformal mapping to be extremal.
(3) The infinitesimal form of Teichmüller's metric.
(4) A theorem which shows how they arise as extremal metrics in a broad class of extremal length problems.
(5) The bijective correspondence between holomorphic quadratic differentials on a given Riemann surface and the space of measure classes of measured foliations.

For the proof of the first minimal norm property we are able to bypass some of the detailed theory of trajectories by means of an averaging device used by Teichmüller. This minimal norm property will be enough to lead us to (1), (2), and (3) above. To obtain (4) and (5) we need the second minimal norm property, which depends on the more detailed description of trajectories.

2.1. TRAJECTORIES OF QUADRATIC DIFFERENTIALS

Throughout this chapter, we assume R is a Riemann surface obtained from a compact surface by removing a finite number of points, that is, a surface of finite analytic type. The removed points are called punctures. We let $\bar{R}$ be the surface R with the punctures filled in. Thus $\bar{R}$ is compact. If g is the genus of R and n is the number of punctures, we assume $3g - 3 + n > 0$, except that when $g = 1$ we allow n to be zero. This is enough to make the vector space $A(R)$ of integrable holomorphic quadratic differentials on R have positive dimension.

For a holomorphic quadratic differential φ the order of a zero of φ at p_0 is the exponent of the first nonvanishing term in the Taylor series expansion for φ centered at p_0 in terms of any local parameter. If z_1 and z_2 are local parameters and φ_1 and φ_2 are expressions for φ in terms of these local parameters valid in neighborhoods N_1 and N_2, both of which contain the point z_0, then the equation

$$\varphi_1(z_1) = \varphi_2(z_2)\left(\frac{dz_2}{dz_1}\right)^2$$

shows that the order of a zero of φ does not depend on the choice of local parameter. The point is that $dz_2/dz_1 \neq 0$ anywhere in the overlap $N_1 \cap N_2$ of the two neighborhoods N_1 and N_2.

The quadratic differential φ has a zero at p_0 if the order of φ at p_0 is one or more.

The quadratic differential is said to be meromorphic if it is holomorphic except at isolated points where it has poles. The order of a pole at p_0 is the exponent of the leading term in the Laurent expansion for φ centered at p_0 in terms of any local parameter. Just as in the case of defining the order of a zero, the order of a pole does not depend on the choice of local parameter. Sometimes, instead of saying φ has a pole or order k, we will say φ has a zero of order $-k$. If the pole is of order one, we call it a simple pole.

We call a point p_0 a critical point of the quadratic differential φ if it is either a zero or a pole of φ. Elements of the vector space $A(R)$ are, by definition, holomorphic on R. They can have poles at punctures of R, but the poles can only be simple. Note that the punctures of R are the points of $\bar{R} - R$. A point p_0 which is not a critical point of φ is called regular.

Let p_0 be a point on $\bar{R}$ and z a local parameter with $z(p_0) = z_0$. Let $\varphi(z)$ be the functional expression for φ in terms of the local parameter z. Assume p_0 is a regular point for φ. We obtain a special kind of local parameter ζ, called a natural parameter, by letting

$$\zeta = \int_{z_0}^{z} \sqrt{\varphi(z)}\, dz. \tag{1}$$

It is clear that if $\zeta_1(z_1(p))$ and $\zeta_2(z_2(p))$ are two natural parameters coming from φ and defined in overlapping coordinate patches U_1 and U_2, then

$$\zeta_1(z_1(p)) = \pm\zeta_2(z_2(p)) + (\text{const.}) \tag{2}$$

for p in $U_1 \cap U_2$.

Notice that $d\zeta^2 = \varphi(z)\, dz^2$ for any natural parameter ζ associated with φ. A parametric curve $\gamma\colon I \to S$ is called a horizontal (vertical) trajectory of φ if, given any local coordinate z defined in a patch overlapping the image of γ, the function $z(\gamma(t))$ satisfies $\varphi(\gamma(t))\gamma'(t)^2 > 0\,(<0)$. This means that in the ζ-plane, where ζ is a natural parameter, the curve $\gamma(t)$ is transformed into a horizontal (vertical) line. Clearly, this notion is independent of the choice of local coordinate. In fact, for any two different natural parameters, the transition mapping is of the form (2) and it is a transition which preserves horizontal and vertical lines.

In an obvious sense, the horizontal and vertical trajectories of φ give two transverse foliations in $\bar{R}$ in a neighborhood of any nonsingular point of φ. With a slight extension of the notion of transversality, we can also include the singular points. Let φ have a zero of order m at p in R. At any such point there will exist a local coordinate z with $z(p) = 0$ such that $\varphi(z)\, dz^2$ takes the form $z^m\, dz^2$. Let $d\zeta = z^{m/2}\, dz$. Although for odd integers m, ζ is not a single valued function of z,

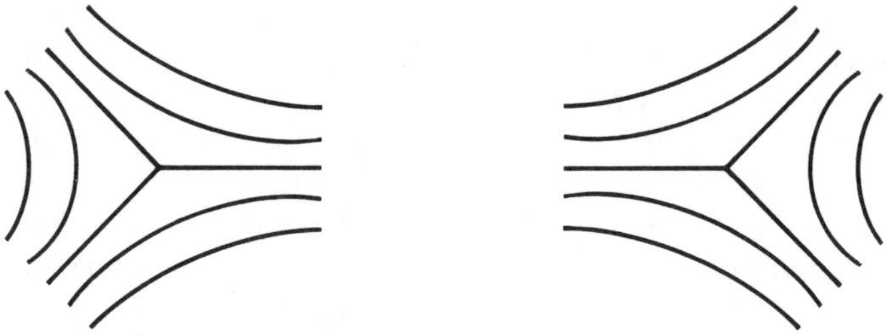

Figure 2.1

for any integer $m \neq -2$ a radial line $t\omega$ (where $t \geqslant 0$ and $|\omega| = 1$), emanating from the origin in the z-plane, will be horizontal if $(t\omega)^m\omega^2 > 0$, that is, if $\omega^{m+2} = 1$. In order for φ to have finite norm, m must always be larger than or equal to -1. The only points where m can be negative are at the punctures because we require the quadratic differential to be holomorphic on the surface R. For the case where $m = 1$, the trajectories in the z-plane have the appearance shown in Figure 2.1.

The directions of the vertical trajectories emanating from the origin come from the equation $\omega^{m+2} = -1$. We postpone the formal definition of measured foliations until Chapter 11. It turns out that the horizontal and vertical trajectories together with the quadratic differential determine two measured foliations which are transversal. Two foliations are transversal at a singular point if they have a C^1-topological structures equivalent to the horizontal and vertical trajectories of $z^m\,dz^2$ for some integer $m \geqslant -1$ and in some neighborhood of the origin in the z-plane.

2.2 INVARIANTS FOR QUADRATIC DIFFERENTIALS

Any nonconstant, holomorphic quadratic differential φ on R carries with it several invariants. First of all, there is the area element

$$dA_\varphi = |\varphi(z)|\,dx\,dy = d\xi\,d\eta,$$

where $z = x + iy$ is any local parameter and where $\zeta = \xi + i\eta$ is any natural parameter. From this, one obtains the norm of φ by letting

$$\|\varphi\| = \iint_R |\varphi(z)|\,dx\,dy.$$

Of course, the area element and the norm are defined even for quadratic differentials that are not holomorphic. If $\|\varphi\| < \infty$ and φ is holomorphic on R, then it is elementary (by switching to polar coordinates) to see that φ can have at most simple poles at the punctures in $\bar{R} - R$. Conversely, if R is of finite analytic type and φ is holomorphic except for at most simple poles at the punctures in $\bar{R} - R$, then $\|\varphi\| < \infty$.

The expression $ds_\varphi = |\varphi|^{1/2}|dz|$ is a line element. The φ-length of a piecewise differentiable arc γ in R is $\ell_\varphi(\gamma) = \int_\gamma ds_\varphi$. Away from the singularities of φ and in terms of a natural parameter $\zeta = \xi + i\eta$, one has $ds_\varphi^2 = d\xi^2 + d\eta^2$, so local geodesics away from singularities are just straight line segments in the ζ-plane. However, at singularities geodesics can have vertices. Although the curvature of the metric ds_φ is not defined at singular points, in an intuitive sense the points where φ is zero contribute to negative curvature. Since we will consider the trajectory structure of quadratic differentials which are holomorphic in R, there is a negative-curvature-like property which forces global geodesics to be unique ([Ab], [Ah1]). In our approach it is unnecessary to consider the notion of curvature or to define what is meant by a geodesic. However, the fact that the quadratic differential φ is holomorphic on R will enter in an essential way.

For our purposes, the most important notion is the height of a curve. By definition, if γ is a differentiable curve on R, its height with respect to φ is given by

$$h_\varphi(\gamma) = \int_\gamma |\mathrm{Im}(\sqrt{\varphi}(z)\, dz)|. \tag{3}$$

Similarly, its width is given by

$$w_\varphi(\gamma) = \int_\gamma |\mathrm{Re}(\sqrt{\varphi}(z)\, dz)|.$$

Obviously the φ-length of a curve is greater than or equal to its width and its height.

We call a trajectory of φ on $\bar{R}$ critical if, when it is continued in either direction, it meets a zero or a pole of φ. Let b_φ be the subset of $\bar{R}$ which consists of the union of all critical vertical trajectories and any trajectory which meets a puncture of the surface. Since there are a finite number of singularities of φ and finitely many punctures on the surface, b_φ consists of finitely many smooth images of intervals, and therefore b_φ has measure zero. (In the generic case, b_φ is a dense subset of $\bar{R}$.)

A vertical segment is the continuous image of an interval which lies on a vertical trajectory. A vertical segment is called regular (or sometimes noncritical) if it does not meet a critical point.

2.3. THE FIRST MINIMAL NORM PROPERTY

An element of $A(R)$, that is, a holomorphic quadratic differential of finite norm, satisfies a minimum norm property subject to side conditions determined by the differential itself. The side conditions are conditions on its heights, and the minimality is relative to quadratic differentials ψ which are not necessarily holomorphic. Such a quadratic differential ψ is merely an assignment of a function ψ^z to each local coordinate z such that $\psi^z\, dz^2 = \psi^\zeta\, d\zeta^2$ in overlapping coordinate neighborhoods.

For the purposes of the first minimal norm property, we assume that ψ is a locally L_1 function with respect to any local parameter. Let $\psi^\zeta = \psi(\xi, \eta)$ be the expression for ψ in terms of a natural parameter $\zeta = \xi + i\eta$. Since $\psi(\xi, \eta)$ is integrable in any coordinate patch, by Fubini's theorem, for almost every ξ, $\psi(\xi, *)$ is an integrable function of η. Thus for almost every vertical segment β the integral

$$h_\psi(\beta) = \int_\beta |\mathrm{Im}(\sqrt{\psi(z)}\, dz)|$$

is well defined. This integral is an unoriented line integral, taken in the positive sense regardless of what orientation is given to β.

Theorem 1 (The first Minimal Norm Property). *Let R be a Riemann surface of finite analytic type. Assume φ is a holomorphic quadratic differential on R with $\|\varphi\| = \iint_R |\varphi|\, dx\, dy < \infty$. Let ψ be another quadratic differential which is locally integrable. Assume there is a constant M such that for almost every noncritical vertical segment β, one has $h_\varphi(\beta) \leqslant h_\psi(\beta) + M$. Then*

$$\|\varphi\| \leqslant \iint_R |\sqrt{\varphi(z)}|\, |\sqrt{\psi(z)}|\, dx\, dy. \tag{4}$$

Proof. Because R is of finite analytic type, every noncritical vertical trajectory of φ can be continued infinitely in both directions. When we say this, we do not exclude the possibility that the trajectory may be closed. The verification of this assertion is given at the beginning of Section 2.5.

We need the following lemma.

Lemma 1. *Let g be a nonnegative function on R and let g be integrable with respect to the area element $|\varphi(z)|\, dx\, dy$ induced by the holomorphic quadratic differential φ. Let $\|\varphi\| < \infty$ and ζ be a natural parameter for φ. Then for real numbers τ the function $h(\zeta) = g(\zeta + i\tau) + g(\zeta - i\tau)$ is well defined on $R - b_\varphi$ and*

$$\iint_R h(\zeta)\, d\xi\, d\eta = 2 \iint_R g(\zeta)\, d\xi\, d\eta. \tag{5}$$

Proof. First note that integrating over R and integrating over $R - b_\varphi$ are equivalent since b_φ is a set of measure zero. Since noncritical trajectories can be continued infinitely in both directions, after choice of orientation, the expression

$g(\zeta + i\tau)$ is well defined for ζ on any particular noncritical trajectory. The idea behind the lemma is that locally the mapping $\zeta \to \zeta + i\tau$ is well defined and has Jacobian identically equal to one. Therefore, integration of $g(\zeta)\,d\xi\,d\eta$ over a coordinate patch near $\zeta + i\tau$ gives the same value as the integration of $g(\zeta + i\tau)\,d\xi\,d\eta$ over a translated coordinate patch near ζ. In order to piece together this argument to make the global statement (5), we consider the Riemann surface $\tilde{R}$ which is a double covering of R constructed by continuation of the square root of the quadratic differential φ. Let $\pi\colon \tilde{R} \to R$ be this covering. Corresponding to a natural parameter ζ on R and the restriction of π to some open subset of $\tilde{R}$ where π is unramified and one-to-one, we obtain a local parameter $\tilde{\zeta}$ satisfying $\tilde{\zeta} = \zeta \circ \pi$. The functions g and h lift to $\tilde{g}$ and $\tilde{h}$ by the formula $\tilde{g} = g \circ \pi$ and $\tilde{h} = h \circ \pi$. The quadratic differential $\varphi(\zeta)\,d\zeta^2$ on R induces $\tilde{\varphi}\,d\tilde{\zeta}^2$ on $\tilde{R}$ by the formula $\tilde{\varphi} = \varphi \circ \pi(\pi'^2)$. Except at the finitely many ramified points, π is a two-to-one mapping. On the surface $\tilde{R}$ the differential $\tilde{\varphi}\,d\tilde{\zeta}^2$ has a global square root. The vertical trajectories β on R lift to trajectories $\tilde{\beta}$ on $\tilde{R}$ which have global orientation. Thus, the mapping $\tilde{\zeta} \to \tilde{\zeta} + i\tau$ is globally defined and one-to-one and onto on $\tilde{R} - \tilde{b}_\varphi$. Obviously, $\iint_{\tilde{R}} \tilde{h}(\tilde{\zeta})\,d\tilde{\xi}\,d\tilde{\eta} = 2\iint_R h(\zeta)\,d\xi\,d\eta$ and an analogous formula holds for g and $\tilde{g}$. The equality (5) is now obvious and Lemma 1 is proved.

To proceed with the proof of the theorem, let p be a point of $R - b_\varphi$ and define a nonnegative function $g(p)$ by

$$g(p) = \int_{\beta_p} |\mathrm{Im}(\sqrt{\psi(s)}\,ds)|, \tag{6}$$

where β_p is the vertical segment for φ with height b and midpoint p. We emphasize that β_p in (6) is an unoriented vertical segment and the integral is a Lebesque integral.

Notice that if an orientation of β_p is selected, then $g(p)$ can be rewritten as

$$\begin{aligned} g(p) &= \int_{-b/2}^{b/2} |\mathrm{Im}(\sqrt{\psi(p + it)}\,i\,dt)| \\ &= \int_0^{b/2} (|\mathrm{Re}\sqrt{\psi(p + it)}| + |\mathrm{Re}\sqrt{\psi(p - it)}|)\,dt \end{aligned}$$

and this formula is valid no matter which orientation is selected. Thus, one finds that

$$\iint_R g(\zeta)\,d\xi\,d\eta = \int_0^{b/2} \iint_R |\mathrm{Re}\sqrt{\psi(p + it)}| + |\mathrm{Re}\sqrt{\psi(p - it)}|\,d\xi\,d\eta\,dt.$$

From Lemma 1, the right-hand side of this equation becomes

$$2\int_0^{b/2}\iint_R |\mathrm{Re}\sqrt{\psi(\zeta)}|\,d\xi\,d\eta\,dt$$

and thus we obtain

$$\iint_R g(\zeta)\,d\xi\,d\eta = b\iint_R |\mathrm{Re}\sqrt{\psi(\zeta)}|\,d\xi\,d\eta. \tag{7}$$

A word concerning the meaning of the integral on the right-hand side of (7) is in order. The variable ζ is assumed to be a natural parameter for the quadratic differential φ. If ζ_1 and ζ_2 are two natural parameters defined in overlapping neighborhoods and if in terms of these parameters the quadratic differential ψ is represented by ψ_1 and ψ_2, then $\psi_1(\zeta_1) = \psi_2(\zeta_2)(d\zeta_2/d\zeta_1)^2$. Since $d\zeta_2/d\zeta_1 = \pm 1$, the expression $|\mathrm{Re}\sqrt{\psi(\zeta)}|\,d\xi\,d\eta$ is defined independently of the choice of natural parameter. However, one may not use a local coordinate z which is not a natural parameter because the integrand in the right-hand side of (7) ceases to be invariant.

From the hypothesis of Theorem 1, we know that

$$b \leqslant \int_{\beta_p} |\mathrm{Im}\sqrt{\psi(z)}\,dz| + M,$$

where β_p is the vertical segment of height b with midpoint p. This means that $b - M \leqslant g(p)$ for all p in $R - b_\varphi$. From (7), this implies

$$(b - M)\iint_R d\xi\,d\eta \leqslant b\iint_R |\mathrm{Re}\sqrt{\psi(\zeta)}|\,d\xi\,d\eta. \tag{8}$$

Dividing both sides by b and taking the limits as b approaches infinity, one obtains

$$\iint_R d\xi\,d\eta \leqslant \iint_R |\mathrm{Re}\sqrt{\psi(\zeta)}|\,d\xi\,d\eta \leqslant \iint_R |\sqrt{\psi(\zeta)}|\,d\xi\,d\eta. \tag{9}$$

Notice that $\varphi(\zeta) \equiv 1$ for any natural parameter ζ, so the integrand on the right-hand side of (9) may be multiplied by $|\sqrt{\varphi(\zeta)}\,|$ without changing it. Of course, the purpose of this is to render it invariant under changes of holomorphic local coordinates. Then (9) becomes

$$\iint_R |\varphi(z)|\,dx\,dy \leqslant \iint_R |\mathrm{Re}\sqrt{\psi(\zeta)}|\,d\xi\,d\eta \leqslant \iint_R |\sqrt{\psi(z)}\sqrt{\varphi(z)}|\,dx\,dy, \tag{10}$$

and this completes the proof of Theorem 1.

Theorem 2. *Let* φ *and* ψ *satisfy the same hypothesis as in Theorem* 1. *Then*

$$\|\varphi\| \leqslant \|\psi\| \tag{11}$$

and, if this inequality is an equality, then $\psi(z) \equiv \varphi(z)$ *for almost every* z.

Proof. Schwarz's inequality gives

$$\iint_R |\sqrt{\psi(z)}\sqrt{\varphi(z)}|\, dx\, dy \leqslant \|\psi\|^{1/2}\|\varphi\|^{1/2}. \tag{12}$$

Substituting this into (10) and dividing both sides by $\|\varphi\|^{1/2}$ yields (11). Moreover, if there is equality in (11), then (10) and (12) yield

$$\|\varphi\| \leqslant \iint_R |\sqrt{\psi(z)}\sqrt{\varphi(z)}|\, dx\, dy \leqslant \|\varphi\|.$$

When an application of Schwarz's inequality yields equality, the two functions must be multiples of one another. Thus $|\sqrt{\psi(z)}| = c|\sqrt{\varphi(z)}\,|$. Since (10) is an equality, one has $c = 1$. Equality in (10) also forces $\mathrm{Re}\sqrt{\psi(\zeta)} = \pm 1$ for any natural parameter ζ. Since $\varphi(\zeta) = 1$ and $|\varphi(\zeta)| = |\psi(\zeta)|$, this obviously forces $\psi(\zeta) = 1$, for any natural parameter ζ. Thus $\psi = \varphi$ almost everywhere and the proof is complete.

2.4. THE REICH–STREBEL INEQUALITY

In order to obtain the inequality of Reich and Strebel from Theorem 1, we need two lemmas. Lemma 2 says that the height of a noncritical vertical segment is minimum among all homotopic arcs with the same endpoints. Lemma 3 concerns the extent to which a quasiconformal self-mapping of R which is homotopic to the identity can distort heights.

Lemma 2. *Let* φ *be a holomorphic quadratic differential on* R *and* $\tilde{R}$ *be the universal covering surface of* R. *Let* β *be a differentiable mapping of a closed interval into a vertical trajectory of* φ *such that a lifting* $\tilde{\beta}$ *of* β *is a one-to-one mapping into* $\tilde{R}$. *Let* γ *be any differentiable mapping from the same interval into* R *which has the same endpoints as* β *and which is homotopic to* β *with fixed endpoints. Then* $h_\varphi(\beta) \leqslant h_\varphi(\gamma)$.

Proof. Our first step is to lift the arcs β and γ and the differential φ to the universal covering surface $\tilde{R}$, where they become $\tilde{\beta}$, $\tilde{\gamma}$, and $\tilde{\varphi}$. Notice that $h_{\tilde{\varphi}}(\tilde{\beta}) = h_\varphi(\beta)$ and $h_{\tilde{\varphi}}(\tilde{\gamma}) = h_\varphi(\gamma)$. We select the liftings of β and γ in such a way that $\tilde{\beta}$ and $\tilde{\gamma}$ have coinciding initial and terminal points and there is a homotopy connecting $\tilde{\beta}$ and $\tilde{\gamma}$ with fixed endpoints in $\tilde{R}$.

The next step is to replace $\tilde{\gamma}$ by a homotopic curve which is a product $\tilde{\alpha}_1\tilde{\beta}_1 \cdots \tilde{\alpha}_n\tilde{\beta}_n$ of a chain of vertical segments $\tilde{\beta}_i$ and horizontal segments $\tilde{\alpha}_i$ for which $h_\varphi(\prod_{i=1}^n \tilde{\alpha}_i\tilde{\beta}_i) \leqslant h_\varphi(\tilde{\gamma})$. To see that this can be done, we cover $\tilde{\gamma}$ by parametric disks parameterized by natural parameters. Then we take a subdivision $\{x_i\}_{i=1}^m$ of the interval such that each $\tilde{\gamma}([x_{i-1}, x_i])$ is contained in one parametric disk. Within each disk it is a simple matter to see that $\tilde{\gamma}([x_{i-1}, x_i])$ can be replaced by one horizontal and one vertical segment, such that the height of the vertical segment is less than or equal to the height of $\tilde{\gamma}([x_{i-1}, x_i])$.

The third step is to observe that we may assume $\tilde{\alpha}_1\tilde{\beta}_1 \cdots \tilde{\alpha}_n\tilde{\beta}_n$ has no self-intersection. The third step is achieved in two stages. First one arranges for the number of points of self-intersection to be finite. The only way they could be infinite is for part of a segment $\tilde{\alpha}_i$ (or $\tilde{\beta}_i$) to coincide with part of segment $\tilde{\alpha}_j$ (or $\tilde{\beta}_j$). If this happens, it is clear that $\tilde{\alpha}_j$ or $\tilde{\beta}_j$ may be shifted slightly to the side in one of the parametric disks without losing the homotopy. The second stage prescribes a way of reducing the number of self-intersections by at least one. You move along the path until you come to the first self-intersection point. Then you proceed along the path, marking as you go in red, until you return to that intersection point. The part marked in red may contain further intersection points. Whether it does or not, you delete from the path the part marked in red. You have reduced by at least one the number of self-intersection points and you have not lost the homotopy because $\tilde{R}$ is simply connected.

The fourth step is to observe that you may assume $\tilde{\alpha}_1\tilde{\beta}_1 \cdots \tilde{\alpha}_n\tilde{\beta}_n$ does not intersect $\tilde{\beta}$ except at the two endpoints. Since the inequality to be proved is $h_\varphi(\tilde{\beta}) \leqslant \sum_{i=1}^n h_\varphi(\tilde{\beta}_i)$, one simply deletes the segments where $\tilde{\beta}$ is common with any of the $\tilde{\beta}_i$ and then one proves the inequality between each successive point of intersection.

The fifth and final step is to treat the case where $\tilde{\beta}$ and $\tilde{\alpha}_1\tilde{\beta}_1 \cdots \tilde{\alpha}_n\tilde{\beta}_n$ joined at the two endpoints make up a simple closed curve C. We will show that there is a measure-preserving injective map from $\tilde{\beta}$ into $\bigcup_{i=1}^n \tilde{\beta}_i$ defined at all but a finite number of points of $\tilde{\beta}$. Given a point on $\tilde{\beta}$, we consider the horizontal trajectory $\tilde{\alpha}$ passing through this point inside the curve C. Since $\tilde{\varphi}$ has only finitely many zeros inside C, by omitting consideration of finitely many points of $\tilde{\beta}$ we can assume the horizontal trajectory is noncritical inside of C. It must be a crosscut, by which we mean $\tilde{\alpha}$ is a simple arc with two endpoints on the curve C. If this were not the case, starting along the trajectory $\tilde{\alpha}$ in one direction from some point p inside C, there would be a sequence of points z_n on $\tilde{\alpha}$ which would converge to a point z inside of C. Let u be the arclength parameter along $\tilde{\alpha}$ initialized so that the distance from p to $\tilde{\alpha}(u)$ is u and oriented so that $\tilde{\alpha}(u_n) = z_n$ with u_n an increasing sequence. Let ζ be a natural parameter with domain in a neighborhood N of z. If z is a critical point, clearly by changing u_n to $u_n + \varepsilon$ and taking a subsequence of $\tilde{\alpha}(u_n + \varepsilon)$ we get a sequence of points on $\tilde{\alpha}$ which converge to a noncritical point z inside of C. We keep the same notation, z_n, for the sequence. Let δ be a vertical arc through z contained in N and containing z in its interior. Since z_n converges to z, $\tilde{\alpha}$ must return to δ. But then we can make a

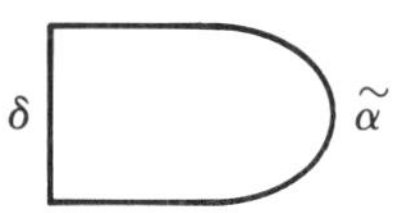

Figure 2.2a

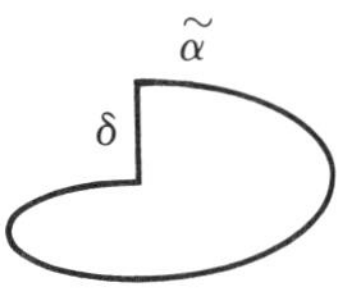

Figure 2.2b

simple closed surve consisting of a subarc of $\tilde{\alpha}$ and a subarc of δ. This simple closed curve C_1 could be essentially of two possible types, shown in Figures 2.2a and 2.2b. By the argument principle, the change in argument of $\varphi(z)$ along a curve C_1 is 2π times the sum of the orders of the zeros of $\tilde{\varphi}$ inside C_1. In particular,

$$\int_{C_1} d \arg \tilde{\varphi}(z) \geqslant 0,$$

assuming the integration along C_1 is in the positive direction. Since $\arg \tilde{\varphi}(z)\, dz^2$ is constant along δ and $\tilde{\alpha}$, we see that $d \arg \tilde{\varphi}(z) = -2d(\arg dz)$ along δ and $\tilde{\alpha}$. Since the vertices are regular points of $\tilde{\varphi}$, the total change in $\arg \tilde{\varphi}$ at the two vertices of Figure 2.2a is π and the total change in $\arg \varphi$ at the two vertices in Figure 2.2b is 0. Thus we find that

$$\int_{C_1} d \arg \tilde{\varphi}(z) = -2\pi + \pi \qquad \text{in Figure 2.2a}$$

and

$$\int_{C_1} d \arg \tilde{\varphi}(z) = -4\pi \qquad \text{in Figure 2.2b.}$$

In either case, we get a negative number and this is a contradiction. Thus we see that $\tilde{\alpha}$ is a crosscut of C with two endpoints on C.

The next important observation is that, while one of the endpoints of $\tilde{\alpha}$ is on $\tilde{\beta}$, the other endpoint must be on one of the $\tilde{\beta}_j$. With a similar use of the argument principle, the proof of this is an easy exercise.

We see that the crosscuts map $\tilde{\beta}$ into $\bigcup \tilde{\beta}_i$. Since the measure $|d\eta|$ is preserved (where $\zeta = \xi + i\eta$ is a natural parameter) as we move along horizontal trajectories, we get

$$h_\varphi(\tilde{\beta}) \leqslant \sum h_\varphi(\tilde{\beta}_i) \leqslant h_\varphi(\gamma).$$

Remark. Lemma 2 can be generalized to measured foliations. Let $|dv|$ be a measured foliation and let β be any curve which is quasitransversal to the leaves of the foliation $|dv|$. (These notions are defined in Chapter 11.) Let γ be any curve which has the same endpoints as β and which is homotopic to β. Let $h_v(\gamma) = \int_\gamma |dv|$. Then $h_v(\beta) \leqslant h_v(\gamma)$. Hence Lemma 2 would follow on letting $|dv| = |\mathrm{Im}\sqrt{\varphi(z)}\, dz|$.

We will not need this stronger form of the lemma. We mention it only to observe that the fact φ is holomorphic is not essential in this lemma. All that is necessary is that its singularities take a form to which the argument principle can be applied.

Lemma 3. *Let φ be a holomorphic quadratic differential on R with $\|\varphi\| < \infty$. Let f be a quasiconformal self-mapping of R which is homotopic to the identity. Then there exists a constant M such that for every noncritical vertical segment β of φ, one has*

$$h_\varphi(\beta) \leqslant h_\varphi(f(\beta)) + M.$$

The constant M depends on φ and f but not on β.

Proof. As before, let $\bar{R}$ be the completion of R with the n punctures filled in. So $\bar{R}$ is compact with no punctures. Since f is quasiconformal, f extends to $\bar{f}$, a quasiconformal self-mapping of $\bar{R}$ and $\bar{f}$ fixes the punctures because f *is homotopic to the identity on* R. The line element $ds_\varphi = |\varphi|^{1/2}|dz|$ determines a finite-valued metric on $\bar{R}$. To see that the distance from a point in R to a puncture is finite, one observes that φ has at most simple poles and so to find the length of a short arc ending at a puncture, one has to calculate an integral of the form $\int_0^a t^{-1/2}\,dt$, and this clearly converges.

Let f_t be the homotopy connecting f to the identity, so $f_0(p) = p$ and $f_1(p) = f(p)$. Let $\ell(p)$ be the infimum of the φ-lengths of all curves which go from p to $f(p)$ and which are homotopic with fixed endpoints to the curve $f_t(p)$. Clearly, $\ell(p)$ is a continuous function on the compact set $\bar{R}$. Let M_1 be the maximum of this function.

Let β be a noncritical vertical segment of φ with endpoints p and q. The segment β and the curve which consists of $f_t(p)$ followed by $f(\beta)$ and then followed by $f_{1-t}(q)$ is clearly homotopic to β with fixed endpoints. By Lemma 2,

$$h_\varphi(\beta) \leqslant h_\varphi(f_t(p)) + h_\varphi(f(\beta)) + h_\varphi(f_{1-t}(q)).$$

Since the φ-length of a curve is greater than its height and the first and third terms in this inequality are bounded by M_1, the lemma is proved if we let $M = 2M_1$.

Theorem 3 (The Inequality of Reich and Strebel). *Let R be a Riemann surface of finite analytic type and let φ be a holomorphic quadratic differential on R with $\|\varphi\| < \infty$. Let f be a quasiconformal self-mapping of R which is homotopic to the identity and let $\mu(z) = f_{\bar{z}}/f_z$ be the Beltrami coefficient of f. Then*

$$\|\varphi\| \leqslant \iint_R |\varphi(z)| \frac{\left|1 - \mu(z)\dfrac{\varphi(z)}{|\varphi(z)|}\right|^2}{1 - |\mu(z)|^2}\,dx\,dy. \tag{13}$$

Proof. Let ψ be defined by

$$\psi(z) = \varphi(f(z))f_z^2(z)\left(1 - \frac{\mu(z)\varphi(z)}{|\varphi(z)|}\right)^2. \tag{14}$$

We will show that ψ satisfies the hypotheses of Theorem 1. An elementary calculation shows that ψ is quadratic differential. From Lemma 3, $h_\varphi(\beta) \leqslant h_\varphi(f(\beta)) + M$ for all noncritical vertical segments β. From the definition of h_φ, we have

$$h_\varphi(f(\beta)) = \int_{f(\beta)} |\mathrm{Im}\sqrt{\varphi(f)}\, df|.$$

Since $df = f_z\, dz + f_{\bar{z}}\, d\bar{z} = f_z(1 + \mu(d\bar{z}/dz))\, dz$, by introducing $\sqrt{\psi}$ from (14), this last integral becomes

$$h_\varphi(f(\beta)) = \int_\beta |\mathrm{Im}\sqrt{\psi(z)}\,(1 + \mu(d\bar{z}/dz))(1 - \mu(\varphi/|\varphi|))^{-1}\, dz|.$$

Since $\varphi(z)\, dz^2 < 0$ along the vertical segment β, one easily sees that $\varphi/|\varphi| = -d\bar{z}/dz$ along β. The final result is that $h_\varphi(f(\beta)) = h_\psi(\beta)$. Hence from Lemma 3, $h_\varphi(\beta) \leqslant h_\psi(\beta) + M$ for all vertical segments β. Theorem 1 tells us that

$$\|\varphi\| \leqslant \iint_R |\sqrt{\psi(z)}\sqrt{\varphi(z)}|\, dx\, dy. \tag{15}$$

Substituting (14) into (15) yields

$$\|\varphi\| \leqslant \iint_R |\varphi(f(z))|^{1/2}|f_z|\,|1 - \mu\varphi/|\varphi||\,|\varphi(z)|^{1/2}\, dx\, dy. \tag{16}$$

Introducing a factor of $(1 - |\mu|^2)^{1/2}$ into the numerator and denominator of (16) and applying Schwarz's inequality yields

$$\|\varphi\| \leqslant \left(\iint |\varphi(f(z))|\,|f_z|^2(1 - |\mu|^2)\, dx\, dy\right)^{1/2}\left(\iint |\varphi|\frac{|1 - \mu\varphi/|\varphi||^2}{1 - |\mu|^2}\, dx\, dy\right)^{1/2}.$$

The first integral on the right-hand side of this expression is simply $\|\varphi\|^{1/2}$, and so we have (13).

Remark 1. If, instead of using the stronger inequality (4), one uses the inequality (11), then one obtains

$$\|\varphi\| \leqslant \iint_R |\varphi(w)|\frac{|1 - \mu(z)\varphi(z)/|\varphi(z)||^2}{1 - |\mu(z)|^2}\, du\, dv$$

where $w = u + iv = f(z)$. This is enough to prove Teichmüller's uniqueness theorem, but it does not yield (13), which is more useful in Teichmüller theory.

Remark 2. Inequality (13) can be proved easily for holomorphic quadratic differentials all of whose noncritical trajectories are closed in the same way that we proved inequality (31) in Section 1.10. If we could appeal to the fact that such differentials are dense (to be proved in Chapter 11), we would immediately obtain (13) in the general case.

2.5. TRAJECTORY STRUCTURE

The remainder of this chapter is devoted to proving the *second minimal norm property.* This property is not used until the last chapter, the chapter on measured foliations, so the reader may prefer to bypass the rest of this chapter until reading Chapter 11.

In order to obtain the second minimal norm property a more detailed analysis of the trajectory of a quadratic differential is required. The analysis has been developed by many people, including Jenkins [Je1, Je2] Strebel [St1], and others. All of the material of this section is contained in greater detail in Strebel's book on quadratic differentials [St4].

We only need that part of the theory of trajectories which relates to quadratic differentials on surfaces of finite analytic type. In effect, this means a quadratic differential induces a decomposition of the Riemann surface into domains of two types, spiral domains and ring domains. In this section we consider primarily horizontal trajectories. Since the vertical trajectories of $-\varphi(z)\,dz^2$ are the same as the horizontal trajectories of $\varphi(z)\,dz^2$, the difference is of no consequence. Also, it is also frequently convenient to use the letter $w = u + iv$ for a natural parameter.

Accordingly, let P_0 be a regular point for $\varphi(z)\,dz^2$ on $\bar{R}$ (recall that $\bar{R}$ is the surface R with the punctures filled in) and let w be a natural parameter centered at P_0. Thus

$$w(P) = \int_{P_0}^{P} \sqrt{\varphi(z)}\,dz$$

and $dw^2 = \varphi(z)\,dz^2$. Let $\Phi(P) = w$ and $\Phi(P_0) = 0$. Then the map Φ^{-1} can be continued along open neighborhoods containing the horizontal trajectory α through P_0. Thus Φ yields a mapping from α to a maximal horizontal interval $(u_{-\infty}, u_{+\infty})$ in the w-plane, where $w = u + iv$. If there are two points u_1 and u_2 in this interval for which $\Phi^{-1}(u_1) = \Phi^{-1}(u_2)$, then α is a closed trajectory. If not, then every closed subinterval $[u_1, u_2]$ is mapped homeomorphically into $\bar{R}$ and is therefore a closed Jordan arc of φ-length $u_2 - u_1$.

Since $\bar{R}$ has no boundary, the maximal horizontal interval in the w-plane along which Φ maps into the trajectory α has a finite endpoint $\Phi^{-1}(u_\infty)$ if, and only if, there is a sequence u_n increasing to $u_{+\infty}$ such that $P_n = \Phi^{-1}(u_n)$ converges to a critical point of φ. By taking a subsequence, we may assume P_n converges to a point P in $\bar{R}$. If $u_{+\infty}$ is finite and P is not a critical point, then there is a nonsingular natural parameter at P and we can continue the map Φ^{-1} beyond the point $u_{+\infty}$, contradicting its maximality. On the other hand, if P_n approaches a singular point, then the maximal horizontal trajectory terminates either at a pole or a zero of φ.

Definition. *Let P_0 be a regular point of φ, let α be the maximal horizontal trajectory through P_0, and let Φ be a continuation of a natural parameter mapping defined on all of α with $\Phi(P_0) = 0$. Then $\Phi([0, u_{+\infty})) = \alpha^+$ is called the positive ray of α with initial point P_0. Similarly, $\Phi((u_{-\infty}, 0]) = \alpha^-$ is called the negative ray of α with initial point P_0.*

We have seen that the ray α^+ has finite length if, and only if, it leads into a critical point. Of course, there may be two points u_1 and u_2 such that $\Phi^{-1}(u_1) = \Phi^{-1}(u_2)$ on the ray α^+. In that case α^+ and α sweep out the same closed trajectory.

Any trajectory α is of the form $\Phi^{-1}((u_{-\infty}, u_{+\infty}))$ for some natural parameter Φ. Choose a closed subinterval $[u_1, u_2]$ in $(u_{-\infty}, u_{+\infty})$ on which the mapping Φ^{-1} is one-to-one. Since φ is regular along $\Phi^{-1}([u_1, u_2])$, by continuation it is possible to construct a rectangle $R(\varepsilon)$ in the w-plane (possibly very narrow) of the form $R(\varepsilon) = \{(u, v): u_1 \leqslant u \leqslant u_2, |v| \leqslant \varepsilon\}$ such that $\Phi^{-1}[R(\varepsilon)]$ is a nonoverlapping strip S on R having $\Phi^{-1}([u_1, u_2])$ as a bisecting horizontal segment.

Ring Domains

We now consider the case that α is a closed trajectory. There exists a largest positive number u_1 such that Φ^{-1} is one-to-one on the half-open interval $[0, u_1)$ and $\Phi^{-1}(0) = \Phi^{-1}(u_1)$. By continuation, it is obvious that there is a rectangle

$$R(\varepsilon_1, \varepsilon_2) = \{(u, v): 0 \leqslant u < u_1, \varepsilon_1 \leqslant v \leqslant \varepsilon_2\}$$

contained in $\bar{R}$ such that Φ^{-1} is one-to-one on $R(\varepsilon_1, \varepsilon_2)$ and extends to a mapping of the closed rectangle on which Φ^{-1} identifies the point iv with the point $u_1 + iv$. Then, one can extend $R(\varepsilon_1, \varepsilon_2)$ to the largest rectangle in which Φ is defined at regular points of the quadratic differential φ and on which Φ^{-1} maps the rectangle

$$R(v_1, v_2) = \{(u, v): 0 \leqslant u \leqslant u_1, v_1 < v < v_2\}$$

onto a ring domain of the Riemann surface with the points iv identified with $u_1 + iv$. If either $v_1 = -\infty$ or $v_2 = +\infty$, then the area of the rectangle would be infinite and φ would have infinite norm. Let A be the ring domain which is so constructed.

Suppose the quadratic differential φ has at least one critical point (this excludes the case of a torus) and also suppose that there is a regular point P on the boundary of the ring domain. Let α_P^+ be a horizontal ray with initial point P. One finds that α_P^+ cannot be closed because it would then lie on the middle line of a ring of closed trajectories some of which would lie in the ring domain A. Thus, A would not be maximal. It follows that α_P^+ leads into a critical point. Moreover, there is no point P_1 on α_P^+ after P with the φ-distance from P to P_1 greater than u_1. The reason is that Φ would be one-to-one on the segment $[P, P_1]$ and we could erect a strip with $[P, P_1]$ as its median line on which Φ is one-to-one and part of this strip would map into the ring domain A. This would contradict the fact that Φ^{-1} identifies the images of the points iv and $u_1 + iv$ in A. Thus we see that α_P^+ must lead into a critical point before it has traveled a distance u_1. Of course, the same argument applies to α_P^-. We have proved the following theorem.

Theorem 4. *Let α be a closed trajectory of a holomorphic quadratic differential φ of finite norm and with at least one critical point. Then α is contained in a maximal ring domain A which is swept out by horizontal trajectories of φ and the boundary of A in R consists either of critical points or of trajectories which lead into critical points in both directions.*

From the maximality it is clear that the ring domain is uniquely determined.

The situation is special if the quadratic differential has no critical points. That can happen only if the Riemann surface is a torus, possibly with some punctures. But since the quadratic differential has no critical points and, in particular, no poles, it can be extended to a quadratic differential on the torus with no punctures. From uniformization, Theorem 1 of Section 1.2, this means it is of the form $\varphi(z)\,dz^2 = c\,dz^2$ on $\mathbb{C}/L$, where L is a lattice and c is a complex constant. It turns out that for φ to have closed horizontal trajectories, c must be a positive multiple of $(m + n\bar{\tau})^2$, where the lattice is generated by 1 and τ and where m and n are integers. Now the closure of the ring domain A will be the whole torus, and it is clear that it is nonunique because it can be shifted upward or downward in the v-coordinate.

Going back to the case where φ is a quadratic differential with critical points, let A_1 and A_2 be two maximal ring domains consisting of closed horizontal trajectories of φ. If they have any point in common, they must obviously be equal. Moreover, we claim that if α_1 is a closed trajectory of A_1 and α_2 is a closed trajectory of A_2, then α_1 and α_2 cannot be homotopic. Moreover, neither of the curves α_1 or α_2 can be homotopically trivial or homotopic to a puncture.

First we will show that α_1 cannot be homotopic to α_2 on R, that is, by a homotopy that does not move across the punctures. If they were homotopic, then by continuation, we could lift φ to a holomorphic quadratic differential $\tilde{\varphi}$ in a doubly connected domain. This doubly connected domain would have $\tilde{\alpha}_1$ and $\tilde{\alpha}_2$ as inner and outer boundary contours where $\tilde{\alpha}_1$ and $\tilde{\alpha}_2$ are lifts of the curves α_1 and α_2. Also $\arg \tilde{\varphi}(z)\,dz^2$ is positive along $\tilde{\alpha}_1$ and $\tilde{\alpha}_2$. Now, by the argument principle,

$$\frac{1}{2\pi}\int_{\tilde{\alpha}_2-\tilde{\alpha}_1} d\arg\tilde{\varphi}(z)$$

is the total number of zeros of $\tilde{\varphi}$ inside the doubly connected domain. But since $d\arg\tilde{\varphi}(z) = -2d\arg dz$ along the trajectories $\tilde{\alpha}_1$ and $\tilde{\alpha}_2$, the integral is obviously zero. This shows that a boundary contour of A_1 does not contain critical points, and we know this cannot be the case.

Now, assume α_1 is homotopically trivial or homotopic to a puncture. Again, by continuation, we could lift α_1 and φ to $\tilde{\alpha}_1$ and $\tilde{\varphi}$, where $\tilde{\alpha}_1$ bounds a disk or a punctured disk and $\tilde{\varphi}$ is holomorphic except for possibly a simple pole at the puncture. On calculating

$$\frac{1}{2\pi}\int_{\tilde{\alpha}_1} d\arg\tilde{\varphi},$$

we find, on the one hand, that it must be bigger than or equal to -1, that is, the sum of the orders of the singular points of $\tilde{\varphi}$ inside the disk. On the other hand, $d\arg\tilde{\varphi} = -2d\arg dz$ along the trajectory $\tilde{\alpha}_1$, and, since we integrate in a counterclockwise direction, the integral is -2. We have proved the following result.

Theorem 5. *Let α_1 and α_2 be two closed trajectories of a holomorphic quadratic differential of finite norm on Riemann surface R of finite analytic type. Suppose the maximal ring domains A_1 and A_2 constructed in Theorem 4 do not coincide. Then they are disjoint and the curves α_1 and α_2 are not homotopic. Moreover, for any ring domain A_1, the trajectory α_1 cannot be homotopically trivial or homotopic to a single puncture.*

Spiral Domains

Next we consider a regular point P_0 for the quadratic differential φ which is the initial point of a ray α^+. Moreover, we assume α^+ is nonclosed and never runs into a critical point. Thus a natural parameter Φ^{-1} can be extended by continuation to a mapping which takes the positive real axis isometrically to the ray α^+ with $\Phi^{-1}(0) = P_0$. Our next goal is to show that such a ray is recurrent in the sense of the next theorem.

Theorem 6. *With the notations just described, let β be a noncritical vertical segment with one of its endpoints at P_0. Then the ray α^+ must return to β, that is, there is a point P after P_0 on α^+ which also lies on β. Moreover, for some point P on α^+, P is on β and α^+ passes through β in the same direction that it emanates from β at P_0.*

Proof. If the ray α^+ passes through β in the same direction that it emanates from β and P_0, we will say that it passes through β in the positive direction.

To begin the proof of the theorem, pick a point P in α^+ not equal to P_0. By shortening β to β_0 we may assume that the interval $[P_0, P]$ on α^+ and β_0 are two sides of an embedded rectangle in R and that no horizontal line through this rectangle is part of a closed trajectory.

Mark every horizontal ray γ^+ with initial point on β_0 and which departs from β_0 in the same direction as α^+ and which leads into a critical point before returning to β_0 in the positive direction.

If two marked rays γ_1^+ and γ_2^+ tend to the same critical point along the same prong, then one of them, say γ_1^+, would have to be a subray of the other, γ_2^+. In that case, γ_2^+ would cross β_0 in the positive direction at the initial point of γ_1^+, and thus γ_2^+ would not be marked. We conclude that only finitely many of the rays are marked since there are only finitely many prongs leading into finitely many critical points. Let β_0' be a subinterval of β_0 with the same initial point P_0 but containing the initial points of none of the marked rays.

If none of the rays γ^+ with initial point on β_0' return to β_0 in the positive direction, then since they are not marked, none of them can lead into a critical point. This means we could construct an embedded infinite strip S with a vertical side on β_0' and one horizontal side along α^+. To see that the strip is embedded, consider Figure 2.3. Let $u_1 + iv_1$ and $u_2 + iv_2$ be coordinates such that $\Phi^{-1}(u_1 + iv_1) = \Phi^{-1}(u_2 + iv_2)$. By continuation we would get $\Phi^{-1}(u_1 - t + iv_1) = \Phi^{-1}(u_2 - t + iv_2)$ and we would then find that the ray starting at the point iv_1 on β_0' would return to β_0' in the positive direction at the point $\Phi^{-1}(iv_2) = \Phi^{-1}(u_2 - u_1 + iv_2)$. This contradicts our assumption that the rays in S do not return.

Now, the strip would have infinite area and this, of course, contradicts the fact that the total area of R measured by the quadratic differential φ is finite. We conclude that one of the rays γ^+ with initial point in β_0' returns to β in the positive direction.

If α^+ is this trajectory, we are done. If α^+ is not this trajectory, then α^+ must return in the manner shown in Figure 2.4. Now apply the same argument to the smaller strip S' bounded by α^+ and the dotted line in Figure 2.4. Since the rays of S' have not returned to β_0 at the point where the rays of S did, we can continue S' further until it meets β_0. When this happens, the ray α^+ will be forced to lead into β_0 because the strip S' has already covered the points along the continuation of β_0 just above the point P_0.

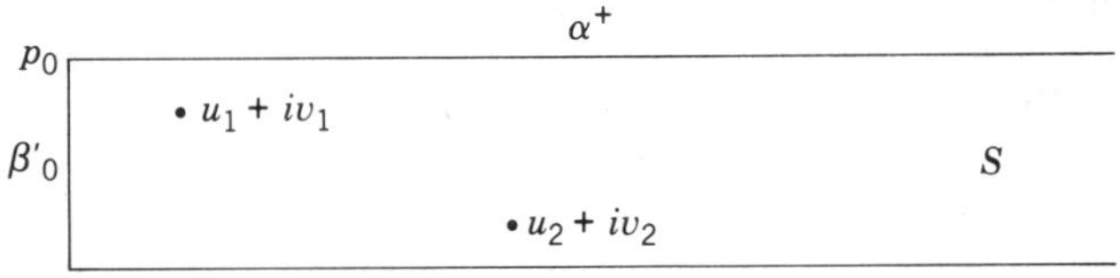

Figure 2.3

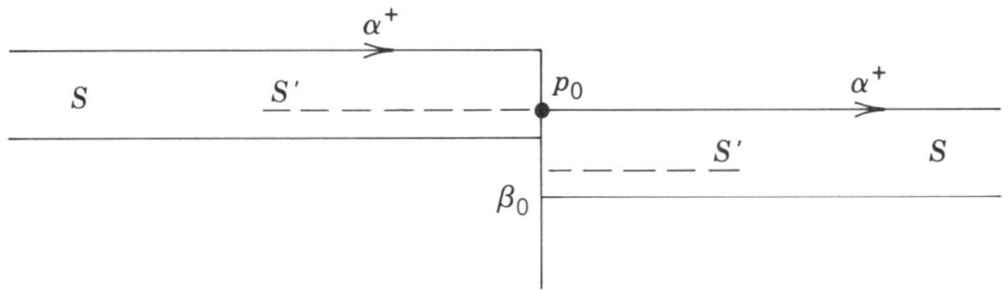

Figure 2.4

Definition. *A trajectory ray is called regular if it does not lead to a critical point.*

Definition. *The limit set A of a regular trajectory ray α^+ is the set of points P in $\bar{R}$ which can be expressed as limits of sequences of the form $\Phi^{-1}(u_n)$, where $u_n \to \infty$.*

Theorem 6 implies that a nonclosed regular trajectory ray α^+ is recurrent, by which we mean that its initial point is in A. Since A does not change if we pick a different point on a horizontal trajectory α as initial point, it is clear that $\alpha \subset A$ and, since A is closed, $\bar{\alpha} \subset A$. On the other hand, A is contained in $\bar{\alpha}$, so we have $\bar{\alpha} = A$ for any nonclosed, regular trajectory α in A.

Theorem 7. *Let α^+ be a regular, nonclosed trajectory ray and A its limit set. Then the interior A^0 of A is a nonempty connected open set, that is, a domain. The boundary $A - A^0$ of A^0, if it is nonempty, consists of critical trajectories of finite length and the critical points that they lead into. Moreover, the limit set of any regular horizontal trajectory ray which has initial point in A is equal to A.*

Proof. First we assume that P is a regular point of φ in A and that the horizontal trajectory γ through P has infinite length. We will show that P is in the interior of a rectangle contained in A. Since α^+ is assumed to be regular, this will show, in particular, that A^0 is not empty. Now since P is a limit point of α^+, it is obvious that any other point on γ is a limit point of α^+ and, therefore, $\bar{\gamma} \subset A$. Since γ has infinite length, one of the rays γ^+ or γ^- with initial point at P must have infinite length and therefore one of these rays is recurrent. Choose a vertical segment β with P as one of its endpoints. Let P_0 be the point of first return of one of the rays, γ^+ or γ^-, to β.

Construct a rectangle U_0 which has as one of its sides a segment of γ with midpoint P and a median vertical segment along β from P to P_0. (See Figure 2.5.) By making U_0 small enough, we may assume φ is regular throughout U_0. If the segment $[P, P_0]$ is contained in A, then so is every horizontal segment of U_0 passing through $[P, P_0]$ and we see that U_0 would be contained in A. If $[P, P_0]$ is not contained in A and a point Q_1 between P and P_0 is not in A, then Q_1 is contained in a largest open subinterval I_1 on β not contained in A. Let P_1 be the endpoint of I_1 nearest to P. P_1 is not equal to P because γ is recurrent. P_1 is itself a regular point of a trajectory γ_1 of finite length. If γ_1 had infinite length, it would be recurrent and would pass through a point on I_1. But clearly γ_1 would then be

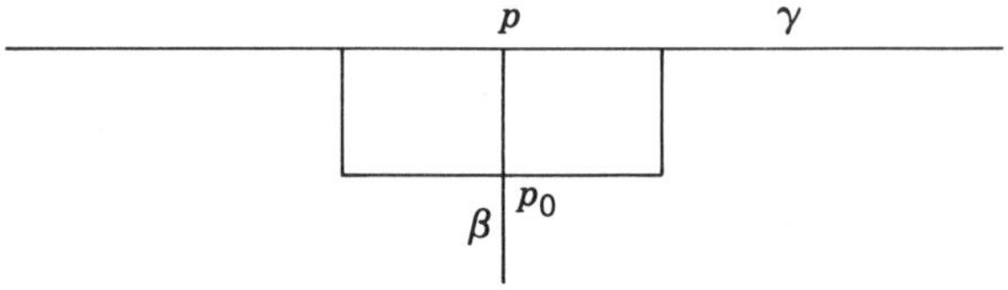

Figure 2.5

contained in A, and this contradicts the definition of I_1. If $[P, P_1]$ is contained in A, then the subrectangle U_1 of U_0 with median vertical segment $[P, P_1]$ would be contained in A and we would be done. If $[P, P_1]$ is not contained in A, we get a point Q_2 between P and P_1 not in A and we can construct a maximal open interval I_2 containing Q_2 on β but not in A and proceed in the same fashion. If this process proceeds indefinitely, we obtain a sequence of trajectories γ_n of finite length running through the upper endpoint of each interval I_n and passing through the rectangle U_0. This would be a contradiction, because the total length of critical trajectories of finite length is finite.

By the same argument, A must contain a rectangle adjacent to γ and lying above the point P, and hence P is in A^0, the interior of A.

Second, assume P in A is a regular point and lies on a trajectory of finite length. The trajectory α^+ must cross arbitrarily close to P on at least one side of γ. Choose a vertical segment β with initial point at P and such that α^+ intersects β in every neighborhood of P. Then on using α instead of γ and on repeating the argument above, we find a rectangle U with one side along a segment of γ with P as its midpoint and U contained in A.

In the last case, P in A could be a critical point. Then one of the sectors determined by the critical trajectories ending at P must contain a sequence of segments of the ray α^+ which contain P in their limit set. By applying the same argument as above, we find that a sufficiently small relatively open neighborhood of P in this section must be contained in A.

The statement in the theorem about the boundary $A - A^0$ is now obvious. Suppose α_1^+ is a regular horizontal trajectory with initial point in A^0. It cannot be closed because then it would be contained in an open ring domain of closed trajectories none of which could be limit points of any other trajectory. Since by assumption it does not lead into a critical point, it must have infinite length and must be recurrent. Let A_1 be the limit set of α_1^+. A_1 is a domain and obviously $\alpha \subset A_1$ and $\alpha_1 \subset A$. Thus $A_1 \subset A$ and $A \subset A_1$.

Definition. *A domain A^0 formed by taking the interior of the limit set of a regular, nonclosed trajectory ray is called a spiral domain.*

Decomposition of a Spiral Domain into Rectangles

Let β be any vertical segment in a spiral domain A. Mark the two sides of β by β^+ and β^-. Mark each point on β^+ with a label y_n^+ if either

(1) it is an endpoint of β^+, or
(2) if the horizontal ray emanating from β^+ at y_n^+ leads to a critical point of φ before returning to β, or
(3) if the horizontal ray emanating from β^+ at y_n^+ leads to an endpoint of β.

Similarly, mark the points y_m^- on β^-. Pick a monotone parametrization of β and let the points y_n^+ and y_m^- have their natural ordering.

Define a ray emanating from β^+ to be of the first kind if the point of its first return to β is on β^-. A ray emanating from β^+ is of the second kind if its first return to β is on β^+. Clearly, every ray emanating from β^+ in the interior of the interval $[y_n^+, y_{n+1}^+]$ is of the same kind. We will call the interval $[y_n^+, y_{n+1}^+]$ a first kind interval if its interior rays are of the first kind. In a parallel manner, we define intervals of the second kind. Clearly, the intervals of the first kind on β^+ are paired with intervals of the first kind on β^-, and two such paired intervals determine a rectangle S_k. Similarly, there are rectangles S_j of the second kind and the total length of all intervals of the second kind on β^- is equal to the total length of all intervals of the second kind on β^+. The union of all of the finite number of closed rectangles S_j and S_k obviously contains the limit set of any recurrent ray passing through β. Therefore, the union of these rectangles is A itself.

This observation holds true no matter how short the originally selected interval β is. In the next section we will let β become arbitrarily small.

We need a final result for ring domains and spiral domains which is parallel to Theorem 5.

Theorem 8. *Let φ be a holomorphic quadratic differential on a surface R of finite analytic type. Let $\bar{R}$ be the surface R with the punctures filled in. Assume φ has at least one critical point and is of finite norm. Then the total number of maximal ring domains and spiral domains associated to φ is not more than $3g - 3 + n$. The closure of the union of these domains is $\bar{R}$.*

Proof. Let $A_1, \ldots, A_m$ be the set of disjoint ring domains and spiral domains. We will construct a set of simple closed curves $\gamma_1, \ldots, \gamma_m$ with each γ_j contained in A_j with the following properties:

(1) No γ_j is homotopic to any γ_k if $j \neq k$.
(2) No γ_j is homotopically trivial or homotopic to a single puncture.

From elementary topology, this will imply $m \leqslant 3g - 3 + n$. (This is proved in Lemma 1 of Chapter 10.)

To construct the curves γ_j, if A_j is a ring domain, let γ_j be any one of the closed trajectories in A_j. If A_j is a spiral domain, select a vertical segment β in the interior of A_j and construct the rectangle decomposition of A_j associated with β. If this contains a strip of the first kind, let γ_j be formed from a horizontal segment along the strip joined at the two ends by a segment of β. If every strip in the decomposition is of the second kind, let S^+ be a strip both of whose ends lie on β^+ and let S^- be a strip both of whose ends lie on β^-. Let α^+ be the horizontal segment joining the two ends of S^+ and let α^- be a horizontal segment joining the two ends of S^- with one endpoint at P. Now, let the curve γ_j consist of α^- followed by α^+ followed by appropriate segments of β so as to make a closed curve homotopic to a simple closed curve.

In every case γ_j has either no vertices or two vertices or four vertices, each one lying on the vertical segment β with the curve entering from one side of β and departing from the opposite side.

The rest of the proof that the curves γ_j have properties (1) and (2) is exactly like the proof of Theorem 5. If γ_j were homotopic to γ_k for some $j \neq k$, then a boundary contour of A_k would not contain any critical points. But we know the boundaries of every A_k consist of critical trajectories of finite length and the critical points they lead into.

Now assume we constructed maximal ring domains and spiral domains $A_1, \ldots, A_k$. If closure of $(\bigcup_{j=1}^{k} A_j)$ is not equal to R, we could construct another ring domain or spiral domain A_{k+1}. Since the process must end with some integer $m \leqslant 3g - 3 + n$, we have the closure of $\bigcup_{k=1}^{m} A_k = \bar{R}$.

2.6. THE SECOND MINIMAL NORM PROPERTY

For a closed curve γ in R, define $h_\psi[\gamma]$ to be the infimum of the values $h_\psi(\gamma')$, where γ' varies over all closed curves in R freely homotopic to γ. Let $\mathscr{S}$ be the set of all simple closed curves in R which are not homotopically trivial and not homotopic to a puncture. In order to define $h_\psi(\gamma)$ we must have a meaningful interpretation of the integral

$$\int_\gamma |\mathrm{Im}(\sqrt{\psi(z)}\, dz)|.$$

Since the path of integration is not necessarily a piece of horizontal or vertical trajectory, Fubini's theorem does not apply as it did for the first minimal norm property. Thus, it is convenient now to make the assumption that ψ be continuous. If $|\mathrm{Im}(\sqrt{\psi}\, dz)|$ is nonzero along a nonrectifiable part of γ, we interpret the integral to be $+\infty$. Thus $h_\psi[\gamma]$ is well defined for every homotopy class $[\gamma]$.

Theorem 9 (The Second Minimal Norm Property). *Assume R is of finite analytic type. Let φ be an integrable holomorphic quadratic differential on R and ψ another quadratic differential, continuous except possibly at the punctures of R. Suppose $h_\varphi[\gamma] \leqslant h_\psi[\gamma]$ for all γ in $\mathscr{S}$. Then*

$$\|\varphi\| \leqslant \iint_R |\sqrt{\psi}\sqrt{\varphi}|\, dx\, dy \leqslant \|\varphi\|^{1/2}\|\psi\|^{1/2} \tag{17}$$

and $\|\varphi\| = \|\psi\|$ only if $\varphi = \psi$.

Remark. The proof is exactly the same as the one given by Marden and Strebel [MarS]. In their version of the theorem ψ is assumed to be harmonic and the conclusion is that $\|\varphi\| \leqslant \|\psi\|$.

Proof. We use the decomposition of R into maximal ring domains and spiral sets, but we use the decomposition which is induced by the *vertical* trajectories of φ.

First let A be a ring domain swept out by closed vertical trajectories of φ. Let α be a horizontal segment connecting the two boundary contours of the ring domain. With respect to a natural parameter $\zeta = \xi + i\eta$ the domain $A - \alpha$ is mapped onto a rectangle $\{0 < \xi < a, 0 < \eta < b\}$ where a is the φ-width of the rectangle and b is the height of any one of the closed vertical trajectories of A. Let β be a simple closed curve moving along one of the closed trajectories and going around the ring once. In Theorem 8, we have seen that β is not homotopically trivial and not homotopic to a puncture. Thus β is in $\mathscr{S}$.

The next observation is that $h_\varphi[\beta] = h_\varphi(\beta)$. That is, the height of β is less than or equal to the height of any homotopic curve γ. This can be viewed as a consequence of Lemma 2 in Section 2.4. In fact let $f_t(s)$ be a homotopy of β to γ. By this we mean $f_t(s)$ is a continuous function from $I \times S^1$ into R with $f_0(s)$ a parameterization of β and $f_1(s)$ a parametrization of γ. Let $n\beta$ and $n\gamma$ be the curve β and γ followed n times around. Clearly, f induces a homotopy of $n\beta$ onto $n\gamma$, which we denote by the same letter f. Let γ_1 be the curve $f_t(0)$ followed by the curve $n\gamma$ followed by the curve $f_{1-t}(1)$. Clearly γ_1 is homotopic to $n\beta$ by a homotopy which fixes the endpoints, and so Lemma 2 implies

$$nb = h_\varphi(n\beta) \leqslant h_\varphi(\gamma_1) = nh_\varphi(\gamma) + (\text{const}).$$

Here, the constant represents the integral $\int |\text{Im}\sqrt{\varphi}\, dz|$ along the curve $f_t(0)$, $0 \leqslant t \leqslant 1$, and the curve $f_{1-t}(1)$, $0 \leqslant t \leqslant 1$.

On dividing both sides by n and taking the limit as $n \to \infty$, we obtain

$$b \leqslant h_\varphi(\gamma)$$

for any curve γ homotopic to β. This shows that $h_\varphi[\beta] = h_\varphi(\beta)$.

Since β is in $\mathscr{S}$, the hypothesis of the theorem tells us that

$$b \leqslant \int_\beta |\text{Im}\sqrt{\psi(\zeta)}\, d\zeta| = \int_\beta |\text{Re}\sqrt{\psi(\zeta)}|\, |d\eta|. \tag{18}$$

The equality on the right holds because ζ is a natural parameter for φ and thus $d\zeta = i\, d\eta$ along a vertical trajectory. Integrating both sides of (18) across the ring domain A from $\xi = 0$ to $\xi = a$, we obtain

$$\iint_A |\varphi(z)|\, dx\, dy \leqslant \iint_A |\text{Re}\sqrt{\psi(\zeta)}|\, d\xi\, d\eta. \tag{19}$$

Now suppose A is a spiral domain. We pick a regular horizontal segment α contained in A and let S_n and S_m be the rectangles of the first and second kind based on α. We use the decomposition described in Section 2.5, except it is based on a horizontal instead of a vertical segment. We further subdivide the rectangles S_m of the second kind so that the ones S_m^+ based on the positive side α^+ of α can be paired with the ones S_m^- based on the negative side α^- of α in such a way that the width of S_m^+ is equal to the width of S_m^-. This can clearly be done because the total length of intervals of the second kind on α^- is equal to the total length of intervals of the second kind on α^+.

For a rectangle S_n of the first kind, let β_n be a vertical segment joining the two ends which meet α on opposite sides. Let α_n be a subsegment of α joining the two endpoints of β_n and let $\beta_n\alpha_n$ be the simple closed curve made up from β_n and α_n.

Now we need a slight improvement of Lemma 2 of Section 2.4. Replace the single vertical segment β in Lemma 2 by two vertical segments β_1 and β_2 and a horizontal segment α_1 connecting an endpoint of β_1 to an endpoint of β_2. Assume that β_1 and β_2 emanate from α_1 on opposite sides of α_1. Then a curve which is homotopic to $\beta_2\alpha_1\beta_1$ which fixes endpoints must have height greater than or equal to the height of β_1 plus the height of β_2. Hence, by the same argument we just used for the case of a ring domain, we see that

$$h_\varphi[\beta_n\alpha_n] = h_\varphi(\beta_n) = b_n.$$

Since $\beta_n\alpha_n$ is in $\mathscr{S}$, the hypothesis tells us that $h_\varphi[\beta_n\alpha_n] \leqslant h_\psi[\beta_n\alpha_n]$, and we find that

$$b_n \leqslant \int_{\beta_n} |\mathrm{Re}\sqrt{\psi(\zeta)}|\,|d\eta| + \int_{\alpha_n} |\mathrm{Im}\sqrt{\psi(\zeta)}|\,|d\xi|.$$

On letting a_n be the width of S_n and integrating with respect to ξ across the width of S_n, we obtain

$$a_n b_n \leqslant \iint_{S_n} |\mathrm{Re}\sqrt{\psi(\zeta)}|\,|d\xi\,d\eta| + a_n \int_{\alpha_n} |\mathrm{Im}\sqrt{\psi(\zeta)}|\,|d\xi|. \tag{20}$$

Now consider two paired rectangles S_m^+ and S_m^- of the second kind, each with width a_m and with heights b_m^+ and b_m^-. Let β_m^+ and β_m^- be vertical segments joining the two ends of S_m^+ and S_m^-, respectively. Let α_m and α_m' be subsegments of α shown in the Figure 2.6 and let the closed curve γ be formed by following β_m^+, then α_m', then β_m^-, and finally α_m. That is, $\gamma = \alpha_m\beta_m^-\alpha_m'\beta_m^+$. Although γ is not a simple curve in Figure 2.6b, obviously by shifting α_m slightly upward and α_m' slightly downward it can be made into a homotopic simple curve with the same height. Again, we see that γ is in $\mathscr{S}$ and that $h_\varphi[\gamma] = h_\varphi(\gamma) = b_m^+ + b_m$.

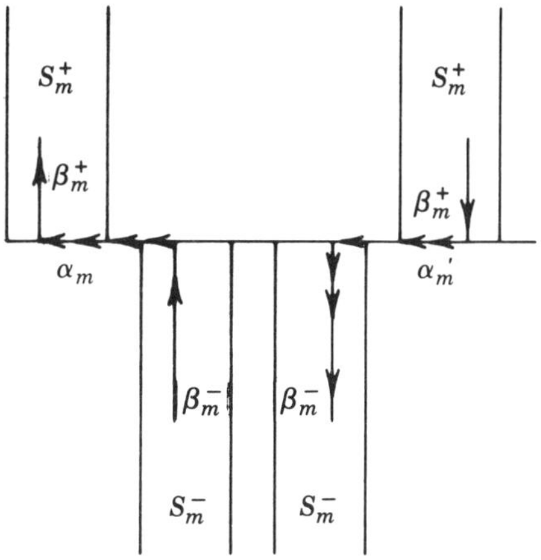

Figure 2.6a

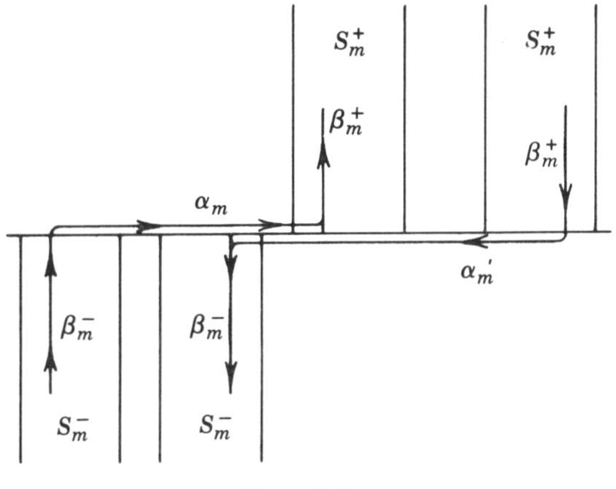

Figure 2.6b

Thus

$$b_m^+ + b_m^- \leqslant \int_{\beta_m^+ + \beta_m^-} |\mathrm{Re}\sqrt{\psi(\zeta)}|\,|d\eta| + \int_{\alpha_m + \alpha_m'} |\mathrm{Im}\sqrt{\psi(\zeta)}|\,|d\xi|$$

and, on integrating both sides of this inequality across the common width of S_m^+ and S_m^-, we obtain

$$a_m(b_m^+ + b_m^-) \leqslant \iint\limits_{S_m^+ \cup S_m^-} |\mathrm{Re}\sqrt{\psi(\zeta)}|\,d\xi\,d\eta + a_m \int_{\alpha_m + \alpha_m'} |\mathrm{Im}\sqrt{\psi(\zeta)}\,|d\xi| \qquad (21)$$

Clearly, $\sum a_n + 2\sum a_m = a =$ the length of α. Thus on adding up the areas of all the rectangles S_n and S_m^+ and S_m^- in the decomposition of A based on α and using inequalities (20) and (21), we see that

$$\iint_A |\varphi(z)|\,dx\,dy \leqslant \iint_A |\mathrm{Re}\sqrt{\psi(\zeta)}|\,d\xi\,d\eta + a\int_\alpha |\mathrm{Im}\sqrt{\psi(\zeta)}|\,|d\xi|. \tag{22}$$

Since ψ is continuous on the closed interval α, the second integral on the right-hand side of (22) is finite. If we allow the length of α, which is a, to approach 0, the right-hand term in the right side of (22) approaches zero.

To complete the proof of the theorem, we sum inequalities (22) and (19) over all ring domains and spiral domains for the quadratic differential φ and obtain

$$\begin{aligned}\iint_R |\varphi(z)|\,dx\,dy &\leqslant \iint_R |\mathrm{Re}\sqrt{\psi(\zeta)}|\,d\xi\,d\eta\\ &\leqslant \iint_R |\sqrt{\psi(\zeta)}|\,d\xi\,d\eta \leqslant \iint |\sqrt{\psi(\zeta)}\sqrt{\varphi(\zeta)}|\,d\xi\,d\eta\\ &\leqslant \iint_R |\sqrt{\psi(z)}\sqrt{\varphi(z)}|\,dx\,dy.\end{aligned}$$

The uniqueness part of the theorem follows in exactly the same way as it did for the first minimal norm property.

Notes. The material of this chapter is based on ideas of Teichmüller [T1], Bers [Ber2], Marden and Strebel [MarS], Reich and Strebel [ReiS1, ReiS2], and others. A minimum norm property very close to the ones given here was proved by Marden and Strebel [MarS]. They stated the property by way of comparison to harmonic quadratic differentials. The property is generalized in two ways here: first, by allowing comparison to measurable quadratic differentials and second by allowing comparison to continuous quadratic differentials. Moreover, the inequality in the conclusion is strengthened to a form involving square roots of quadratic differentials. The idea of looking at heights instead of lengths of curves is due to Thurston. The averaging device in Lemma 1 is due to Teichmüller [T1, T2] and reformulated by Bers in [Ber2]. It also appears in Abikoff's book [Ab]. The strengthened minimal norm properties and the use of the first minimal norm property to prove the Reich–Strebel inequality is due to the author.

The recurrence property (Theorem 6 of Section 2.5) leads to applications of the theory of quadratic differentials to interval exchange transformations [Masu4].

The detailed analysis of the trajectory structure given in Section 2.5 is due to Jenkins [Je2, Je3] and Strebel [St1, St4].

EXERCISES

SECTION 2.1

1. Draw a picture of the horizontal and vertical trajectories of $z^{-1}dz^2$ in a neighborhood of $z = 0$.
2. Draw a picture of the horizontal and vertical trajectories of $z^{-2}dz^2$.

SECTION 2.2

3. Let $\gamma(t) = t$, $0 < t < 1$. Find the width and height of γ with respect to the quadratic differential $\varphi(z) = z^{-1}dz^2$.
4. Let $\gamma(t) = (t + 1)e^{\pi it/2}$, $0 < t < 1$. Find the width and height of γ with respect to the quadratic differential $\varphi(z) = z^{-2}dz^{-2}$.

SECTION 2.4

5. Show that the Reich–Strebel inequality (13) is equivalent to

$$\operatorname{Re} \iint_R \frac{\mu\varphi}{1 - |\mu|^2}\, dx\, dy \leqslant \iint_R \frac{|\mu|^2|\varphi|}{1 - |\mu|^2}\, dx\, dy.$$

6. Let f be a quasiconformal mapping from R to $f(R)$ and f_1 be quasiconformal from $f(R)$ to R. Assume that $f_1 \circ f$ is homotopic to the identity. Let $\mu(z) = f_{\bar{z}}/f_z$ and

$$\mu_1(w) = \frac{\partial f_1/\partial \bar{w}}{\partial f_1/\partial w}$$

and $p = f_z$. From inequality (13) derive

$$\iint_R |\varphi|\, dx\, dy \leqslant \iint_R |\varphi| \frac{|1 - \mu\varphi/|\varphi||}{1 - |\mu|^2} \cdot \frac{|1 - \mu_1\theta\varphi/|\varphi||^2}{1 - |\mu_1|^2}\, dx\, dy,$$

where

$$\theta = \frac{\bar{p}}{p}(1 - \overline{\mu\varphi/|\varphi|})(1 - \mu\varphi/|\varphi|)^{-1}.$$

Remark. The result of Exercise 6 is called the "main inequality" by Reich and Strebel.

SECTION 2.6

7. Let $\alpha > 2$ and $\varphi(z)\,dz^2 = -dz^2/[z(z-1)(z-\alpha)]$. Let $\gamma(t) = \frac{1}{2} + e^{2\pi i t}$, $0 < t < 1$. Write an integral formula for the height of the homotopy class of γ with respect to φ on the surface $\mathbb{C} - \{0, 1, \alpha\}$.

8. Find the decomposition of $\mathbb{C} - \{0, 1, \alpha\}$ into ring domains (and possibly spiral domains) corresponding to the quadratic differential in Exercise 7.

9. Explain why the second minimum norm principle can be stated in terms of just one homotopy class in $(\mathbb{C} - \{0, 1, \alpha\})$ for the quadratic differential φ in Exercise 7.

10. Draw a picture which makes it clear why there are infinitely many homotopy classes in $\mathscr{S}(\mathbb{C} - \{0, 1, \alpha\})$.

3

THE REICH–STREBEL INEQUALITY FOR FUCHSIAN GROUPS

Theorem 3 of the previous chapter gives the inequality

$$\iint_R |\varphi(z)|\, dx\, dy \leqslant \iint_R |\varphi(z)| \frac{|1 - \mu(z)\varphi(z)/|\varphi(z)||^2}{1 - |\mu(z)|^2}\, dx\, dy. \tag{1}$$

This was proved for all holomorphic quadratic differentials φ in $A(R)$, where R is a Riemann surface of finite analytic type. The Beltrami coefficient μ was assumed to be of the form $\mu = f_{\bar{z}}/f_z$, where f is a quasiconformal self-mapping of R which is homotopic to the identity.

Inequality (1) is essential to our development of Teichmüller theory. The efforts of this chapter and the next are directed to proving (1) in a more general setting. In this chapter we prove (1) in the setting of a Fuchsian group Γ and under assumptions which imply that the relevant space of holomorphic quadratic differential forms is finite dimensional. In the next chapter we generalize to infinite dimensional cases.

In the generalized version of inequality (1), the region of integration R is replaced by a fundamental domain ω for Γ. The holomorphic quadratic differential φ is replaced by a holomorphic quadratic differential form on $\mathbb{H}$ which takes real values on the part of the real axis complementary to a Γ-invariant closed subset C of $\hat{\mathbb{R}}$ which contains the limit set of Γ. The Beltrami coefficient μ is of the form $\mu = w_{\bar{z}}/w_z$, where w is a quasiconformal self-mapping of $\mathbb{H}$ for which $w \circ A = A \circ w$ for all A in Γ and $w(x) = x$ for all x in C.

3.1. INTEGRABLE CUSP FORMS

For a Fuchsian group Γ acting on the upper half plane $\mathbb{H}$ we have seen in Chapter 1 that $\mathbb{H}/\Gamma$ has a natural Riemann surface structure even when Γ has elliptic fixed points. We consider this surface as being punctured at the elliptic fixed points.

It is convenient to delete the preimages of the elliptic punctures from $\mathbb{H}$. Accordingly, let $\mathbb{H}_\Gamma$ be $\mathbb{H}$ with all of the fixed points of elliptic elements of Γ removed and let $R = \mathbb{H}_\Gamma/\Gamma$. When Γ is finitely generated and of the first kind, R has finite genus g and n punctures. The total number of punctures n is $n_1 + n_2$, where n_1 is the number of punctures on R coming from parabolic conjugacy classes in Γ and n_2 is the number of punctures on R coming from elliptic conjugacy classes. When Γ is finitely generated and of the second kind, the surface R also has holes.

If Γ is of the second kind, we let C be a closed Γ-invariant subset of the extended real axis which contains the limit set Λ of Γ. If σ is a closed subset of the border of $R = \mathbb{H}_\Gamma/\Gamma$, then such a subset C is obtained by letting $C = \Lambda \cup \pi^{-1}(\sigma)$, where π is the covering mapping of $\mathbb{H}_\Gamma$ onto R extended to the border of R.

We now define $A_s(R, \sigma)$, the space of symmetric holomorphic quadratic differentials on R relative to σ, and $A_s(\Gamma, C)$ the space of symmetric holomorphic quadratic differential forms for Γ relative to C.

Definition. *Let σ be a closed subset of the border of R. Then $A_s(R, \sigma)$ is the Banach space of all holomorphic quadratic differentials φ satisfying*

(a) *φ has at most simple poles at the punctures of R;*
(b) *φ is real valued with respect to boundary uniformizers on the part of the border complementary to σ; and*
(c) *$\|\varphi\| = \iint_R |\varphi(z)|\, dx\, dy$ is finite.*

Definition. *Let C be a Γ-invariant subset of the extended real axis. Then $A_s(\Gamma, C)$ is the Banach space of all holomorphic functions $\tilde{\varphi}$ satisfying*

(i) *$\tilde{\varphi}(A(z))A'(z)^2 = \tilde{\varphi}(z)$ for all A in Γ;*
(ii) *$\tilde{\varphi}$ is real valued on $\hat{\mathbb{R}} - C$; and*
(iii) *$\|\tilde{\varphi}\| = \iint_\omega |\tilde{\varphi}(z)|\, dx\, dy$ is finite, where ω is a fundamental domain for Γ in $\mathbb{H}$.*

Remark. The subscript s in the notations $A_s(R, \sigma)$ and $A_s(\Gamma, C)$ is meant to refer to the symmetry properties (b) and (ii) in the definitions.

Theorem 1. *Let $R = \mathbb{H}_\Gamma/\Gamma$ and let π be the unique extension of the covering mapping $\mathbb{H}_\Gamma \to R$ to the part of the real axis where Γ acts discontinuously. Let σ be a closed subset of the border of R and let $C = \pi^{-1}(\sigma) \cup \Lambda$. Then π induces an isometric isomorphism of $A_s(R, \sigma)$ onto $A_s(\Gamma, C)$ by the formula $\varphi \to \tilde{\varphi} = (\varphi \circ \pi)\pi'^2$.*

Remark. If Γ is of the first kind, then $\Lambda = C = \hat{\mathbb{R}}$ and σ is empty. If Γ is of the second kind, the border of $R = \mathbb{H}_\Gamma/\Gamma$ is obtained by taking the quotient by Γ of the part of the real axis where Γ acts discontinuously.

Proof. It is obvious that $\|\varphi\|$ given in (c) is equal to $\|\tilde{\varphi}\|$ given in (iii) because π is a one-to-one mapping of ω onto a subset of R whose complement has measure zero and the Jacobian of π is $|\pi'(z)|^2$.

It is also obvious that condition (b) on φ corresponds to condition (ii) on $\tilde{\varphi}$. The fact that φ is a quadratic differential converts into the relation in (i) for $\tilde{\varphi}$ because $\pi \circ A = \pi$ for all A in Γ.

We must show that φ has at most a simple pole at $\pi(p_0)$, where p_0 is an elliptic fixed point of an element A in Γ if, and only if, $\tilde{\varphi}$ is holomorphic at p_0. By mapping $\mathbb{H}$ to the unit disk and p_0 to the origin, we may assume that $A(z) = \alpha z$, where α is a primitive nth root of unity and n is the order of the elliptic element A. If we let $\zeta = z^n$, then ζ will be a local coordinate at $\pi(p_0)$ on R. Moreover

$$\varphi(\zeta)\left(\frac{d\zeta}{dz}\right)^2 = \tilde{\varphi}(z). \tag{2}$$

Thus, if M is the order of φ at $\zeta = 0$ and m is the order of $\tilde{\varphi}$ at $z = 0$, by equating the orders in the variable z of the two sides of equation (2), we have $Mn + 2(n-1) = m$. The condition $m \geqslant 0$ is equivalent to $M \geqslant -2(1 - 1/n)$. Since $n \geqslant 2$ and M is an integer, we see that $m \geqslant 0$ if, and only if, $M \geqslant -1$.

Next, we must show that φ may have at most a simple pole at a parabolic puncture. Such a puncture corresponds to a primitive parabolic element A of Γ and, by conjugation, we may assume $A(z) = z + 1$. A local coordinate ζ at the puncture is given by the formula $\zeta = e^{2\pi i z}$. There always exists a fundamental domain which contains $\{z: 0 < x < 1, y > 1\}$. (This is proved in Lemma 1 of Section 8.4.) Hence the condition $\|\tilde{\varphi}\| < \infty$ implies that

$$\int_1^\infty \int_0^1 |\tilde{\varphi}(z)|\, dx\, dy \tag{3}$$

is finite. Changing to integration in the ζ-plane and using (2), (3) becomes

$$\iint_{|\zeta| < e^{-2\pi}} |\varphi(\zeta)|\, d\xi\, d\eta. \tag{4}$$

In order for (4) to be finite the function φ, which is assumed to be holomorphic in the punctured disk, may have at most a simple pole at the origin. The theorem is now proved.

3.2. TRIVIAL MAPPINGS FOR FUCHSIAN GROUPS OF THE FIRST KIND

In Theorem 3 of Section 2.4 we used the notion of a trivial self-mapping of a surface R. A trivial self-mapping of R is merely a homeomorphism f of R which is homotopic to the identity. We need the corresponding notion for a Fuchsian group.

Definition. *Let Γ be a Fuchsian group of the first kind acting on $\mathbb{H}$ and let w be a quasiconformal self-mapping of $\mathbb{H}$. Then w is trivial for Γ if $w \circ A \circ w^{-1} = A$ for all A in Γ.*

First, recall from Section 1.8 that any quasiconformal homeomorphism w of $\mathbb{H}$ has a unique continuous extension to the real axis, which we again denote by w. We now show that if such a mapping w is a trivial mapping for a group of the first kind, then its extension fixes all points of the real axis. If x is an attracting fixed point of a hyperbolic element and z is in $\mathbb{H}$, then $w(x) = \lim_{n\to\infty} w(A^n(z)) = \lim_{n\to\infty} A^n(w(z)) = x$. Since such points are dense in $\mathbb{R}$, the mapping w must fix all points of $\mathbb{R}$. Moreover, if p is an elliptic fixed point of an element A of Γ, then $w(p) = w(A(p)) = A(w(p))$, so $w(p)$ is also a fixed point of A. Since an elliptic element has only one fixed point in $\mathbb{H}$, we see that $w(p) = p$. On the other hand, if $w(x) = x$ for all x in $\mathbb{R}$ and if $w \circ A \circ w^{-1}$ equals a Möbius transformation, then obviously $w \circ A \circ w^{-1} = A$. We have proved the following proposition.

Proposition 1. *Let w be a quasiconformal self-mapping of $\mathbb{H}$ which conjugates a group Γ of the first kind into a group of Möbius transformations. Then w is trivial if and only if $w(x) = x$ for all x in $\mathbb{R}$. Moreover, a trivial mapping w automatically fixes any elliptic fixed points of Γ.*

Proposition 2. *Let Γ be a Fuchsian group of the first kind acting on $\mathbb{H}$ and let $R = \mathbb{H}_\Gamma/\Gamma$. Then a trivial mapping f of R naturally induces a trivial mapping w for Γ.*

Proof. Let $\pi\colon \mathbb{H}_\Gamma \to \mathbb{H}_\Gamma/\Gamma$ be the natural mapping and $R = \mathbb{H}_\Gamma/\Gamma$ the quotient Riemann surface. Let $\pi_1\colon \mathbb{H} \to \mathbb{H}_\Gamma$ be the universal covering of $\mathbb{H}_\Gamma$ with covering group Γ_1. (If Γ has no elliptic elements, then $\Gamma_1 = \{\text{identity}\}$.) Then $\tilde{\pi} = \pi \circ \pi_1$ is the universal covering mapping from $\mathbb{H}$ to R and it has covering group $\tilde{\Gamma}$. Now $\tilde{\Gamma}$ consists of all self-mappings A of $\mathbb{H}$ for which there exists an element B of Γ such that $\pi_1 \circ A = B \circ \pi_1$. Moreover, Γ_1 is a normal subgroup of $\tilde{\Gamma}$ and $\tilde{\Gamma}/\Gamma_1$ is isomorphic to Γ.

Let $f\colon R \to R$ be a quasiconformal mapping. Then there exists a quasiconformal mapping $\tilde{w}\colon \mathbb{H} \to \mathbb{H}$ such that $\tilde{\pi} \circ \tilde{w} = f \circ \tilde{\pi}$. If $\tilde{w}_1$ is another lifting of f (satisfying $\tilde{\pi} \circ \tilde{w}_1 = f \circ \tilde{\pi}$), then $\tilde{w}_1 = B \circ \tilde{w}$ for some B in $\tilde{\Gamma}$. Clearly, for any A in $\tilde{\Gamma}$, $\tilde{w} \circ A$ is a lifting of f. Therefore, there exists B in $\tilde{\Gamma}$ such that $\tilde{w} \circ A = B \circ \tilde{w}$.

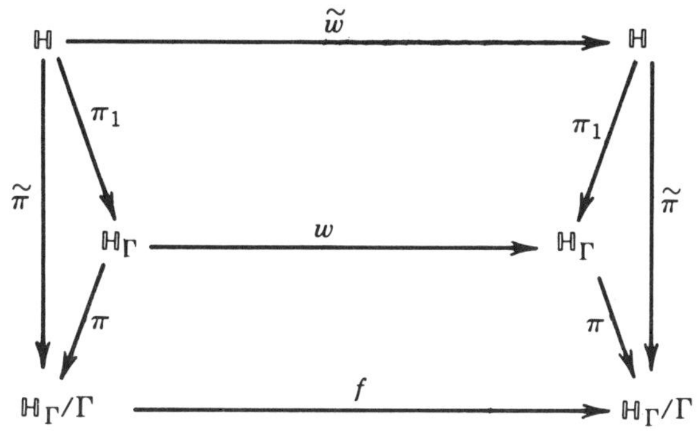

Figure 3.1

Now f is homotopic to the identity on R. Since homotopic maps lift to homotopic maps, the lifting $\tilde{w}$ is homotopic to the identity by a homotopy $\tilde{w}_t$ with $\tilde{w}_1 = \tilde{w}$ and $\tilde{w}_0 = \text{identity}$ (see Figure 3.1). This homotopy gives a continuous curve of homomorphisms $\chi_t: \tilde{\Gamma} \to \tilde{\Gamma}$ determined by

$$\tilde{w}_t \circ A = \chi_t(A) \circ \tilde{w}_t.$$

Then $\chi_0(A) = A$ and $\chi_t(A)$ is a continuous mapping into the discrete group $\tilde{\Gamma}$ and, therefore, $\chi_t(A) = A$ for $0 \leqslant t \leqslant 1$. We conclude that $\tilde{w} \circ A \circ \tilde{w}^{-1} = A$ for all A in $\tilde{\Gamma}$.

The mappings described above are summarized in the diagram in Figure 3.1. We have added in one new mapping, namely, w. It is uniquely determined by the condition that the diagram in Figure 3.1 commutes.

To define $w(z)$ we pick a point p in $\mathbb{H}$ such that $\pi_1(p) = z$. Then let $w(z) = \pi_1 \circ \tilde{w}(p)$. To show this is well defined, suppose $\tilde{p}$ is another point for which $\pi_1(\tilde{p}) = z$. Then there exists A_1 in Γ_1 for which $A_1(p) = \tilde{p}$. Then $\tilde{w}(\tilde{p}) = \tilde{w} \circ A_1(p) = A_1 \circ \tilde{w}(p)$ and

$$\pi_1 \circ \tilde{w}(\tilde{p}) = \pi_1 \circ A_1 \circ \tilde{w}(p) = \pi_1 \circ \tilde{w}(p).$$

We claim that $\tilde{w} \circ A = A \circ \tilde{w}$ for all A in $\tilde{\Gamma}$ implies $w \circ B = B \circ w$ for all B in Γ. Let B be in Γ and let A in $\tilde{\Gamma}$ satisfy $\pi_1 \circ A = B \circ \pi_1$. Then

$$\begin{aligned} w \circ B \circ \pi_1 &= w \circ \pi_1 \circ A = \pi_1 \circ \tilde{w} \circ A \\ &= \pi_1 \circ A \circ \tilde{w} = B \circ \pi_1 \circ \tilde{w} = B \circ w \circ \pi_1. \end{aligned}$$

This shows $w \circ B = B \circ w$ and thus that w is a trivial mapping for Γ.

Proposition 3. *Suppose that Γ is a Fuchsian group of the first kind and that Γ has no elliptic elements. If the mapping w which is the lifting of the mapping f in Figure 3.1 is trivial, then f is also trivial.*

Proof (due to Ahlfors [Ah4]). Suppose $w(x) = x$ for all x in $\mathbb{R}$ and w is the lift of a quasiconformal mapping $f: \mathbb{H}/\Gamma \to \mathbb{H}/\Gamma$. We must construct a homotopy which connects f to the identity. Let ρ be the noneuclidean metric for $\mathbb{H}$ and define $w_t(z)$ to be the point in $\mathbb{H}$ which lies on the noneuclidean segment joining z to $w(z)$ such that the noneuclidean distance from z to $w_t(z)$ and the noneuclidean distance from $w_t(z)$ to $w(z)$ are in the proportion $t:(1-t)$. Then $w_t(z)$ is obviously continuous in both t and z and $w_t \circ A(z) = A \circ w_t(z)$ because $w \circ A = A \circ w$ and the transformation A is an isometry in the noneuclidean metric. Thus w_t projects to a homotopy f_t which connects f to the identity.

Remark. By a theorem of Marden [Mar], Proposition 3 is true even when Γ contains elliptic elements. That is, if w is induced by f and if w is trivial, then there is a homotopy $f_t: \mathbb{H}_\Gamma/\Gamma \to \mathbb{H}_\Gamma/\Gamma$ which connects f to the identity. But this is more difficult to prove. The argument used above does not work because, for a pair of points z and $w(z)$, the noneuclidean segment from z to $w(z)$ may pass through an elliptic point and, therefore, the homotopy would not project to the punctured surface $\mathbb{H}_\Gamma/\Gamma$. Ultimately in Theorems 1 and 1′ of Section 9.1 we prove a theorem of Bers and Greenberg [BerGr] which implies Proposition 3 for Fuchsian groups with elliptic elements. That theorem was also proved in the finitely generated case by Earle and Kra [EK2]. Further references for this fact are given in the bibliographical notes to Chapter 9. It is even true that the homotopy for f in Proposition 3 may be replaced by "isotopy through quasiconformal mappings." This follows from the paper of Reich [Rei1] (see the notes for Chapter 6). It has also been proved in a recent paper of Earle and McMullen [EM] by using the extension theorem of Douady and Earle [DE].

3.3. THE REICH–STREBEL INEQUALITY FOR FINITELY GENERATED FUCHSIAN GROUPS OF THE FIRST KIND

We have already introduced in Section 3.1 the space of symmetric integrable holomorphic quadratic differential forms, $A_s(\Gamma, C)$. When Γ is of the first kind there is only one choice for the invariant closed set C, namely, $C = \Lambda = \hat{\mathbb{R}}$. Hence, we simply write $A(\Gamma)$ instead of $A_s(\Gamma, C)$. Note that whereas $A_s(\Gamma, C)$ is a real vector space, $A(\Gamma)$ is a complex vector space.

Theorem 2. *Let Γ be a finitely generated Fuchsian group of the first kind acting on $\mathbb{H}$ with fundamental domain ω. Let $M_0(\Gamma)$ be the space of all trivial Beltrami differentials μ for Γ. That is, $\mu = w_{\bar{z}}/w_z$, where w is a quasiconformal homeomorphism of $\mathbb{H}$ for which $w \circ A \circ w^{-1} = A$ for all A in Γ. Then for every μ in $M_0(\Gamma)$ and every φ in $A(\Gamma)$, one has*

$$\iint_\omega |\varphi(z)|\, dx\, dy \leqslant \iint_\omega \frac{|1 - \mu\varphi/|\varphi||^2}{1 - |\mu|^2} |\varphi|\, dx\, dy. \tag{5}$$

Proof. When Γ is finitely generated and torsion free, the theorem follows from Theorem 3 of the previous chapter. It is only necessary to observe that $A(R)$ and $A(\Gamma)$ are isometric by Theorem 1, where $R = \mathbb{H}/\Gamma$ and trivial maps from R into R correspond to self-mappings w of $\mathbb{H}$ for which $w \circ A \circ w^{-1} = A$, by Propositions 2 and 3. Then inequality (1) translates into inequality (5). Now suppose Γ contains elliptic elements. We need an algebraic theorem which says that any finitely generated Fuchsian group Γ contains a subgroup of finite index Γ_0 which is torsion free. The proof of this result is algebraic and not contained in this book. It can be found in Fox [Fox], Mennicke [Men], or Selberg [Se]. Since Γ is of the first kind, the subgroup Γ_0 of finite index must also be of the first kind. If μ is trivial for Γ, it is obviously trivial for Γ_0. Moreover, $A(\Gamma) \subset A(\Gamma_0)$. Therefore, we have inequality (5) with ω replaced by ω_0, a fundamental domain for Γ_0. If n is the index of Γ_0 in Γ, we can obtain ω_0 by taking the interior of the closures of n iterates of ω (see Section 1.6). Moreover, the integrals in (5) over each iterate of ω will be equal. Therefore, (5) follows by dividing through by n the corresponding inequality for ω_0.

The remainder of this chapter, as well as all of Chapter 4, is devoted to proving (5) for arbitrary Fuchsian groups Γ and arbitrary spaces $A_s(\Gamma, C)$. The reader who wishes to use Teichmüller theory only for Riemann surfaces of finite analytic type may go on to Chapter 5.

3.4. TRIVIAL MAPPINGS FOR GROUPS OF THE SECOND KIND

The objective of this section is to extend the notion of a trivial mapping to groups of the second kind and to surfaces with boundary and to prove propositions analogous to Proposition 2 and 3 for finitely generated groups of the second kind. As before, we are able to show that the two concepts of triviality coincide under the natural isomorphism only in the case that the group has no elliptic elements. To show they coincide in the general case would take us too far afield. However, we are still able to obtain the Reich–Strebel inequality for groups with elliptic elements.

As before, let C be a closed subset of $\hat{\mathbb{R}}$ which is invariant under Γ and which contains the limit set Λ of Γ. Let $\sigma = (C - \Lambda)/\Gamma$ be a distinguished subset of the border of the Riemann surface $R = \mathbb{H}_\Gamma/\Gamma$. Throughout this section we assume Γ is nonelementary, by which we mean Λ contains at least three points.

Definition. *A quasiconformal self-mapping w of $\mathbb{H}$ is C-trivial for the group Γ if $w \circ A \circ w^{-1} = A$ for every A in Γ and if $w(x) = x$ for every x in C.*

Definition. *A quasiconformal mapping $f: R \to R$ is σ-trivial if there is a continuous curve of continuous mappings $f_t: R \to R$ such that*

(i) *f_0 is the identity on R;*
(ii) *f_t extends continuously to the border of R and $f_t(p) = p$ for p in σ and $0 \leqslant t \leqslant 1$; and*
(iii) *$f_1 = f$.*

All of the mappings in Figure 3.1 are defined in the same way as in Section 3.2 and we have the following result.

Proposition 4. *If f is σ-trivial and if w is the corresponding lifting in Figure 3.1, then w is C-trivial, where $\sigma = (C - \Lambda)/\Gamma$. Moreover, if Γ contains no elliptic elements and w is C-trivial, then f is σ-trivial.*

Proof. Just as in Proposition 2, we can show that if f is σ-homotopic to the identity, then $w \circ B = B \circ w$ for all B in Γ. It follows that $w(x) = x$ for every x in Λ. Let w_t be a lifting of the homotopy f_t. Let J be a component of $\hat{\mathbb{R}} - \Lambda$. Now J is an open interval and $w(J)$ must be the same interval since w fixes the endpoints of J, which are in the limit set Λ. For p in $C \cap J$ we know that $w(p) = B(p)$, where B is in the subgroup of Γ which fixes J. If B is not the identity, then the continuous curve $w_t(p)$ which joins p to $B(p)$ would project to a continuous curve $f_t(\pi(p))$ which winds around the boundary component determined by J. But $f_t(\pi(p)) = \pi(p)$ for each t, $0 \leqslant t \leqslant 1$ and $\pi(p)$ in σ. Thus B is the identity and we see that $w(x) = x$ for all x in C.

Conversely, suppose $w(x) = x$ for all x in C and suppose that Γ has no elliptic elements. Let ρ be the noneuclidean metric for the region $\hat{\mathbb{C}} - \Lambda$. This differs from the proof of Lemma 1 where we took the noneuclidean metric for $\mathbb{H}$. Define $w_t(z)$ to be the point in $\mathbb{H}$ on the noneuclidean segment (with respect to ρ) joining z to $w(z)$ such that the noneuclidean distance from z to $w_t(z)$ and the noneuclidean distance from $w_t(z)$ to $w(z)$ are in the proportion $t:(1 - t)$. Then $w_t(Az) = A \circ w_t(z)$ for A in Γ and $w_t(p) = p$ for p in C. Therefore, w_t determines a σ-homotopy f_t which joins f to the identity.

3.5. FINITELY GENERATED GROUPS OF THE SECOND KIND

Let Γ be a finitely generated Fuchsian group of the second kind acting on $\mathbb{H}$ and let C be a Γ-invariant closed subset of $\hat{\mathbb{R}}$ containing the limit set of Γ. Assume $\sigma = (\Lambda - C)/\Gamma$ is a finite set. Let $M_0(\Gamma, C)$ be the set of all functions $\mu(z)$ defined on $\mathbb{H}$ of the form $\mu = w_{\bar{z}}/w_z$, where w is a quasiconformal self-mapping of $\mathbb{H}$ satisfying $w \circ A \circ w^{-1} = A$ for all A in Γ and $w(x) = x$ for all x in C.

Theorem 3. *In the above setting, for all φ in $A_s(\Gamma, C)$ and all μ in $M_0(\Gamma, C)$, one has inequality* (5).

Proof. For this case we consider the surface $\hat{R} = (\hat{\mathbb{C}} - C)/\Gamma$, which is a surface of finite analytic type. A trivial Beltrami coefficient μ in $M_0(\Gamma, C)$ extends to a Beltrami coefficient $\hat{\mu}(z)$ equal to μ in the upper half plane and $\overline{\mu(\bar{z})}$ in the lower half plane. If Γ is torsion free, then the Ahlfors's homotopy (Propositions 3 and 4) for μ extends to a homotopy for the self-mapping of $\hat{R}$ with Beltrami coefficient $\hat{\mu}$. Inequality (5) follows even for the case where Γ has elliptic elements by the same device used to prove Theorem 2.

Remark. From the results of the next chapter, we will see that Theorem 3 is true whether or not Γ is finitely generated and whether or not $\sigma = (C - \Lambda)/\Gamma$ is a finite set.

Notes. Formulation of inequality (1) in the setting of Fuchsian groups does not appear in the literature. The isometry between the Banach spaces $A(R, \sigma)$ and $A(\Gamma, C)$ in Section 3.1 is well known and can be found in the book by Kra [Kr2] or the papers of Ahlfors and Bers referred to there. The notion of trivial mappings and Ahlfors's homotopy for torsion free Fuchsian groups is treated in Ahlfors [Ah4]. Bers [Ber10] considers the same problem of extending to general Fuchsian groups an inequality analogous to but different from (1).

The technique we use to generalize the Reich–Strebel inequality to Fuchsian groups which contain elliptic elements relies in an essential way on an algebraic theorem. The theorem says that any finitely generated Fuchsian group contains a torsion free normal subgroup of finite index. References for this result are given at the end of the proof of Theorem 2 in Section 3.3. A recent paper of R. S. Kulkarni [Ku] investigates the minimum index of a torsion free normal subgroup of a $\{p, q, r\}$-triangle group. In our application it is not necessary for the subgroup to be normal.

4

DENSITY THEOREMS FOR QUADRATIC DIFFERENTIALS

In Chapter 3, the inequality of Reich and Strebel is proved for any finitely generated Fuchsian group of the first kind, or a finitely generated Fuchsian group of the second kind and a finite closed set on the border of the associated Riemann surface. This is enough to treat the theory of finite dimensional Teichmüller spaces, and if one does not wish to develop the theory for infinite dimensional spaces, most of the results to be proved in this chapter are unnecessary.

In order to extend the inequality, we need to look at three topics. The first is a density theorem for rational functions with simple poles in a closed set C in the space of all integrable holomorphic functions on $\hat{\mathbb{C}} - C$. The second is a theorem on the surjectivity of the Poincaré theta series operator. The third is a theorem on the existence of reproducing kernel functions for connected plane domains with more than two boundary points. Finally, using these topics, we can give the exhaustion process which accomplishes the desired generalization of the Reich–Strebel inequality.

Except for the exhaustion process, all of the above theorems are presented in greater generality in Kra [Kr2]. The exposition given here is more brief and adapted to our special needs.

A similar but different exhaustion process is given by Bers [Ber10].

4.1. BERS'S APPROXIMATION THEOREM

Let C be a closed subset of $\hat{\mathbb{C}}$ and C_0 a dense subset of C. Let $R(C_0)$ be the set of rational functions $r(z)$ which are holomorphic except for at most simple poles at points of C_0 and for which

$$\|r\| = \iint_{\mathbb{C}} |r(z)|\, dx\, dy < \infty.$$

Let $A(C)$ be the set of all functions φ holomorphic in $\mathbb{C} - C$ for which $|\varphi(z)| = 0(|z|^{-4})$ as $z \to \infty$ when $\infty \notin C$ and for which

$$\|\varphi\| = \iint_{\mathbb{C}-C} |\varphi(z)|\, dx\, dy < \infty.$$

Theorem 1. *If C_0 is dense in C, then $R(C_0)$ is dense in $A(C)$ in the L_1-norm.*

Proof. If C is a finite set, then the hypothesis implies $C_0 = C$ and the theorem is obvious.

If C is infinite, obviously C_0 is infinite and by using a conjugation by a Möbius transformation we may assume 0, 1, and ∞ are points of C_0. Suppose $\mu(z)$ is an L_∞ complex valued function which satisfies the orthogonality condition

$$\iint \mu(z) r(z)\, dx\, dy = 0$$

for all rational functions $r(z)$ in $R(C_0)$. From the Hahn–Banach theorem and the Riesz theorem which says that L_∞ is the dual space to L_1, Theorem 1 will follow if we show that this orthogonality condition implies $\iint_{\mu\varphi} dx\, dy = 0$ for all φ in $A(C)$.

We form the potential function

$$F(z) = -\frac{z(z-1)}{\pi} \iint_{\mathbb{C}} \frac{\mu(\zeta)\, d\xi\, d\eta}{\zeta(\zeta-1)(\zeta-z)}. \tag{1}$$

We will show that the potential function F in (1) satisfies

$$\frac{\partial}{\partial \bar{z}} F(z) = \mu(z),$$

where the derivative is taken in the distributional sense. Notice that part of the integrand in (1),

$$r(\zeta) = \frac{z(z-1)}{\zeta(\zeta-1)(\zeta-z)}, \tag{2}$$

is a rational function with only simple poles in $R(C_0)$, so long as z is in C_0.

We can now outline a heuristic proof of the theorem. We will see that $F(z)$ defined in (1) is continuous and so the hypotheses that μ is orthogonal to $R(C_0)$ and that C_0 is dense in C tell us that $F(z) = 0$ for z in C. Let φ be in $A(C)$. Then, where the double integrals are taken over the domain $\mathbb{C} - C$, we have

$$\iint \mu(z)\varphi(z)\,dx\,dy = \iint F_{\bar{z}}(z)\varphi(z)\,dx\,dy$$

$$= -\frac{1}{2i}\iint \frac{\partial}{\partial\bar{z}}(F(z)\varphi(z))\,dz \wedge d\bar{z}$$

$$= +\frac{1}{2i}\iint d(F(z)\varphi(z)\,dz)$$

$$= \frac{1}{2i}\int F(z)\varphi(z)\,dz = 0.$$

The last integral is a line integral along the boundary of the domain of integration, where F is identically zero. Of course, $\varphi(z)$ is not continuous on this boundary and the boundary certainly need not be a smooth curve. Thus the line integral is not defined.

In order to make this heuristic proof valid, we need to establish continuity properties of the potential function F and to use a device known as Ahlfors's mollifier.

Lemma 1. *The potential function $F(z)$ defined in* (1) *has the following properties:*

(i) $F(z) = 0$ *for all z in C;*
(ii) $|F(z) - F(t)| \leqslant (\text{const})|z - t|\log|z - t|^{-1}$ *for* $|z|$ *and* $|t| < R$ *and* $|z - t| < \frac{1}{2}$;
(iii) $|F(z)| \leqslant (\text{const})|z|\log|z|$ *for* $|z| > 3$; *and*
(iv) $F_{\bar{z}}(z) = \mu(z)$ *in the sense of distributions.*

The constants in (ii) *and* (iii) *depend only on R and* $\|\mu\|_\infty$.

Proof. Since C_0 is dense in C and (ii) implies $F(z)$ is continuous, (i) follows from the hypothesis that $\mu(\zeta)$ is orthogonal to the rational function of ζ given in (2). Notice that (2) has the partial fractions decomposition

$$\frac{z(z-1)}{\zeta(\zeta-1)(\zeta-z)} = \frac{1}{\zeta - z} - \frac{z}{\zeta - 1} + \frac{z-1}{\zeta}. \tag{3}$$

It is therefore obvious that for $|z| \leqslant R$, the function

$$F_1(z) = -\frac{1}{\pi} \iint\limits_{|\zeta|<2R} \frac{\mu(\zeta)}{\zeta - z}\, d\xi\, d\eta$$

differs from $F(z)$ by a holomorphic function. Thus it suffices to show the continuity condition (ii) for the function $F_1(z)$. Clearly

$$F_1(z) - F_1(t) = -\frac{1}{\pi} \iint\limits_{|\zeta|<2R} \frac{\mu(\zeta)(z - t)}{(\zeta - z)(\zeta - t)}\, d\xi\, d\eta$$

and on making the substitution $(z - t)w = \zeta - z$, one obtains

$$|F_1(z) - F_1(t)| \leqslant \frac{1}{\pi} |z - t|\, \|\mu\|_\infty \iint \left| \frac{1}{w(w + 1)} \right| du\, dv,$$

where the integral is over the region $|w| < 3R/|z - t|$. This integral can be divided into two parts: the part where $|w| < 3$ and the part where $3 < |w| < 3R/|z - t|$. The first part is bounded by a constant times $|z - t|$ and the second is bounded by a constant times

$$|z - t| \iint \frac{1}{|w|^2}\, du\, dv, \tag{4}$$

where the integral (4) is taken over the domain $3 < |w| < 3R/|z - t|$. This clearly works out to condition (ii).

To verify (iii), write

$$\frac{\pi |F(z)|}{\|\mu\|_\infty |z(z - 1)|} \leqslant \iint\limits_{|\zeta|<2} \frac{d\xi\, d\eta}{|\zeta(\zeta - 1)(\zeta - z)|} + \iint\limits_{|\zeta|>2} \frac{d\xi\, d\eta}{|\zeta(\zeta - 1)(\zeta - z)|}.$$

The first integral is obviously less than a constant times $|z|^{-1}$ for $|z| > 3$. Thus we must show that the second integral gives a contribution of order $0(|z|^{-1} \log|z|)$ for $|z| > 3$. The second integral is less than

$$2 \iint\limits_{|\zeta|>2} \frac{d\xi\, d\eta}{|\zeta - z|\, |\zeta|^2}. \tag{5}$$

Let $\zeta = zw$ and $d\xi\, d\eta = |z|^2\, du\, dv$, where $\zeta = \xi + i\eta$ and $w = u + iv$. Then (5) is less than a constant times $|z|^{-1}$ times

$$\iint\limits_{\frac{2}{|z|}<|w|} \frac{du\,dv}{|w-1|\,|w|^2}. \tag{6}$$

If (6) is broken down into the sum of the integral over the region $2|z|^{-1} < |w| < \frac{1}{2}$ and the integral over the region $\frac{1}{2} < |w|$, the first integral contributes a term of order $\log|z|$ and the second a term which is constant. Putting all this together, the estimate (iii) follows.

To prove (iv), first suppose μ is C^2 and has compact support. Then from (3) it is obvious that $F(z)$ defined in (1) differs from

$$G(z) = -\frac{1}{\pi}\iint \frac{\mu(\zeta)\,d\xi\,d\eta}{\zeta - z}$$

by a holomorphic function and so $F_{\bar{z}} = G_{\bar{z}}$. Notice that

$$G(z) = -\frac{1}{\pi}\iint \frac{\mu(\zeta + z)}{\zeta}\,d\xi\,d\eta$$

and so

$$G_{\bar{z}}(z) = -\frac{1}{\pi}\iint \frac{\mu_{\bar{\zeta}}(\zeta)\,d\xi\,d\eta}{\zeta - z}.$$

Since μ has compact support, Stoke's theorem yields

$$G_{\bar{z}}(z) = -\frac{1}{2\pi i}\iint \frac{\mu_{\bar{\zeta}}\,\overline{d\zeta} \wedge d\zeta}{\zeta - z} = -\frac{1}{2\pi i}\lim_{\varepsilon\to 0}\int_{\gamma_\varepsilon} \frac{\mu(\zeta)}{(\zeta - z)}\,d\zeta = \mu(z), \tag{7}$$

where γ_ε is a circle of radius ε about z traced out in the clockwise direction. One sees that $F_{\bar{z}}(z) = \mu(z)$. To finish the proof of part (iv), we must show that

$$\iint F(z)\varphi_{\bar{z}}\,dx\,dy = -\iint \mu(\zeta)\varphi(\zeta)\,d\xi\,d\eta \tag{8}$$

for every C^2 function φ which has compact support. But

$$\iint F\varphi_{\bar{z}}\,dx\,dy = -\frac{1}{\pi}\iint \mu(\zeta)\iint \varphi_{\bar{z}}\frac{z(z-1)}{\zeta(\zeta-1)(\zeta-z)}\,dx\,dy\,d\xi\,d\eta$$

and, by the same argument used to show (7), for the integral on the inside of the right-hand side, we have

$$-\frac{1}{\pi}\iint \varphi_{\bar{z}}(z)\frac{z(z-1)}{\zeta(\zeta-1)(\zeta-z)}\,dx\,dy = -\varphi(\zeta).$$

The end result is (8).

For the next step in the proof of Theorem 1 we introduce Ahlfors's mollifier. Let $j(t)$ be a real valued smooth function such that $j(t) \equiv 0$ for $t \leqslant 1$ and $j(t) \equiv 1$ for $t \geqslant 2$ and form

$$m_n(z) = j\left(\frac{n}{\log\log \delta^{-1}(z)}\right), \tag{9}$$

where $\delta(z)$ is the minimum of e^{-2} and the distance from z to the closed set C.

Lemma 2. *For positive integers n, the function $m_n(z)$ defined in* (9) *has the following properties:*

(i) $0 \leqslant m_n(z) \leqslant 1$;
(ii) *for each z, there exists n_0 such that $m_n(z) = 1$ for n larger than n_0;*
(iii) $m_n(z) \equiv 0$ *in a neighborhood of C; and*

(iv) $\left|\frac{\partial m_n}{\partial \bar{z}}\right| \leqslant (\text{const})\,\frac{1}{n}\cdot\frac{1}{\delta\log\delta^{-1}}.$

Proof. The first three properties are obvious. Notice that from the definition of $\delta(z)$ it is obvious that

$$|\delta(z) - \delta(w)| \leqslant |z - w|$$

and from this inequality one can show that the partial derivatives of δ with respect to z and $\bar{z}$ exist in the sense of distributions and they are both bounded by 1. A calculation gives

$$\frac{\partial m_n}{\partial \bar{z}} = j'(\;)\frac{-n}{(\log\log\delta^{-1})^2}\cdot\frac{1}{\delta\log\delta^{-1}}\cdot\frac{\partial\delta}{\partial\bar{z}}. \tag{10}$$

Here, the argument of j' is the same as the argument of j in (9), and since $j'(t) \equiv 0$ unless $1 < t < 2$, in (10) we may assume that

$$1 < \frac{n}{\log\log\delta^{-1}} < 2.$$

Therefore, (10) yields

$$\left|\frac{\partial m_n}{\partial \bar{z}}\right| \leqslant (\text{const}) \frac{1}{n} \frac{1}{\delta \log \delta^{-1}}, \tag{11}$$

which proves Lemma 2.

To proceed with the proof of Theorem 1 recall properties (i) and (ii) of Lemma 1. They tell us that for z with distance less than $\frac{1}{2}$ to C and for $|z| < R$

$$|F(z)| \leqslant C(R)\delta(z) \log \delta^{-1}(z). \tag{12}$$

Since $\delta(z)$ is never more than e^{-2}, this inequality obviously holds throughout $|z| < R$ for a constant $C(R)$ depending only on R and not on z.

For any function φ holomorphic in the complement of C, we have

$$\begin{aligned} \iint\limits_{|z|<R} \mu\varphi m_n \, d\bar{z}\, dz &= \iint\limits_{|z|<R} m_n \, d(F\varphi \, dz) \\ &= \int_{|z|=R} m_n F\varphi \, dz + \iint\limits_{|z|<R} F\varphi \frac{\partial m_n}{\partial \bar{z}} \, d\bar{z}\, dz. \end{aligned} \tag{13}$$

From (12) and property (iv) of Lemma 2, the double integral on the right in (13) is less than a constant times $(1/n) \iint_{|z|<R} |\varphi| \, dx\, dy$ and, if φ is integrable, this approaches 0 as $n \to \infty$. Taking the limit as $n \to \infty$ in (13), we arrive at

$$\left|\iint\limits_{|z|<R} \mu\varphi \, dz\, d\bar{z}\right| \leqslant \int_{|z|=R} |F\varphi| \, |dz| \leqslant (\text{const})R \log R \int_{|z|=R} |\varphi(z)| \, |dz|. \tag{14}$$

But it is not possible for there to exist a constant $C > 0$ for which

$$\int_{|z|=R} |\varphi(z)| \, |dz| \geqslant \frac{C}{R \log R}. \tag{15}$$

If (15) were true, then

$$\begin{aligned} \iint |\varphi(z)| \, dx\, dy &\geqslant \int_2^\infty \int_{|z|=R} |\varphi(z)| \, |dz| \, dR \\ &\geqslant C \int_2^\infty \frac{dR}{R \log R} = \infty, \end{aligned}$$

which contradicts the assumption that φ is integrable, and the proof of Theorem 1 is now complete.

Remark. It is not always necessary for C_0 to be dense in C. The potential function F defined in (1) is holomorphic on any open subset of the complement of the support of μ. We only need a subset C_0 of C which is sufficiently large that if F vanishes on C_0, then it vanishes on the boundary of the complement of C (see [Ber4]). This remark is of no use in our situation because we will apply the theorem for sets C with no interior.

4.2. A DENSITY THEOREM FOR FUCHSIAN GROUPS

Let C be a closed Γ-invariant set for the Fuchsian group Γ and assume $\Lambda \subseteq C \subseteq \hat{\mathbb{R}}$. Let ω be a fundamental domain for Γ in $\hat{\mathbb{C}} - \Lambda$. Obviously, we may choose ω so that it is symmetric about $\mathbb{R}$; that is, $\omega = \overline{\omega}$. If Γ is of the first kind, then ω is necessarily disconnected, but we may assume the component in each half plane is noneuclidean convex. If Γ is of the second kind, we may assume ω is connected and bounded by circular arcs and vertices.

The space $A(\Gamma, C)$ consists of all functions φ holomorphic in $\hat{\mathbb{C}} - C$ for which

(i) $\varphi(B(z))B'(z)^2 = \varphi(z)$ for all B in Γ; and
(ii) $\|\varphi\| = \iint_\omega |\varphi(z)|\, dx\, dy < \infty$.

The space $R(\Gamma, C)$ is the subset of $A(\Gamma, C)$ whose elements are holomorphic in $\mathbb{C} - \Lambda$ except for at most isolated simple poles on $C - \Lambda$ which are finite in number on the orbit space $(C - \Lambda)/\Gamma$. Note that by reflection the space $A_s(\Gamma, C)$ introduced in the previous chapter can be regarded as a subspace of $A(\Gamma, C)$.

Theorem 2. *Suppose $\Lambda \subseteq C_0 \subseteq C \subseteq \hat{\mathbb{R}}$ and C_0 and C are invariant sets for the Fuchsian group Γ with limit set Λ. Then $R(\Gamma, C_0)$ is dense in $A(\Gamma, C)$ if C_0 is dense in C.*

Proof. It is easy to check that $R(B\Gamma B^{-1}, B(C_0)) \cong R(\Gamma, C)$ and $A(B\Gamma B^{-1}, B(C)) \cong A(\Gamma, C)$ for any Möbius transformation B. Since C must have at least three points [or else $A(\Gamma, C)$ is empty], we may assume, after a conjugation, that 0, 1, and ∞ are points in C_0. Form the summation

$$r(\zeta) = \sum \frac{z(z-1)B'(\zeta)^2}{B(\zeta)(B(\zeta)-1)(B(\zeta)-z)}, \tag{16}$$

where z is in C_0 and the sum is over all B in Γ. The series in (16) converges absolutely and uniformly on compact subsets of $\mathbb{C} - C_0$. It has simple poles at the points of $\{B(z) | B \text{ in } \Gamma\}$. These points form a discrete set on the real axis with the limit set removed since Γ acts properly discontinuously at points not in its limit set. The convergence of the series (16) follows from the inequalities

$$\iint_{\omega} |r(\zeta)|\, d\xi\, d\eta \leqslant \sum_{B} \iint_{B(\omega)} \left| \frac{z(z-1)}{\zeta(\zeta-1)(\zeta-z)} \right| d\xi\, d\eta$$

$$= \iint_{\mathbb{C}} \frac{|z(z-1)|}{|\zeta(\zeta-1)(\zeta-z)|}\, d\xi\, d\eta < \infty,$$

where the summation runs over all B in Γ. It is also easy to see that $r(B(\zeta))B'(\zeta)^2 = r(\zeta)$ for B in Γ. Therefore, $r(\zeta)$ is in $R(\Gamma, C_0)$.

To prove the theorem we suppose μ is an L_∞-function on ω for which $\iint_\omega \mu r = 0$ for all r in $R(\Gamma, C_0)$ and we must show that $\iint_\omega \mu\varphi = 0$ for all φ in $A(\Gamma, \varphi)$. First, extend the function μ to all of $\mathbb{C}$ by stipulating

$$\mu(B(z)) = \frac{\mu(z)B'(z)}{\overline{B'(z)}} \qquad \text{for } z \text{ in } \omega \text{ and } B \text{ in } \Gamma. \tag{17}$$

This defines μ up to a set of measure zero because $\mathbb{C} - \Lambda$ is the disjoint union, except for a countable union of analytic arcs, of the sets $B(\omega)$ for B in Γ. For the function $r(\zeta)$ in (16), we have

$$\iint_{\omega} r(\zeta)\mu(\zeta)\, d\xi\, d\eta = \iint_{\mathbb{C}} \frac{z(z-1)}{\zeta(\zeta-1)(\zeta-z)}\, \mu(\zeta)\, d\xi\, d\eta. \tag{18}$$

Therefore, the hypothesis that $\iint_\omega r(\zeta)\mu(\zeta)\, d\xi\, d\eta = 0$ for all r in $R(\Gamma, C_0)$ tells us that $F(z)$ defined in (1) is zero for z in C_0. By the continuity of F, this means $F(z) \equiv 0$ for z in C and the proof of Theorem 1 tells us that

$$\iint_{\mathbb{C}} \Phi(\zeta)\mu(\zeta)\, d\xi\, d\eta = 0 \tag{19}$$

for all Φ in $A(C)$. This then implies

$$\iint_{\omega} \{\sum \Phi(B(\zeta))B'(\zeta)^2\}\mu(\zeta)\, d\xi\, d\eta = 0 \tag{20}$$

for all Φ in $A(C)$, where the summation in the curly brackets in (20) is taken over all B in Γ. To complete the proof of the theorem, we need to know that the summation inside the integral in (20) sweeps out a dense subspace of $A(\Gamma, C)$ as Φ varies in $A(C)$. In the next two sections we will show that this summation gives a surjective mapping.

4.3. POINCARÉ THETA SERIES

Let Θ be the linear operator from $A(C)$ into $A(\Gamma, C)$ defined by

$$\Theta(F)(z) = \sum F(B(z))B'(z)^2, \tag{21}$$

where the summation is over all B in a Fuchsian group Γ. Assume that 0, 1, and ∞ are in C.

Theorem 3. *The series in* (21) *converges absolutely and uniformly on compact subsets of* $\mathbb{C} - C$. *The series* (21) *defines a linear operator* Θ *from* $A(C)$ *to* $A(\Gamma, C)$ *which is surjective and has norm less than or equal to* 1. *Moreover,* Θ *maps the unit ball in* $A(C)$ *onto the ball of radius* $\frac{1}{3}$ *in* $A(\Gamma, C)$.

Proof. The statements about the convergence of the series and the fact that $\|\Theta\| \leqslant 1$ follows immediately from the following string of equalities and inequalities:

$$\|\Theta F\|_{\omega} \leqslant \iint_{\omega} \sum |F(B(z))B'(z)^2| \, dx \, dy$$

$$= \iint_{\bigcup B^{-1}(\omega)} |F(z)| \, dx \, dy = \iint_{\mathbb{C}} |F(z)| dx \, dy = \|F\|,$$

where the summation and union are over all B in Γ.

For the proof of the rest of the Theorem 3, we need to prove the existence of a reproducing kernel function for the domain $D = \mathbb{C} - C$, which is the topic of the next section. At the end of the next section, we finish the proof of Theorem 3.

4.4 KERNEL FUNCTIONS FOR PLANE DOMAINS

Lemma 3. *For any plane domain D with three or more boundary points in* $\hat{\mathbb{C}}$, *there is a kernel function* $K(z, \zeta)$ *defined for* (z, ζ) *in* $D \times D$ *which is holomorphic in z and antiholomorphic in* ζ *which has the following properties:*

(i) $K(z, \zeta) = \overline{K(\zeta, z)}$;
(ii) *for every conformal self-mapping A of D,* $K(Az, A\zeta)A'(z)^2\overline{A'(\zeta)}^2 = K(z, \zeta)$;
(iii) $\iint |K(z, \zeta)| \, dx \, dy \leqslant \pi\rho^2(\zeta)$;
(iv) *for every integrable holomorphic function* φ *on* D, $\varphi(z) = (3/\pi) \iint \rho^{-2}(\zeta)K(z, \zeta)\varphi(\zeta) \, d\xi \, d\eta$; *and*
(v) *for each fixed* ζ *in D,* $\sup_z |K(z, \zeta)|\rho^{-2}(z) < \infty$.

In the above formulas and integrals, D is the domain of integration and $\rho(z)$ is the Poincaré metric for the domain D. If confusion can arise because of discussion of different domains, we will sometimes write ρ_D and K_D.

We first discuss the case where D is the unit disk Δ. Then K and ρ are given explicitly by the following formulas:

$$K_\Delta(z, \zeta) = (1 - z\bar{\zeta})^{-4},$$
$$\rho_\Delta(z) = (1 - |z|^2)^{-1}. \tag{22}$$

In this case, property (i) of the lemma is completely obvious and the verification of (ii) is an elementary exercise. To verify (iii), let

$$g(\zeta) = \iint_\Delta |K_\Delta(z, \zeta)|\, dx\, dy$$

and note that by using the change of variable $\zeta \mapsto A(\zeta)$ we get $g(A(\zeta))|A'(\zeta)|^2 = g(\zeta)$ for any Möbius transformation A which preserves Δ. Thus we can determine $g(\zeta)$ by calculating $g(0)$. Obviously, $g(0) = \pi$ and we get property (iii) with equality.

To prove (iv), observe that for $r < 1$ the mean value property for harmonic functions says

$$\varphi(0) = \frac{1}{2\pi} \int_0^{2\pi} \varphi(re^{i\theta})\, d\theta.$$

Thus,

$$\varphi(0) \int_0^1 (1 - r^2)^2 r\, dr = \frac{1}{2\pi} \int_0^{2\pi} \int_0^1 (1 - r^2)^2 \varphi(re^{i\theta}) r\, dr\, d\theta,$$

which holds as long as $\varphi(z)$ is integrable in Δ. This can be rewritten as

$$\varphi(0) = \frac{3}{\pi} \iint_\Delta \rho_\Delta(\zeta)^{-2} K_\Delta(0, \zeta) \varphi(\zeta)\, d\xi\, d\eta. \tag{23}$$

A simple calculation using the invariance properties of ρ_Δ and K_Δ now yields (iv). Property (v) for K_Δ and ρ_Δ is obvious.

Now we treat the case where D is a connected open set with at least three boundary points in $\hat{\mathbb{C}}$. This hypothesis implies that there is a universal covering mapping $p\colon \Delta \to D$ with a Fuchsian covering group G for which $\Delta/G \cong D$. The noneuclidean metric ρ for the domain D is defined by the relation $\rho(p(z))|p'(z)| = \rho_\Delta(z)$.

To obtain the kernel function, we first form the series

$$F(z, \zeta) = \sum_{B \in G} K_{\Delta}(B(z), \zeta) B'(z)^2 \tag{24}$$

defined for z and ζ in Δ and then let the kernel function be determined by the equation

$$K(p(z), p(\zeta)) p'(z)^2 \overline{p'(\zeta)^2} = F(z, \zeta). \tag{25}$$

Ultimately, we will show that K in (25) is well defined and satisfies all the properties of Lemma 3. First, we need to establish analogous properties for $F(z, \zeta)$.

Lemma 4. *The series for $F(z, \zeta)$ in* (24) *converges absolutely and uniformly on compact subsets of Δ to a function which is holomorphic in z and antiholomorphic in ζ and which satisfies*

(i) $F(z, \zeta) = \overline{F(\zeta, z)}$;
(ii) *for every B in G, $F(Bz, \zeta) B'(z)^2 = F(z, \zeta)$;*
(iii) $F(Az, A\zeta) A'(z)^2 \overline{A'(\zeta)^2} = F(z, \zeta)$ *for all A in the normalizer of G;*
(iv) *for every G-automorphic quadratic differential ψ for which $\iint_{\Delta/G} |\psi(z)| \, dx \, dy < \infty$,*

$$\psi(z) = \frac{3}{\pi} \iint_{\Delta/G} \rho_{\Delta}^{-2}(\zeta) F(z, \zeta) \psi(\zeta) \, d\xi \, d\eta;$$

and
(v) *for fixed $|\zeta| < 1$, $\sup |F(z, \zeta) \rho_{\Delta}^{-2}(z)| < \infty$.*

Proof. For fixed ζ, the function $K_{\Delta}(z, \zeta)$ is integrable in the variable z and the statement about (24) converging absolutely and uniformly on compact subsets of Δ follows in the same way we proved the corresponding statement for the series in (16). This observation also shows property (ii). The fact that $F(z, \zeta)$ is holomorphic in z and antiholomorphic in ζ is obvious. To verify (i) it is necessary to use the symmetry property for K_{Δ} and the fact that $K_{\Delta}(A(z), A(\zeta)) A'(z)^2 \overline{A'(\zeta)^2} = K_{\Delta}(z, \zeta)$.

To prove (iii), let A be in the normalizer of G. Then

$$\begin{aligned} F(Az, A\zeta) A'(z)^2 \overline{A'(\zeta)^2} &= \sum_{B \in G} K(BAz, A\zeta) B'(Az)^2 A'(z)^2 \overline{A'(\zeta)^2} \\ &= \sum K(BAz, A\zeta)(BA)'(z)^2 \overline{A'(\zeta)^2} \\ &= \sum K(AA^{-1}BAz, A\zeta)(AA^{-1}BA)'(z)^2 \overline{A'(\zeta)^2} \\ &= \sum K(A^{-1}BAz, \zeta)(A^{-1}BA)'(z)^2 = F(z, \zeta). \end{aligned}$$

Using the invariance property of $F(z, \zeta)$ in (ii) and the invariance of ρ_Δ, the formula in (iv) reduces to the formula in part (iv) of Lemma 3 for the case $D = \Delta$, a case which we have already proved.

To prove (v) it suffices to show that

$$\sum_{A \in G} |A'(z)|^2 \leqslant (\text{const})\rho_\Delta^2(z), \tag{26}$$

where the constant depends only on the group G. To see this, form the function

$$h(z) = \frac{1}{\pi} \iint_\omega \frac{d\xi\, d\eta}{|1 - z\bar{\zeta}|^4}, \tag{27}$$

where ω is a fundamental domain for G. Note that

$$\sum_{A \in G} h(Az)|A'(z)|^2 = \frac{1}{\pi} \iint_\Delta \frac{d\xi\, d\eta}{|1 - z\bar{\zeta}|^4} = \frac{1}{(1 - |z|^2)^2}. \tag{28}$$

The denominator of the integrand in (27) is clearly less than $2^4 = 16$, and thus

$$h(z) \geqslant \frac{1}{16\pi} \times (\text{Euclidean area of } \omega). \tag{29}$$

From (28) and (29), we see that (26) is true with the constant equal to $16\pi/(\text{Euclidean area of } \omega)$.

Now we return to the kernel function K defined in (25). The fact that K is well defined follows from property (iii) of Lemma 4. To see that K satisfies property (iii) of Lemma 3, we calculate

$$\begin{aligned} \iint_D |K(p(z), \zeta)|\, d\xi\, d\eta &= \iint_\omega |K(p(z), p(\zeta))|\, |p'(\zeta)|^2\, d\xi\, d\eta \\ &= \iint_\omega |F(z, \zeta)|\, d\xi\, d\eta |p'(z)|^{-2} \\ &\leqslant |p'(z)|^{-2} \iint_\Delta |K_\Delta(z, \zeta)|\, d\xi\, d\eta \\ &= \pi |p'(z)^{-2}| \rho_\Delta(z)^2 = \pi \rho_D(p(z))^2 \end{aligned}$$

The verification of the rest of the properties for the kernel function K stated in Lemma 3 follow easily from the analogous properties of the function F and we leave them to the reader.

We can now show that the theta series operator from $A(C)$ to $A(\Gamma, C)$ is surjective. Let $\varphi \in A(\Gamma, C)$ and let ρ and K be the noneuclidean metric and the kernel function described in Lemma 3 for the domain $D = \hat{\mathbb{C}} - C$. Let ω be a fundamental domain for Γ in D and χ_ω be the characteristic function of ω. Let

$$\Phi(z) = \frac{3}{\pi} \iint_D K(z, \zeta)\chi_\omega(\zeta)\varphi(\zeta)\rho^{-2}(\zeta)\, d\xi\, d\eta. \tag{30}$$

By properties (i) and (v) of Lemma 3, we know that

$$\sup |K(z, \zeta)|\rho^{-2}(\zeta) < \infty.$$

Since $\varphi(\zeta)$ is integrable over ω, clearly $\chi_\omega(\zeta)\varphi(\zeta)$ is integrable over D, and thus the integrand in (30) is integrable for each fixed z and gives a holomorphic function of z. Moreover, from property (iii) in Lemma 3, we see that

$$\iint_D |\Phi(z)|\, dx\, dy \leqslant \frac{3}{\pi} \iint \iint |K(z, \zeta)|\, dx\, dy |\chi_\omega \varphi \rho^{-2}|\, d\xi\, d\eta$$

$$\leqslant 3 \iint_\omega |\varphi|\, d\xi\, d\eta. \tag{31}$$

Finally, on using properties (ii) and (iv), we see that

$$\Theta\Phi(z) = \frac{3}{\pi} \sum \iint_\omega K(Az, \zeta)A'(z)^2 \varphi(\zeta)\rho^{-2}(\zeta)\, d\xi\, d\eta$$

$$= \frac{3}{\pi} \sum \iint_{A^{-1}(\omega)} K(Az, A\zeta)A'(z)^2 \varphi(A\zeta)\rho^{-2}(A\zeta)|A'(\zeta)|^2\, d\xi\, d\eta$$

$$= \frac{3}{\pi} \iint_D K(z, \zeta)\varphi(\zeta)\rho^{-2}(\zeta)\, d\xi\, d\eta = \varphi(\zeta).$$

This shows that Θ is surjective and inequality (31) shows that it maps the ball of radius 1 onto the ball of radius $\frac{1}{3}$. Theorem 3 is now completely proved.

4.5. THE INEQUALITY OF REICH AND STREBEL FOR ARBITRARY FUCHSIAN GROUPS

Let Γ be a Fuchsian group with at least three points in its limit set Λ and assume C is a closed Γ-invariant set with $\Lambda \subseteq C \subseteq \hat{\mathbb{R}}$. In this section, let ω be a fundamental domain for Γ in $\mathbb{H}$. Let w be a quasiconformal self-mapping of $\hat{\mathbb{C}}$ such that

$$\begin{aligned} &\text{(i)} \quad w(\bar{z}) = \overline{w(z)}, \\ &\text{(ii)} \quad w \circ A \circ w^{-1} = A \qquad \text{for every } A \text{ in } \Gamma, \text{ and} \\ &\text{(iii)} \quad w(x) = x \qquad \text{for every } x \text{ in } C. \end{aligned} \tag{32}$$

Let μ be the Beltrami coefficient of w, that is, $w_{\bar{z}}/w_z = \mu$. Such a Beltrami coefficient μ coming from a quasiconformal mapping w satisfying (32) is called a trivial Beltrami coefficient.

Let φ be a quadratic differential in $A_s(\Gamma, C)$, by which we mean φ satisfies

$$\begin{aligned} &\text{(i)} \quad \varphi \text{ is holomorphic in } \mathbb{C} - C, \\ &\text{(ii)} \quad \varphi(Bz)B'(z)^2 = \varphi(z) \qquad \text{for all } B \text{ in } \Gamma, \\ &\text{(iii)} \quad \|\varphi\| = \iint_{\omega} |\varphi(z)|\, dx\, dy < \infty, \text{ and} \\ &\text{(iv)} \quad \varphi(z) \text{ is real valued if } z \text{ is in } \mathbb{R} - C. \end{aligned} \tag{33}$$

Theorem 4. *For μ and φ satisfying the above conditions, one has the following inequality:*

$$\|\varphi\| \leqslant \iint_{\omega} |\varphi(z)| \frac{|1 - \mu(z)\varphi(z)/|\varphi(z)||^2}{1 - |\mu|^2}\, dx\, dy. \tag{34}$$

Proof. Clearly, if φ_n is a sequence in $A_s(\Gamma, C)$ for which $\|\varphi_n - \varphi\| \to 0$ and φ_n satisfies (34), then φ also satisfies (34). Now let Γ_1 be a finitely generated subgroup of Γ with limit set Λ_1. Clearly $\Lambda_1 \subseteq \Lambda \subseteq C \subseteq \hat{\mathbb{R}}$. From Chapter 3, Theorem 3, we know the inequality holds for the finitely generated group Γ_1 and any Γ_1-invariant closed set C_1 with $\Lambda_1 \subseteq C_1 \subseteq \hat{\mathbb{R}}$ and for which $(C_1 - \Lambda_1)/\Gamma_1$ is a finite set. Clearly, holding Γ_1 fixed, we can let C_1 become a larger and larger subset of C, and, by Theorem 2 of this chapter, we obtain (34) for the closed set C and any finitely generated subgroup of Γ.

For the next step, we take an increasing sequence Γ_n of finitely generated subgroups of Γ whose union is Γ. Let Θ and Θ_n be the theta series operators for

Γ and Γ_n. Thus we have surjective mappings:

$$\Theta\colon A(C) \to A(\Gamma, C) \quad \text{and} \quad \Theta_n\colon A(C) \to A(\Gamma_n, C).$$

From the methods of construction of fundamental domains described in Section 1.6, we can take fundamental domains ω for Γ and ω_n for Γ_n in $\mathbb{H}$ with $\omega \subset \omega_n$ such that $\bigcap_{n=1}^{\infty} \bar{\omega}_n = \bar{\omega}$. Moreover, we have the following equality, which is true except for a set which is a countable union of analytic arcs:

$$\bigcup_{B \in \Gamma_n} B(\omega_n - \omega) = \mathbb{H} - \bigcup_{B \in \Gamma_n} B(\omega).$$

If we let $D_n = \bigcup_{B \in \Gamma_n} B(\omega)$, then D_n is an increasing sequence of open sets whose union is equal to $\mathbb{H}$ except for a set of measure zero.

Before proving inequality (34) it is convenient to rewrite it in the equivalent form given in the next lemma.

Lemma 5. *Inequality* (34) *is true for a given* μ *and a given* φ *if, and only if,*

$$\operatorname{Re} \iint_{\omega} \frac{\mu\varphi}{1 - |\mu|^2}\, dx\, dy \leqslant \iint_{\omega} \frac{|\mu|^2|\varphi|}{1 - |\mu|^2}\, dx\, dy. \tag{35}$$

Proof. Simply expand the numerator in the integrand on the right-hand side of (34). We get

$$|\varphi|\{1 - |\mu|^2 - 2 \operatorname{Re} \mu\varphi/|\varphi| + 2|\mu|^2\}.$$

The term involving $1 - |\mu|^2$ can be divided out and then we can subtract $\|\varphi\|$ from both sides of the inequality. Obviously (35) follows and this reasoning is reversible.

Our task is to prove (35). Let $\varphi = \Theta F$ and $\varphi_n = \Theta_n F$. Since μ is a Beltrami coefficient for Γ and $\Gamma_n \subset \Gamma$, it is easy to show that

$$\iint_{\omega_n} \frac{\mu\varphi_n}{1 - |\mu|^2}\, dx\, dy = \iint_{\omega} \frac{\mu\varphi}{1 - |\mu|^2}\, dx\, dy.$$

Since we have proved the finitely generated case, we know that

$$\operatorname{Re} \iint_{\omega} \frac{\mu\varphi\, dx\, dy}{1 - |\mu|^2} \leqslant \iint_{\omega_n} \frac{|\mu|^2|\Theta_n F|}{1 - |\mu|^2}\, dx\, dy. \tag{36}$$

Now the idea is to take the limit in (36) as n approaches ∞. Since $|\mu|^2/(1 - |\mu|^2)$ is a bounded function, the fact that we obtain (35) in the limit follows from the following lemma.

Lemma 6

$$\lim_{n\to\infty}\left|\iint_{\omega_n}|\Theta_n F| - \iint_{\omega}|\Theta F|\right| \to 0.$$

Proof. We must estimate two terms,

$$\left|\iint_{\omega_n}|\Theta_n F| - \iint_{\omega}|\Theta_n F|\right| \quad\text{and}\quad \left|\iint_{\omega}|\Theta_n F| - \iint_{\omega}|\Theta F|\right|.$$

The first term is bounded by $\iint_{\mathbb{H}-D_n}|F|$, where D_n is the union of the sets $B(\omega)$ for B in Γ_n. Since Γ_n increases to Γ, D_n increases to $\mathbb{H}$. The fact that F is integrable implies that this limit approaches zero. The second term is bounded by

$$\iint_{\omega}\sum|F(B(z))|\,|B'(z)^2|\,dx\,dy,$$

where the sum is over B in $\Gamma - \Gamma_n$. This term is equal to $\iint_{\mathbb{H}-D_n}|F|$, which, as we just mentioned, converges to zero.

Notes. Bers's approximation theorem appears in [Be4]. The density theorem in Section 4.2 is due to Ahlfors [Ah3] and Bers [Ber4], as are the theorems on the surjectivity of Poincaré theta series in Section 4.3 ([Ah3], [Ber5]). The proof of convergence of the series (26) is adapted from a paper of Earle [E1] and a paper of Godement [Go]. Much of the organization and the clever proof of surjectivity of the theta series operator given at the end of Section 4.4 is due to Kra[Kr2]. The proof of the existence of the kernel functions F and K are also adapted from Kra [Kr2]. The proof of the extension of the Reich–Strebel inequality to general Fuchsian groups given in Section 4.5 is due to this author. Bers proves a similar result in [Ber10]. Strebel [St3] generalizes the inequality to arbitrary Riemann surfaces by different methods.

EXERCISES

SECTION 4.1.

1. Let $J_\varepsilon(z)$ be a nonnegative real valued function with support in $\{z\colon |z| \leqslant \varepsilon\}$. Assume J_ε is C^2 and $\iint_{\mathbb{C}} J_\varepsilon(z)\,dx\,dy = 1$. For any L_1-function $f(z)$ defined for z in $\mathbb{C}$, let

$$f_\varepsilon(z) = \iint_{\mathbb{C}} J_\varepsilon(z-\zeta)f(\zeta)\,d\xi\,d\eta,$$

where $\zeta = \xi + i\eta$. Show that $\iint |f_\varepsilon|\,dx\,dy \leqslant \iint |f|\,dx\,dy$, where the integrals are over $\mathbb{C}$.

Remark. Here f_ε is called the ε-mollification of f.

2. Suppose f satisfies a Lipschitz condition, that is, $|f(z) - f(w)| \leqslant K|z - w|$. Show that f_ε satisfies a Lipschitz condition with the same constant K. Conclude that $|(\partial/\partial z)f_\varepsilon|$ and $|(\partial/\partial \bar{z})f_\varepsilon|$ are both bounded by K.
3. For the f in Exercise 2 show that f_ε converges to f uniformly in $\mathbb{C}$.
4. For the f in Exercise 2, show there is an L_∞-function g such that

$$\iint f\varphi_{\bar{z}}\,dx\,dy = \iint g\varphi\,dx\,dy$$

for all C^1 functions φ with compact support. [In this case, the function $-g$ is called the generalized $(\partial/\partial\bar{z})$-derivative of f.]

Hint: Show that the linear functional $\ell(\varphi) = \iint f\varphi_{\bar{z}}\,dx\,dy$ is bounded in the sense that $|\ell(\varphi)| \leqslant K \iint |\varphi|\,dx\,dy$. Then use the Hahn–Banach and Riesz representation theorems to find g.

SECTION 4.4

5. Consider the inner product of functions defined in the unit disk given by

$$\langle f(z), g(z)\rangle = \frac{1}{\pi}\iint_{|z|<1} \rho^{-2}(z) f(z)\overline{g(z)}\,dx\,dy.$$

(a) Show that the holomorphic functions $f(z)$ for which $\|f\|^2 = \langle f(z), f(z)\rangle < \infty$ is a Hilbert space.
(b) Show that the set of functions $\{c_n z^n\}$ makes an orthonormal set with the proper choice of constant c_n.
(c) Show that the function

$$K(z, \zeta) = \sum_{n=0}^{\infty} c_n^2 z^n (\bar{\zeta})^n$$

is a constant multiple of the kernel function $K_\Delta(z, \zeta)$ defined in (22).

6. Let

$$R(z) = \frac{\lambda - 1}{z(z-1)(z-\lambda)}$$

and let Γ be the cyclic group generated by $z \to \lambda z$, $\lambda \neq 0$, and $|\lambda| \neq 1$. Show that

$$\Theta R(z) = \begin{cases} +z^{-2} & \text{if } |\lambda| > 1, \\ -z^{-2} & \text{if } |\lambda| < 1, \end{cases}$$

where Θ is the theta series for the group Γ.

5

TEICHMÜLLER THEORY

This chapter gives the definitions of Teichmüller space of a Riemann surface and of a Fuchsian group and tells under what conditions they are equivalent. It also gives the manifold structure for Teichmüller space, which is a system of coordinate charts modeled on a Banach space. For many cases and, in particular, for the case of surfaces of finite analytic type, the coordinate mappings are complex analytic and the Banach space is finite dimensional. The chapter also contains part of the infinitesimal theory and, in particular, a theorem on the existence of trivial curves with given infinitesimally trivial tangent vector.

We now give a heuristic argument which shows why the space of different complex analytic structures on a surface of genus g has complex dimension $3g - 3$, when $g > 1$ and dimension 1 when $g = 1$. By the end of the chapter, a rigorous argument is provided.

From the uniformization theorem, Theorem 1 of Chapter 1, the compact Riemann surfaces of genus $\geqslant 1$ are realized by the complex plane factored by a lattice or the upper half plane factored by a torsion free Fuchsian group. We count the number of free parameters needed to determine the conjugacy class of the covering group of the surface. When the genus is one, there is a lattice L generated by two translations $z \to z + \omega_1$ and $z \to z + \omega_2$, where ω_1 and ω_2 are independent over $\mathbb{R}$. Conjugation by the transformation $T(z) = \omega_1^{-1}z$ changes the lattice L to the lattice TLT^{-1} generated by $z \to z + 1$ and $z + z + \tau$, where $\tau = \omega_2/\omega_1$. The complex parameter τ determines the one complex dimensional family of complex structures on a torus.

In the case where the genus is more than one, the usual way to select a set of generators for a fundamental group is to take loops α_j and β_j, $1 \leqslant j \leqslant g$, where α_j and β_j go around the jth handle in the way indicated in Figure 5.1. On choosing a basepoint for the surface and connecting each of these loops to the basepoint, they become a basis for the fundamental group satisfying the commutator relation

$$\alpha_1\beta_1\alpha_1^{-1}\beta_1^{-1}\cdots\alpha_g\beta_g\alpha_g^{-1}\beta_g^{-1} = 1.$$

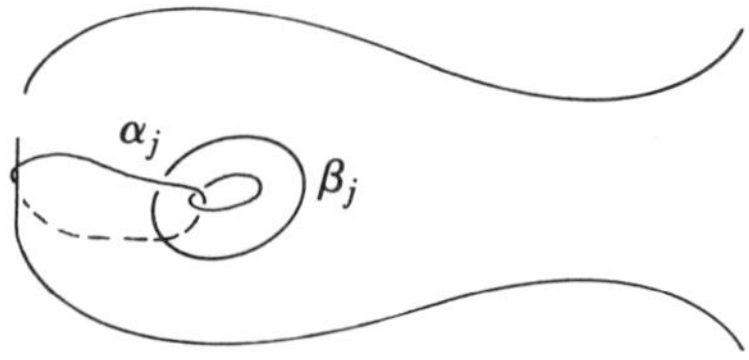

Figure 5.1

The liftings of these loops to the Fuchsian covering group yield Möbius transformations A_j and B_j satisfying the commutator relation

$$A_1 B_1 A_1^{-1} B_1^{-1} \cdots A_g B_g A_g^{-1} B_g^{-1} = 1.$$

Since each Möbius transformation is determined by three real parameters, the $2g$ transformations, $A_1, B_1, \ldots, A_g, B_g$, are determined by $6g$ real parameters. The commutator relation takes away three degrees of freedom and the fact that the group Γ and any conjugate of it, $B\Gamma B^{-1}$, determine the same complex structure takes away three more degrees of freedom. We are left with $6g - 6$ real parameters.

In our setup for Teichmüller space, the definitions include surfaces much more general than compact surfaces of finite genus. The surfaces can have infinite genus or infinite connectivity or both. They can also be bordered surfaces with certain points of the border fixed. An interesting example is obtained by letting the surface be the extended complex plane with a Cantor subset of the unit interval removed. This particular example has genus zero, no border, and infinite connectivity.

We have already used the notion of a bordered Riemann surface in Section 3.4. We remind the reader of the following facts. If a Riemann surface R is represented by $\mathbb{H}/\Gamma$, where Γ is the universal covering group of R, the border of R is the set $(\hat{\mathbb{R}} - \Lambda)/\Gamma$. Each component I of $\hat{\mathbb{R}} - \Lambda$ is an open interval. It is easy to see that if Γ has three or more limit points, then there is a cyclic group $\Gamma(I)$ which contains all of the elements of Γ which preserve I. Now $\Gamma(I)$ is either trivial or is generated by a hyperbolic transformation whose fixed points are the two endpoints of I. The part of the border of the surface corresponding to I is represented by $I/\Gamma(I)$, and the inverse of the universal covering maps gives a chart which realizes the border as a real analytic curve.

5.1. TEICHMÜLLER SPACE OF A RIEMANN SURFACE

Let R be a Riemann surface, possibly with border, and σ a closed subset of the border. In this section we define $T(R, \sigma)$, the Teichmüller space of R relative to σ. In the case where R is the unit disk and σ its circumference, $T(R, \sigma)$ is universal Teichmüller space. If σ is the whole border of R, then $T(R, \sigma)$ is called unreduced

Teichmüller space. If the border is nonempty but σ is empty, $T(R, \sigma)$ is called reduced Teichmüller space. For surfaces of finite analytic type there is no border and these distinctions do not exist.

$T(R, \sigma)$ as a Deformation Space

Consider the set $\mathrm{Def}(R, \sigma)$ of all pairs (f, R^*), where f is a quasiconformal homeomorphism of the surface R onto a surface R^*. Two pairs (f_0, R_0) and (f_1, R_1) in $\mathrm{Def}(R, \sigma)$ are equivalent if there is a conformal map $c: R_0 \to R_1$ such that $c \circ f_0$ is homotopic to f_1 in the following sense: There is a continuous curve of continuous functions $g_t: R \to R_1$ which extend continuously to the border of R such that

(a) $g_0(z) = c \circ f_0(z)$ and $g_1(z) = f_1(z)$ for z in R; and
(b) $g_t(p) = c \circ f_0(p) = f_1(p)$ for every p in σ and every t with $0 \leqslant t \leqslant 1$.

From the theory of quasiconformal mapping, f_0 and f_1 extend continuously to the border of R in a unique way.

By definition, $T(R, \sigma)$ is $\mathrm{Def}(R, \sigma)$ factored by this equivalence relation.

$T(R, \sigma)$ as an Orbit Space

Let $M(R)$ be the open unit ball in the Banach space $L_\infty(R)$. By $L_\infty(R)$ we mean the space of all Beltrami differentials μ. Such a differential μ is an assignment of a measurable complex valued function μ^z to each local parameter z on R such that

(a) $\mu^z(z)(d\bar{z}/dz) = \mu^\zeta(\zeta)(d\bar{\zeta}/d\zeta)$ for any two parameters z and ζ with overlapping domains; and
(b) $\|\mu\|_\infty = \sup\{\|\mu^z(z)\|_\infty \text{ for all parameters } z\} < \infty$.

It follows from the theory of the Beltrami equation (see Section 1.8) that for any μ in $M(R)$ there is a quasiconformal homeomorphism w from R onto another Riemann surface R^μ satisfying

$$w_{\bar{z}} = \mu w_z, \tag{1}$$

and which extends continuously and uniquely to the border of R. Let $D_0(R, \sigma)$ be the group of quasiconformal homeomorphic self-mappings of R which are homotopic to the identity in the sense described above. Specifically, h is in $D_0(R, \sigma)$ if there exists a continuous curve of continuous self-mappings g_t of R extending continuously to the border of R such that

$$g_0(z) = z \quad \text{and} \quad g_1(z) = h(z) \qquad \text{for } z \text{ in } R$$

and

$$g_t(p) = h(p) = p \qquad \text{for } p \text{ in } \sigma \quad \text{and} \quad 0 \leqslant t \leqslant 1.$$

The group operation for $D_0(R, \sigma)$ is composition and obviously there is a group action $D_0(R, \sigma) \times M(R) \to M(R)$ given by $(h, \mu) \mapsto h^*(\mu)$, where $h^*(\mu) = (w \circ h)_{\bar{z}}/(w \circ h)_z$ and w satisfies (1). If ν is the Beltrami coefficient for h, that is, if $\nu = h_{\bar{z}}/h_z$, then the Beltrami coefficient $h^*(\mu)$ is

$$h^*(\mu) = \frac{\nu(z) + \mu(h(z))\theta(z)}{1 + \overline{\nu(z)}\mu(h(z))\theta(z)}$$

where $\theta(z) = \overline{(h_z)}/h_z$.

Obviously $\|h^*(\mu)\|_\infty < 1$ since it is the Beltrami coefficient of a quasiconformal mapping. Teichmüller space can now be defined as the set of orbits in $M(R)$ under this group action. Since it is a different definition from the one already given, we denote it by $T_1(R, \sigma)$.

Proposition 1. *$T(R, \sigma)$ and $T_1(R, \sigma)$ are naturally isomorphic.*

Proof. There is a natural map α: $T(R, \sigma)$ to $T_1(R, \sigma)$. Given an equivalence class $[f, R^*]$ in $T(R, \sigma)$ and a local parameter g on R^*, we form

$$\mu = (g \circ f)_{\bar{z}}/(g \circ f)_z. \tag{2}$$

Notice that μ is defined independently of g and $\mu d\bar{z}/dz$ is invariant on R. We define $\alpha([f, R^*]) = \mu$, where μ is given by (2). To show that α is well defined assume $[f_0, R_0] \sim [f_1, R_1]$. Then there is a conformal mapping c: $R_0 \to R_1$ such that $c \circ f_0$ and f_1 are homotopic as maps from R to R_1 and $c \circ f_0$ and f_1 coincide on σ as does the homotopy g_t which connects them. Let $h_t = f_1^{-1} \circ g_t$. Then h_t: $R \to R$ and h_1 is the identity and $h_0 = f_1^{-1} \circ g_0 = f_1^{-1} \circ c \circ f_0$. Let μ_i be the Beltrami coefficient of f_i for $i = 0$ and 1. Now $h_1^*(\mu_1) = \mu_1$ and $h_0^*(\mu_1)$ is the Beltrami coefficient of $f_1 \circ h_0 = c \circ f_0$ and so $h_0^*(\mu_1) = \mu_0$. Therefore, μ_1 and μ_0 are in the same orbit under $D_0(R, \sigma)$.

This same argument applies in reverse and shows that if μ_0 and μ_1 are in the same orbit under $D_0(R, \sigma)$, then there exists a qc-homeomorphism h of (R, σ) homotopic to the identity for which $f_1 \circ h$ and f_0 have the same Beltrami coefficient. This implies the existence of a conformal mapping c: $R_0 \to R_1$ for which $f_1 \circ h = c \circ f_0$. Finally, the surjectivity of α follows from the existence of quasiconformal homeomorphisms w satisfying (1) for given μ in $M(R)$.

5.2. TEICHMÜLLER SPACE OF A FUCHSIAN GROUP

Now suppose the universal covering surface $\tilde{R}$ of the Riemann surface is conformally isomorphic to the upper half plane $\mathbb{H}$. The cover transformations of $\tilde{R}$ over R are represented by Möbius transformations with no fixed points in $\mathbb{H}$.

The Riemann surface R is realized as the orbit space $\mathbb{H}/\Gamma$ and the canonical mapping π: $\mathbb{H} \to \mathbb{H}/\Gamma$ is a complex analytic projection (see Section 1.4). If

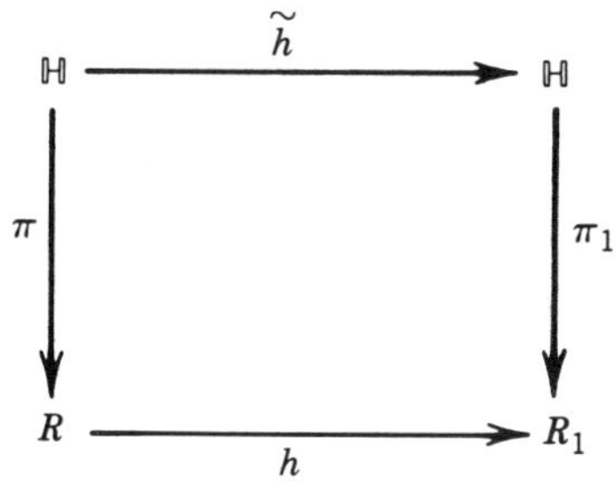

Figure 5.2

$\Gamma_1 = A \circ \Gamma \circ A^{-1}$ for A in PSL(2, $\mathbb{R}$), then $z \to A(z)$ gives a one-to-one correspondence from the orbits of Γ to the orbits of Γ_1 and so yields a bianalytic isomorphism of $\mathbb{H}/\Gamma$ onto $\mathbb{H}/\Gamma_1$. Conversely, if h is a quasiconformal homeomorphism of R onto R_1, it can be lifted to a qc-homeomorphism $\tilde{h}$ of $\tilde{R}$ onto $\tilde{R}_1$ which satisfies $\pi_1 \circ \tilde{h} = h \circ \pi$ (see Figure 5.2). If h is conformal, so is $\tilde{h}$ and therefore $\tilde{h}$ is a Möbius transformation. Thus, the groups Γ and Γ_1 are conjugate, $\tilde{h}\Gamma\tilde{h}^{-1} = \Gamma_1$. This equality holds even if $\tilde{h}$ is only quasiconformal but $\tilde{h}$ is no longer a Möbius transformation. It is this freedom which enables one, by use of the quasiconformal mapping $\tilde{h}$, to deform Γ into nonconjugate subgroups of PSL(2, $\mathbb{R}$).

Let Γ be any Fuchsian group, possibly with elliptic elements, and define $M(\Gamma)$ to be the set of Beltrami coefficients μ with support in $\mathbb{H}$ and which are compatible with Γ in the sense that $\mu(B(z))\overline{B'(z)} = \mu(z)B'(z)$ for all B in Γ. We assume that μ is measurable and complex valued and $\|\mu\|_\infty < 1$. Let w_μ be the unique qc-homeomorphism of $\mathbb{H}$ whose extension to the real axis fixes 0, 1, and ∞ and which satisfies (1). Let C be any closed subset of $\hat{\mathbb{R}}$ which contains the limit set of Γ and which is invariant under Γ.

Definition. *Two elements μ and v of $M(\Gamma)$ are C-equivalent if $w_\mu(x) = w_v(x)$ for all x in C.*

Definition. *$T(\Gamma, C)$, the Teichmüller space of the Fuchsian group Γ relative to the closed set C, consists of equivalence classes of elements of $M(\Gamma)$, where $\mu \sim v$ if $w_\mu(x) = w_v(x)$ for all x in C.*

Lemma 1. *If $\mu \in M(\Gamma)$, then $w_\mu \circ A \circ w_\mu^{-1}$ is a Möbius transformation for every A in Γ.*

Proof. The self-mappings of $\mathbb{H}$ given by w_μ and $w_\mu \circ A$ have Beltrami coefficients μ and $\mu(A(z))\overline{A'(z)}/A'(z)$, respectively. Since μ is in $M(\Gamma)$, these Beltrami coefficients are identical. Choose a Möbius transformation A_μ such that $A_\mu \circ w_\mu$ and $w_\mu \circ A$ have the same values at 0, 1, and ∞. Notice that $A_\mu \circ w$ and $w_\mu \circ A$ are normalized solutions of the Beltrami equation with the same Beltrami coefficient. Therefore, by uniqueness, $A_\mu \circ w_\mu = w_\mu \circ A$.

Lemma 2. *Let Γ be a torsion free, discontinuous group of Möbius transformations and C a closed Γ-invariant subset of $\hat{\mathbb{R}}$ which contains the limit set Λ of Γ. Let $R = \mathbb{H}/\Gamma$ and $\sigma = (C - \Lambda)/\Gamma$. Then $T(R, \sigma)$ is canonically isomorphic to $T(\Gamma, C)$.*

Proof. Recall that $T(R, \sigma) = T_1(R, \sigma)$ by Proposition 1 of Section 5.1. Given an element $[\mu]$ in $T_1(R, \sigma)$, we lift μ to a Beltrami coefficient $\tilde{\mu}$ defined on $\mathbb{H}$ and satisfying $\tilde{\mu}(Bz)\overline{B'(z)} = \tilde{\mu}(z)B'(z)$ for every B in Γ. We must show that μ_0 and μ_1 are equivalent in $T_1(R, \sigma)$ if and only if $\tilde{\mu}_0$ and $\tilde{\mu}_1$ are equivalent in $T(\Gamma, C)$. Let w_i be the normalized qc-mapping of $\mathbb{H}$ with Beltrami coefficient $\tilde{\mu}_i$ and let f_i be a qc-mapping of R onto R_1 with Beltrami coefficient μ_i, $i = 0$ or 1. We claim the homotopy from f_0 to f_1 fixing σ lifts to a homotopy from w_0 to w_1 fixing C. The condition that f_0 is homotopic to $c \circ f_1$ for some conformal mapping $c\colon R_0 \to R_1$ forces $w_0 \Gamma w_0^{-1} = w_1 \Gamma w_1^{-1}$ since w_0 and w_1 are normalised. Moreover, for each A in Γ, the homotopy w_t from w_0 to w_1 gives a continuous curve of homomorphisms χ_t from Γ to the discrete group $w_0 \circ \Gamma \circ w_0^{-1}$ determined by $\chi_t(A) \circ w_t = w_t \circ A$. Thus, we see that $\chi_t(A)$ is constant for $0 \leqslant t \leqslant 1$ and therefore $(w_0^{-1} \circ w_t) \circ A^n = A^n \circ (w_0^{-1} \circ w_t)$ for all A in Γ. By letting $n \to \infty$ it follows that w_0 and w_t are identical on the attracting fixed points of elements of Γ and hence on the whole limit set. (Notice that we do not take the inverse of the mapping w_t in this argument. In fact, we are not permitted this freedom because w_t is not necessarily a homeomorphism.) Since f_0 and f_t are identical on σ, one sees that for every point p in C and not in the limit set of Γ, there is an element A in Γ for which $w_0(p) = w_t(A(p))$. Let J be the open interval of $\hat{\mathbb{R}} - \Lambda$ containing p. Since $w_0(x) = w_t(x)$ for x in Λ and, in particular, for the two endpoints of J, we see that $A(J) = J$. If A were not the identity, then $f_0^{-1} \circ f_t$ applied to projection of p would be a loop winding around the boundary component of R corresponding to J and $f_0^{-1} \circ f_t$ would not fix the points of σ. We conclude that $w_0(p) = w_t(p)$ for p in C and that $\tilde{\mu}_0$ and $\tilde{\mu}_1$ are equivalent in $T(\Gamma, C)$.

To show the converse, we start by assuming $\tilde{\mu}_0$ and $\tilde{\mu}_1$ are equivalent. We must show there is a homotopy between f_0 and f_1 which maps $\mathbb{H}/\Gamma$ onto $\mathbb{H}/\Gamma^*$, where $\Gamma^* = w_0 \Gamma w_0^{-1} = w_1 \Gamma w_1^{-1}$. Let ρ and ρ^* be the noneuclidean metrics for the domains $\hat{\mathbb{C}} - \Lambda$ and $\hat{\mathbb{C}} - \Lambda^*$, where Λ and Λ^* are the limit sets of the groups Γ and Γ^*. Given any two points in the upper half plane $\mathbb{H}$, there is a unique line segment connecting them in either of the metrics ρ or ρ^*. The same statement is true if the two points are in a component of $\hat{\mathbb{R}} - \Lambda$ or $\hat{\mathbb{R}} - \Lambda^*$. Notice that for a component I of $\hat{\mathbb{R}} - \Lambda$, one has $w_0(I) = w_1(I)$ because w_0 and w_1 are identical on Λ.

Now the groups Γ and Γ^* consist of isometries of the domains $\hat{\mathbb{C}} - \Lambda$ and $\hat{\mathbb{C}} - \Lambda^*$ in the metrics ρ and ρ^*, respectively. Define $w_t(z)$ to be the point on the ρ^*-line segment between $w_0(z)$ and $w_1(z)$ such that the ρ^*-distances from $w_0(z)$ to $w_t(z)$ and from $w_t(z)$ to $w_1(z)$ are in the ratio $t\colon (1 - t)$. The mapping w_t is not necessarily quasiconformal, but it is clearly continuous from $\mathbb{H} \cup (\hat{\mathbb{R}} - \Lambda)$ to $\mathbb{H} \cup (\hat{\mathbb{R}} - \Lambda^*)$. Since $w_0(Az) = \tilde{A} \circ w_0(z)$ for A in Γ and $\tilde{A}$ in Γ^* and $w_1(Az) = \tilde{A} \circ w_1(z)$, it follows that $w_t(Az) = \tilde{A} \circ w_t(z)$. Moreover, from the definition of w_t, it agrees with w_0 and w_1 at any points x for which $w_0(x) = w_1(x)$. Thus w_t factors to a homotopy f_t from f_0 to f_1 which fixes points of σ.

Remark. The idea of using noneuclidean line segments to construct the homotopy comes from Ahlfors [Ah4]. We have altered his construction by using noneuclidean metrics for the domains $\hat{\mathbb{C}} - \Lambda$ and $\hat{\mathbb{C}} - \Lambda^*$. This is necessary in order to get a homotopy which extends to C and keeps points of C fixed.

5.3. TEICHMÜLLER'S METRIC

Given two points $[\mu]$ and $[\nu]$ in the Teichmüller space $T(\Gamma, C)$ define the Teichmüller's metric d by

$$d([\mu], [\nu]) = \tfrac{1}{2} \inf \log K(w_{\tilde{\mu}} \circ w_{\tilde{\nu}}^{-1}), \tag{3}$$

where $w_{\tilde{\mu}}$ and $w_{\tilde{\nu}}$ are quasiconformal mappings with Beltrami coefficients $\tilde{\mu}$ and $\tilde{\nu}$ and the infinum is taken over all $\tilde{\mu}$ and $\tilde{\nu}$ in the same Teichmüller classes as μ and ν, respectively. From the dilatation properties $K(f) = K(f^{-1})$ and $K(f \circ g) \leqslant K(f)K(g)$, one easily sees that d is symmetric and satisfies the triangle inequality.

To see that d is nondegenerate, suppose $d(0, [\mu]) = 0$. Then there is a sequence of normalized quasiconformal mappings w_n with Beltrami coefficients μ_n for which $K(w_n) \to 1$ and each μ_n is in the same Teichmüller class as μ. This means $w_n \circ A \circ w_n^{-1} = w_\mu \circ A \circ w_\mu^{-1}$ for each n and each A in Γ and $w_n(x) = w_\mu(x)$ for each x in C. By the convergence principle for quasiconformal mappings, Lemma 6 in Section 1.8, we know that w_n must have a subsequence which converges uniformly on compact subsets of $\mathbb{C}$ to some limit w with $K(w) = 1$. We see that $w(z) = z$ for all z in $\mathbb{H} \cup \mathbb{R}$ and, therefore, $w_\mu(x) = x$ for each x in C and $w_\mu \circ A \circ w_\mu^{-1} = A$ for each A in Γ. This shows that $[\mu]$ is the trivial Teichmüller class.

The Teichmüller metric for the Teichmüller space of a Riemann surface R with a specified closed subset σ of its border is defined analogously. Given two points $[f_1]$ and $[f_2]$ in $T(\Gamma, \sigma)$, we define

$$d[(f_1], [f_2]) = \tfrac{1}{2} \inf \log K(\tilde{f}_2 \circ \tilde{f}_1^{-1}), \tag{4}$$

where the infinum is taken over all quasiconformal mappings $\tilde{f}_1$ and $\tilde{f}_2$ in the classes of f_1 and f_2, respectively.

5.4. THE BERS EMBEDDING OF TEICHMÜLLER SPACE

The Bers embedding of $T(\Gamma, C)$ applies when $C = \hat{\mathbb{R}}$. This can happen either when Γ is of the first kind or when σ is the whole border of $\mathbb{H}/\Gamma$. An important example is the case where Γ is the universal covering group of a surface R of finite analytic type. Such a group is necessarily of the first kind and thus $\Lambda = C = \hat{\mathbb{R}}$. By Lemma 2 of the previous section, we then have $T(\Gamma, C) = T(R)$.

Whenever $C = \hat{\mathbb{R}}$, the equivalence relation on $M(\Gamma)$ which determines $T(\Gamma, C)$ takes a particularly simple form. Two Beltrami coefficients μ and ν are equivalent if and only if $w_\mu(x) = w_\nu(x)$ for all x in $\mathbb{R}$. Here, w_μ is a quasiconformal self-mapping of the upper half plane normalized to fix 0, 1, and ∞ with Beltrami coefficient μ. The mapping w_ν has the same definition with μ replaced by ν.

Let w^μ be the unique qc-homeomorphism of $\hat{\mathbb{C}}$ normalized to fix 0, 1, and ∞ with Beltrami coefficient μ in the upper half plane and 0 in the lower half plane.

Lemma 3. *The following conditions are equivalent:*

(i) $w_\mu(x) = w_\nu(x)$ *for all* x *in* $\mathbb{R}$;
(ii) $w^\mu(x) = w^\nu(x)$ *for all* x *in* $\mathbb{R}$; *and*
(iii) $w^\mu(z) = w^\nu(z)$ *for all* z *in the lower half plane.*

Proof. Let f^μ be the unique conformal mapping from $\mathbb{H}^\mu = w^\mu(\mathbb{H})$ to $\mathbb{H}$ normalized to fix 0, 1, and ∞. Since $f^\mu \circ w^\mu | \mathbb{H}$ has the same Beltrami coefficient as w_μ, is normalized at 0, 1, and ∞, and maps $\mathbb{H}$ to $\mathbb{H}$, we have $w_\mu(z) = f^\mu \circ w^\mu(z)$ for z in $\mathbb{H}$. In fact, f^μ is determined by the set $w^\mu(\mathbb{R})$, since a normalized Riemann mapping is determined by the domain. Hence, if $w^\mu(x) = w^\nu(x)$ for every x in $\mathbb{R}$, then $f^\mu(z) = f^\nu(z)$ for z in $\mathbb{H}^\mu$ and this implies $w_\mu(x) = w_\nu(x)$ for all x in $\mathbb{R}$. This shows that (ii) implies (i).

For the opposite implication, suppose $w_\mu(x) = w_\nu(x)$ and form

$$h = \begin{cases} (f^\nu)^{-1} \circ f^\mu & \text{in } \mathbb{H}^\mu \\ w^\nu \circ (w^\mu)^{-1} & \text{elsewhere.} \end{cases}$$

Then h extends continuously to $w^\mu(\mathbb{R})$. It is quasiconformal in the whole plane and conformal off the set $w^\mu(\mathbb{R})$ and it is normalized. Since a quasiconformal image of the real axis is a removable set for analytic mappings, we conclude that h is the identity. Thus $w^\mu = w^\nu$ on $\mathbb{R}$ and $f^\mu = f^\nu$ on $w^\mu(\mathbb{R})$ and we get $w_\mu(x) = w_\nu(x)$.

It is obvious that (ii) is equivalent to (iii) because both mappings w^μ and w^ν are holomorphic in the lower half plane and continuous on its closure.

The next lemma is the basis for the existence of quasi-Fuchsian groups.

Lemma 4. *Given* μ *in* $M(\Gamma)$, *then for every* B *in* Γ, $w^\mu \circ B \circ (w^\mu)^{-1} = B^\mu$ *is a linear fractional transformation which leaves invariant the region* $w^\mu(\mathbb{H}) = \mathbb{H}^\mu$. *That is,* $B^\mu(z) = (az + b)/(cz + d)$ *with* $a, b, c,$ *and* d *complex and* $ad - bc = 1$.

Proof. $w^\mu \circ B$ and w^μ have identical Beltrami coefficients because $\mu(Bz)\overline{B'(z)}/B'(z) = \mu(z)$. Thus by selecting a Möbius transformation B^μ (with complex coefficients) so that $B^\mu \circ w^\mu$ and $w^\mu \circ B$ are identical on the points 0, 1, and ∞, one finds that $w^\mu \circ B = B^\mu \circ w^\mu$ because of the uniqueness of normalized solutions to the Beltrami equation. Since $B^\mu = w^\mu \circ B \circ (w^\mu)^{-1}$, $B^\mu(\mathbb{H}^\mu) = \mathbb{H}^\mu$.

The Schwarzian derivative is defined by $\{f, z\} = f'''/f' - \frac{3}{2}(f''/f')^2$. One easily verifies the Cayley identity: $\{f \circ g, z\} = \{f, g\}g'(z)^2 + \{g, z\}$. Moreover $\{g, z\} \equiv 0$ if and only if g is a Möbius transformation.

Lemma 5. *If μ is in $M(\Gamma)$, then $\{w^\mu, z\} = \varphi$ is a holomorphic quadratic differential for Γ in the lower half plane $\mathbb{H}^*$.*

Proof. Note that w^μ is holomorphic in the lower half plane. On taking the Schwarzian derivative of both sides of the equation $w^\mu \circ B = B^\mu \circ w^\mu$ and using the Cayley identity, we see that $\varphi(Bz)B'(z)^2 = \varphi(z)$ for B in Γ.

Lemma 6 (Nehari–Kraus [Krau, Ne]). *If f is holomorphic and univalent (one-to-one) in the lower half plane, then $|\{f, z\}y^2| \leqslant \frac{3}{2}$.*

Proof. This inequality comes from Bieberbach's area inequality, which says that if $F(\zeta)$ is univalent in $|\zeta| > 1$ and $F(\zeta) = \zeta + b_1/\zeta + b_2/\zeta^2 + \cdots$, then $\sum_1^\infty n|b_n|^2 \leqslant 1$. The area inequality follows from the observation that $(1/2i)\int_{|\zeta|=r} \bar{F}\, dF$ measures the area enclosed by the Jordan curve which is the image under F of the circle $|\zeta| = r$ and hence is nonnegative. Thus

$$\frac{1}{2i}\int_{|\zeta|=r} \bar{F}\, dF = \frac{1}{2i}\int \overline{\left(\zeta + \frac{b_1}{\zeta} + \cdots\right)}\left(1 - \frac{b_1}{\zeta^2} - \frac{2b_2}{\zeta^3} - \cdots\right) d\zeta$$

$$= \pi\left(r^2 - \frac{|b_1^2|}{r^2} - \frac{2|b_2|^2}{r^4} - \cdots\right) \geqslant 0. \tag{5}$$

On letting r decrease to 1, one gets the area inequality. In particular, $|b_1| \leqslant 1$.

Now $\{F, \zeta\} = -(6b_1/\zeta^4) + \cdots$. Let $\zeta = Tz = (z - \bar{z}_0)/(z - z_0)$, where $z_0 = x_0 + iy_0$ and $y_0 < 0$ and let $F(Tz) = f(z)$. As usual, let $\mathbb{H}^*$ be the lower half plane. Then $T(\mathbb{H}^*) = \{$the exterior of the unit disk$\}$ and $T(z_0) = \infty$. Also $\{f, z\} = \{F, \zeta\}\zeta'(z)^2$ and $\zeta'(z) = -2y_0/(z - z_0)^2$. Hence

$$\{f, z\} = \left(-6b_1 + \text{powers of } \frac{1}{\zeta}\right)\left(\frac{1}{\zeta^4}\right)\left(\frac{-4y_0^2}{(z - z_0)^4}\right)$$

$$= \left(-6b_1 + \text{powers of } \frac{1}{\zeta}\right)\left(\frac{-4y_0^2}{(z - \bar{z}_0)^4}\right). \tag{6}$$

Taking the limit as $\zeta \to \infty$ and $z \to z_0$, we get

$$\{f, z_0\} = \frac{6b_1}{4y_0^2}. \tag{7}$$

Since $|b_1| \leqslant 1$, the lemma follows.

Lemma 7 (Ahlfors–Weill). *Suppose φ is a holomorphic function in the lower half plane $\mathbb{H}^*$ which satisfies $\|\varphi(z)y^2\|_\infty \leqslant \frac{1}{2}$. Let $\mu(z) = -2y^2\varphi(\bar{z})$ for z in $\mathbb{H}$ and $\mu(z) \equiv 0$ for z in $\mathbb{H}^*$. Then $\{w^\mu, z\} = \varphi$. Moreover, if φ is a quadratic differential for Γ, then μ is a Beltrami coefficient in $M(\Gamma)$.*

Proof. We begin with the assumption that φ is holomorphic in a neighborhood of the lower half plane $\mathbb{H}^*$, that is, in the exterior of a circle contained in $\mathbb{H}$. We also assume $|\varphi(z)| = O(|z|^{-4})$ as $z \to \infty$. For two linearly independent solutions η_1 and η_2 of $\eta'' = -\frac{1}{2}\varphi\eta$ normalized by $\eta'_1\eta_2 - \eta'_2\eta_1 \equiv 1$, form

$$\hat{f}(z) = \begin{cases} \dfrac{\eta_1(\bar{z}) + (z - \bar{z})\eta'_1(\bar{z})}{\eta_2(\bar{z}) + (z - \bar{z})\eta'_2(\bar{z})} & \text{for } z \text{ in } \mathbb{H} \\[2ex] \dfrac{\eta_1(z)}{\eta_2(z)} & \text{for } z \text{ in } \mathbb{H}^*. \end{cases} \tag{8}$$

Notice that since the solutions η_1 and η_2 are holomorphic along $\mathbb{R}$, the two definitions are continuous and coincide for z in $\mathbb{R}$. Also, in both formulas the numerator and denominator do not vanish simultaneously because $\eta'_1\eta_2 - \eta'_2\eta_1 = 1$. Simple calculations yield $\hat{f}_{\bar{z}}/\hat{f}_z = -2y^2\varphi(\bar{z})$ for z in $\mathbb{H}$ and $\hat{f}_{\bar{z}} = 0$ for z in $\mathbb{H}^*$. Since $\hat{f}$ has Beltrami coefficient μ with $\|\mu\|_\infty < 1$, we know that $\hat{f}$ is quasiconformal.

To show that $\hat{f}$ is a homeomorphism of $\hat{\mathbb{C}}$ onto $\hat{\mathbb{C}}$, from topology it suffices to show it is a local homeomorphism at each point. Since $Jac(\hat{f}) = |\hat{f}_z|^2(1 - |\mu|^2)$ and $\|\mu\|_\infty < 1$, it suffices to show $\hat{f}_z \neq 0$. Now $\hat{f}_z = (\eta_2(z))^{-2}$ in $\mathbb{H}^*$ and since η_2 is holomorphic in $\mathbb{H}^*$, $\hat{f}_z$ cannot be zero. If $\eta_2(z) = 0$, then we consider instead the mapping $1/\hat{f}$. Now $(\partial/\partial z)(1/\hat{f}) = (-\hat{f}^{-1})\hat{f}_z = (\eta_1(z))^{-2}$ in $\mathbb{H}^*$. Since η_1 and η_2 cannot be simultaneously zero, we see that either $\hat{f}$ or $1/\hat{f}$ is a local homeomorphism at every point and, therefore, so is $\hat{f}$ a local homeomorphism at every point in $\mathbb{H}^*$.

If z is in $\mathbb{H}$ we get $\hat{f}_z = (\eta_2(\bar{z}) + (z - \bar{z})\eta'_2(\bar{z}))^{-2}$. Thus $\hat{f}_z$ cannot be zero and we take care of the case when it is ∞ by considering $(1/\hat{f})$ just as we did before. Since η_1 and η_2 are holomorphic on $\mathbb{R}$ and the two formulas for $\hat{f}_z$ match along $\mathbb{R}$, we see that $f_z \neq 0$ on $\mathbb{H} \cup \mathbb{R} \cup \mathbb{H}^*$.

It remains to show that $\hat{f}$ is a local homeomorphism at ∞. Let $t = 1/z$ be a local parameter at ∞. We must show that $\hat{f}_t = (\hat{f}_z)(-1/t^2)$ approaches a nonzero limit as $t \to 0$. Since $|\varphi| = O(|z|^{-4})$ as $z \to \infty$, the solution $\eta_2(z)$ to the equation $\eta'' = -\frac{1}{2}\varphi\eta$ will be of the form $a_2 z + b_2 + O(|z|^{-1})$. Hence $\eta'_2(z) = a_2 + O(|z|^{-2})$ and $\hat{f}_z = (a_2\bar{z} + b_2 + a_2 z - a_2\bar{z} + O(|z|^{-1}))^{-2}$ and $\hat{f}_t = -(a_2 + b_2 t + O(t))^{-2}$ for z in H. One obtains the same formula if z is in $\mathbb{H}^*$. Thus $\lim_{t\to 0} \hat{f}_t = -a_2^{-2}$. This shows $\hat{f}$ is a local homeomorphism at ∞ so long as $a_2 \neq 0$. If $a_2 = 0$, we consider $1/\hat{f}$, and since this has the effect of inverting the formulas in the definition of $\hat{f}$, we get $\lim_{t\to 0}(1/\hat{f})_t = -a_1^{-2}$, where $\eta_1(z) \doteq a_1 z + b_1 + O(|z|^{-1})$. But $\eta'_1\eta_2 - \eta'_2\eta_1 = 1$ gives $a_1 b_2 - a_2 b_1 = 1$ and so not both a_1 and a_2 are zero. Hence $\hat{f}$ is a local homeomorphism in both cases.

To remove the hypothesis that φ is analytic on $\mathbb{R}$ with fourth order zero at ∞, we take a linear fractional transformation S_n such that the closure of $S_n\mathbb{H}^*$ is compact in $\mathbb{H}^*$ and such that $S_n z \to z$ as $n \to \infty$. Form $\varphi_n(z) = \varphi(S_n z)(S_n'(z))^2$ Clearly $|\varphi_n(z)| = O|z|^{-4})$ as $z \to \infty$ and φ_n is holomorphic in a neighborhood of $\mathbb{R}$. Also since $S_n\mathbb{H}^* \subset \mathbb{H}^*$, $\rho(S_n(z))|S_n'(z)| \leqslant \rho(z)$, where $\rho(z)|dz|$ is the noneuclidean metric for $\mathbb{H}^*$. Hence $|S_n'(z)|\, y < |\mathrm{Im}\, S_n(z)|$ and one sees that

$$|y^2\varphi_n(z)| = |\varphi(S_n z)S_n'(z)^2|\, y^2 < |\varphi(S_n z)|\, |\mathrm{Im}\, S_1(z)|^2 \tag{9}$$

and so

$$\| y^2\varphi_n(z)\|_\infty \leqslant \| y^2\varphi(z)\|_\infty .$$

Now $\varphi_n \to \varphi$ normally in $\mathbb{H}^*$ and hence normalized solutions to $\eta'' = -\frac{1}{2}\varphi_n\eta$ will converge normally in $\mathbb{H}^*$ to normalized solutions of $\eta'' = -\frac{1}{2}\varphi\eta$. One forms $\hat{f}_n$ corresponding to φ_n. Each $\hat{f}_n$ is a quasiconformal homeomorphism of $\hat{\mathbb{C}}$ normalized at 0, 1, and ∞ and with dilatation bounded by $\| y^2\varphi(z)\|_\infty < 1$. By the existence theorem for the Beltrami equation, we can form $\hat{f}$, a normalized, quasiconformal homeomorphism of $\hat{\mathbb{C}}$ with Beltrami coefficient $-\frac{1}{2}y^2\varphi(\bar{z})$ in $\mathbb{H}$ and 0 in $\mathbb{H}^*$. Since $y^2\varphi_n(\bar{z})$ converges in the bounded pointwise sense to $y^2\varphi(z)$, we know that $\hat{f}_n$ converges normally to $\hat{f}$ (see Lemma 5 of Section 1.8). Moreover, $\{\hat{f}_n, z\} = \varphi_n$ converges normally to $\{f, z\}$ in $\mathbb{H}^*$ and hence $\{\hat{f}, z\} = \varphi$.

It is trivial to show that if φ is a quadratic differential for Γ, then $y^2\varphi(\bar{z})$ is a Beltrami differential, and this completes the proof of Lemma 7.

Beltrami differentials of the form $y^2\varphi(\bar{z})$ are called harmonic Beltrami differentials.

For $\rho = |dz|/2y$, let $B(\Gamma)$ be the space of holomorphic quadratic differential for Γ defined in the lower half plane $\mathbb{H}^*$ with norm $\|\varphi\|_B = \|\rho^{-2}\varphi\|_\infty$. The condition in Lemma 7 that $\| y^2\varphi(z)\|_\infty < \frac{1}{2}$ is obviously equivalent to $\|\varphi\|_B < 2$.

Theorem 1. *Let Γ be a Fuchsian group acting on $\mathbb{H}$. Then the mapping $\Phi\colon M(\Gamma) \to B(\Gamma)$ defined by $\Phi(\mu) = \{w^\mu, z\}$ induces a one-to-one mapping $\bar{\Phi}\colon T(\Gamma) \to B(\Gamma)$ whose image in $B(\Gamma)$ is contained in the ball of radius* 6 *and contains the ball of radius* 2.

Proof. Lemma 3 implies $\bar{\Phi}$ is well defined and one-to-one and Lemma 6 yields $\|\Phi(\mu)\|_B \leqslant 6$. Lemma 5 tells us that $\{w^\mu, z\}$ is a quadratic differential for Γ and the Ahlfors–Weill extension lemma (Lemma 7) says that the image of Φ contains the open ball of radius 2.

Remark 1. Ahlfors and Bers have shown that $\Phi(M(\Gamma))$ is an open set in $B(\Gamma)$. This fact depends on Ahlfors's theorem concerning the existence of a Lipschitz continuous anti-quasiconformal reflection about a quasicircle $w^\mu(\hat{\mathbb{R}})$ [Ah2, Ah4]. In cases where $B(\Gamma)$ is finite dimensional the fact that $\Phi(M(\Gamma))$ is open follows in an elementary way from the theorem on invariance of domain (see Theorem 4). In the general case (even without the deep theorem on anti-quasiconformal reflections), we can use the above theorem to introduce local coordinates which make $T(\Gamma)$ into a complex manifold. We pursue this topic in the next two sections.

Remark 2. $w^{\mu}(\hat{\mathbb{R}})$ is a quasicircle, that is, the image under a quasiconformal mapping of $\hat{\mathbb{R}}$. In [Ah2], Ahlfors gives a geometric characterization of quasicircles. The books by Ahlfors [Ah4] and Lehto and Virtanen [LehtV] contain the same theorem. Gehring [Ge3] gives further characterizations of quasicircles. In the case where Γ is the identity and $C = \hat{\mathbb{R}}$, the universal Teichmüller space T is $\Phi(M(\Gamma))$. By Ahlfors [Ah4] and Bers [Ber6], T is an open subset of the Banach space $B(\Gamma)$ contained in the closed set S of Schwarzian derivatives of all univalent functions. Gehring showed that when Γ is the identity group, T is in interior of the closure of T [Ge1] and that there are points in S not in the closure of T. Thurston [Th3] showed that S contains uncountably many isolated points.

Remark 3. There is a geometric characterization of Jordan arcs which are images of intervals under quasiconformal mappings (see Rickman [Ri]).

5.5. TRANSLATION MAPPINGS BETWEEN TEICHMÜLLER SPACES

Let $[\mu] \in T(\Gamma, C)$ and $w = w_{\mu}$. Then $\Gamma_1 = w \circ \Gamma \circ w^{-1}$ is also a Fuchsian group and $C_1 = w(C)$ is a closed subset of the extended real axis which contains the limit set Λ_1 of Γ_1. Let d be the Teichmüller metric for $T(\Gamma, C)$ and d_1 be the metric for $T(\Gamma_1, C_1)$.

The mapping w induces a mapping β from $T(\Gamma_1, C_1)$ to $T(\Gamma, C)$. For $[v]$ in $T(\Gamma_1)$ we let $\beta[v]$ be the Beltrami coefficient of $w_v \circ w_{\mu}$. It is obvious that β is an isometry with respect to the metrics d and d_1.

In the case where $C_1 = C = \hat{\mathbb{R}}$, Theorem 1 applies to the group Γ_1 as well as the group Γ and so we have the following diagram:

$$\begin{array}{ccc} T(\Gamma_1) & \xrightarrow{\ \beta\ } & T(\Gamma) \\ \bar{\Phi}_1 \downarrow & & \downarrow \bar{\Phi} \\ B(\Gamma_1) & \longrightarrow & B(\Gamma). \end{array}$$

The next step is to give an explicit description of the mapping $\bar{\Phi} \circ \beta \circ \bar{\Phi}_1^{-1}$ in terms of the quasifuchsian group $\Gamma^{\mu} = w^{\mu} \circ \Gamma \circ (w^{\mu})^{-1}$ and its invariant domains $\Omega = w^{\mu}(\mathbb{H})$ and $\Omega^* = w^{\mu}(\mathbb{H}^*)$. Let f^{μ} be the unique conformal mapping from Ω to $\mathbb{H}$ normalized at 0, 1, and ∞. Note that $f^{\mu} \circ w^{\mu}(z) = w_{\mu}(z)$ for z in $\mathbb{H} \cup \hat{\mathbb{R}}$.

Let $B(\Gamma^{\mu})$ be the Banach space of holomorphic quadratic differentials φ in Ω

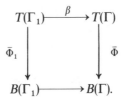

with norm given by

$$\|\varphi\| = \sup_{z\in\Omega} |\varphi(z)\rho_\mu(z)^{-2}|, \tag{10}$$

where ρ_μ is the noneuclidean metric for Ω. Thus $\rho_\mu(z) = \rho(f^\mu(z))|f^{\mu'}(z)|$, where $\rho(\zeta) = |d\zeta|/|\bar{\zeta} - \zeta|$. Similarly, let $B_*(\Gamma^\mu)$ be the Banach space of holomorphic quadratic differentials φ in Ω^* with norm given by

$$\|\varphi\| = \sup_{z\in\Omega^*} |\varphi(z)\rho_*^\mu(z)^{-2}|, \tag{11}$$

where $\rho(z) = \rho_*^\mu(w^\mu(z))|w^{\mu'}(z)|$. Let $j(w) = \bar{w}$. Obviously, the map $\varphi \to \varphi(w^\mu(z))w^{\mu'}(z)^2$ from $B_*(\Gamma^\mu)$ to $B(\Gamma)$ is an isometry. Similarly, if $j(w) \equiv \bar{w}$, then the mapping $\varphi \mapsto j\varphi j(f^\mu(z))f^{\mu'}(z)^2$ from $B(\Gamma)$ to $B(\Gamma^\mu)$ is an isometry.

Given a Beltrami coefficient τ supported in Ω, let w^τ be the unique normalized qc-mapping which is conformal in Ω^* and has Beltrami coefficient τ in Ω.

Lemma 8. *$\{w^\tau, z\}$ has norm $\leqslant 12$ in $B_*(\Gamma^\mu)$.*

Proof. Let g be the unique normalized conformal mapping from $w^\tau(\Omega^*)$ to the lower half plane $\mathbb{H}^*$. Then $g \circ w^\tau \circ w^\nu(z) = z$. Taking the Schwarzian derivative of both sides of this equation, we get

$$\{g, w\}(g^{-1})'(z)^2 + \{w^\tau, w^\nu\}w^{\nu'}(z)^2 + \{w^\nu, z\} = 0. \tag{12}$$

Since $g^{-1} \circ g(z) = z$, $\{g, w\}(g^{-1})'(z)^2 = -\{g^{-1}, z\}$. But from Lemma 6, $\|\{g^{-1}, z\}\| \leqslant 6$ and $\|\{w^\tau, z\}\| \leqslant 6$. Therefore, the norm of $\{w^\tau, z\}$ in $B_*(\Gamma^\mu)$ is $\leqslant 12$.

Theorem 2. *Let $\varphi \in B(\Gamma_1)$ and let $\hat{\varphi} = j\varphi j(f^\mu(z))f^{\mu'}(z)^2$. Let $\hat{\nu} = \rho_\mu^{-2}(\zeta)\overline{\hat{\varphi}(\zeta)}$, where ρ_μ is the noneuclidean metric for the domain $\Omega = w^\mu(\mathbb{H})$. Then the mapping $\bar{\Phi} \circ \beta \circ \bar{\Phi}_1^{-1}$ takes φ into $\{w^{\hat{\nu}}, w\}w^{\mu'}(z)^2 + \{w^\mu, z\}$. Hence, $\bar{\Phi} \circ \beta \circ \bar{\Phi}_1^{-1}$: $B(\Gamma) \to B(\Gamma_1)$ is continuous and holomorphic in a neighborhood of the origin.*

Proof. The formula for $\bar{\Phi} \circ \beta \circ \bar{\Phi}_1^{-1}(\varphi)$ is a routine calculation using the Cayley identity. From the formula, it is clear that the mapping has a bounded Frechét derivative and, in particular, the intermediate mapping $\hat{\nu} \to \{w^{\hat{\nu}}, w\}$ from $M(\Gamma^\mu)$ to $B_*(\Gamma^\mu)$ has derivative at the origin bounded by 12 (from Lemma 8). Moreover, the intermediate mapping $\hat{\nu} \to \{w^{\hat{\nu}}, w\}$ is differentiable at any point where $\|\hat{\nu}\|_\infty < 1$.

5.6. THE MANIFOLD STRUCTURE ON TEICHMÜLLER SPACE

Just as for the case of Lemma 7 and Theorems 1 and 2, throughout this section we again assume $C = \hat{\mathbb{R}}$ and write $T(\Gamma)$ instead of $T(\Gamma, C)$. We use the mapping $\bar{\Phi}$ of Theorem 1 as a local coordinate for $T(\Gamma)$ at the origin. If $[\mu] \in T(\Gamma)$ and

$\Gamma_1 = w_\mu \circ \Gamma \circ (w_\mu)^{-1}$, we let $\bar{\Phi}_1 \circ \beta^{-1}$ be a local coordinate at $[\mu]$. In order to use these mappings as local coordinates, we must show that, restricted to a suitably small neighborhood of the origin, the map $\bar{\Phi}_1$ is a homeomorphism and that the transition mapping $\bar{\Phi} \circ \beta \circ \bar{\Phi}_1^{-1}$ is holomorphic. That the transition mapping is holomorphic follows from the explicit description given in Theorem 2. The next theorem shows that $\bar{\Phi}$ and $\bar{\Phi}_1$ are homeomorphisms when restricted to suitably small neighborhoods of the origin.

Theorem 3. *Let $U = \{[\mu] \in T(\Gamma)\colon d(0, \mu) < \frac{1}{2}\log 2\}$. Then $\bar{\Phi}|U$ is a homeomorphism onto an open set contained in the ball of radius 2 and containing the ball of radius $\frac{2}{3}$ in $B(\Gamma)$.*

Proof. The topology on $T(\Gamma)$ is the quotient topology from $M(\Gamma)$ under the equivalence relation which defines $T(\Gamma)$. Therefore, the continuity of $\bar{\Phi}$ follows from the continuity of Φ. Since $M(\Gamma)$ is the unit ball in a Banach space and since Φ is obviously holomorphic (in the next section we will compute its complex derivative), by Schwarz's lemma $\|\Phi(\mu)\| \leqslant 6\|\mu\|_\infty$. Here we use the fact that Lemma 6 tells us that the image of Φ is contained in the ball of radius 6. Now assume $d(0, [\mu_0]) < \frac{1}{2}\log 2$. From the definition of Teichmüller's metric (3), this implies $[\mu_0]$ is represented by a Beltrami differential μ_0 for which $\|\mu_0\|_\infty < \frac{1}{3}$, and hence $\|\Phi(\mu_0)\| < 2$. Suppose $\varphi_0 = \Phi(\mu_0)$ and $\|\varphi_n - \varphi_0\|$ converges to 0. Then for sufficiently large n, $\|\varphi_n\| < 2$ and $\Phi(-2y^2\varphi_n(\bar{z})) = \varphi_n$ and $\mu_0 \sim -2y^2\varphi_0(\bar{z})$. The hypothesis that $\|\varphi_n - \varphi_0\| \to 0$ then implies that

$$d(-2y^2\varphi_n(\bar{z}), -2y^2\varphi_0(\bar{z})) \to 0 \tag{13}$$

and therefore $\bar{\Phi}^{-1}$ is continuous on $\bar{\Phi}(U)$.

Finally, suppose $\|\varphi\|_B < \frac{2}{3}$. Then $\mu(z) = -2y^2\varphi(\bar{z})$ has L_∞-norm less than $\frac{1}{3}$ and $\Phi(\mu) = \varphi$. Hence,

$$\begin{aligned} d(0, \mu) &< \tfrac{1}{2}\log\frac{1+\frac{1}{3}}{1-\frac{1}{3}} \\ &= \tfrac{1}{2}\log 2. \end{aligned}$$

Thus $\bar{\Phi}(U)$ contains the ball of radius $\frac{2}{3}$.

Theorem 4. *The Teichmüller space $T(\Gamma)$ is a complex manifold modeled on the Banach space $B(\Gamma)$. When $B(\Gamma)$ is finite dimensional, the mapping $\bar{\Phi}\colon T(\Gamma) \to B(\Gamma)$ is a homeomorphism of $T(\Gamma)$ onto a bounded open subset of $B(\Gamma)$.*

Remark. We wish to emphasize that the Bers embedding $\bar{\Phi}$ realizes $T(\Gamma)$ as a bounded open subset of the complex Banach space $B(\Gamma)$ even in the infinite dimensional cases. This fact is not proved in this book. The proof is in Ahlfors's book [Ah4] and depends on establishing geometric properties of quasicircles.

Proof. For any point $[\mu]$ in $T(\Gamma)$, let $\Gamma_1 = w_\mu \circ \Gamma \circ w_\mu^{-1}$. The Banach spaces $B(\Gamma_1)$ and $B(\Gamma)$ are isomorphic. If $\|\mu\|_\infty < \frac{1}{3}$, this fact is obvious by looking at the derivative of the mapping $\bar{\Phi} \circ \beta \circ \bar{\Phi}_1^{-1}$ discussed in Theorem 2. On the other hand, for any μ in $M(\Gamma)$ we can express w_μ in the form $w_\mu = w_{\mu_n} \circ w_{\mu_{n-1}} \circ \cdots \circ w_{\mu_1}$, where $\Gamma_{k+1} = w_{\mu_k} \circ \Gamma_k \circ (w_{\mu_k})^{-1}$ and each $\|\mu_k\|_\infty < \frac{1}{3}$. The mappings $\bar{\Phi}$ and $\bar{\Phi}_1$ give local coordinates if we restrict them to Teichmüller neighborhoods of Teichmüller distance less than $\frac{1}{2}\log 2$ from the origin in the respective spaces $T(\Gamma)$ and $T(\Gamma_1)$. By Theorem 2, the transition mappings are holomorphic.

To prove the second part of the theorem, note that $\bar{\Phi}$ is one-to-one and continuous from $T(\Gamma)$ onto a subset of $B(\Gamma)$. If $B(\Gamma)$ is finite dimensional, then $\bar{\Phi}$ is one-to-one and continuous from $T(\Gamma)$ onto a subset of $B(\Gamma)$ and $T(\Gamma)$ has a manifold structure of the same dimension as $B(\Gamma)$. From the topological theorem on invariance of domain, $\bar{\Phi}$ is therefore a homeomorphism of $T(\Gamma)$ onto a domain in $B(\Gamma)$.

5.7. THE INFINITESIMAL THEORY

From the theory of quasiconformal mapping, $w^\mu(z)$ has a Frechét derivative with respect to μ. Let $F(z) = \lim_{t\to 0} (1/t)(w^{t\mu}(z) - z)$. This limit converges uniformly for z in compact subsets of $\mathbb{C}$ and

$$F(z) = -\frac{\zeta(\zeta-1)}{\pi} \iint_{\mathbb{H}} \frac{\mu(\zeta)\, d\xi\, d\eta}{\zeta(\zeta-1)(\zeta-z)}. \tag{14}$$

We turn next to the function $\Phi(t\mu) = \{w^{t\mu}, z\}$. Since $w^{t\mu}(z) = z + tF(z) + O(t^2)$ uniformly on compact subsets of $\mathbb{C}$, upon taking the Schwarzian derivative in $\mathbb{H}^*$ we get $\{w^{t\mu}, z\} = tF'''(z) + O(t^2)$ uniformly on subsets of $\mathbb{H}^*$. But

$$F'''(z) = -\frac{6}{\pi} \iint_{\mathbb{H}} \frac{\mu(\zeta)\, d\xi\, d\eta}{(\zeta - z)^4}. \tag{15}$$

From this integral formula, it follows that $F'''(Az)\overline{A'(z)}^2 = F'''(z)$ for A in Γ. The calculation depends on the relation $\mu(Az)\overline{A'(z)} = \mu(z)A'(z)$.

We use the notation $\dot{\Phi}[\mu] = \lim_{t\to 0}(1/t)\Phi(t\mu)$. From the above discussion, $\dot{\Phi}[\mu](z) = F'''(z)$.

Theorem 5. *There exists a constant C such that $|\Phi(t\mu)(z) - t\dot{\Phi}[\mu](z)| \leqslant Cy^{-2}t^2$ for $|t|$ less than some positive number δ and where $z = x + iy$. In other words, $\dot{\Phi}$ is the derivative of Φ with respect to the norm for the Banach space $B(\Gamma)$.*

Proof. From Lemma 6 we know that $|\Phi(t\mu)(z)y^2| \leqslant \frac{3}{2}$ for $t < 1/\|\mu\|_\infty$. Now $\Phi(t\mu)(z)y^2$ is a holomorphic function of t, and thus there is a bound on its second derivative for $|t| < \delta$ which depends only on $\frac{3}{2}$ and the amount by which δ is less than $1/\|\mu\|_\infty$. The inequality of the theorem therefore follows.

Corollary. *For any function φ in $B(\Gamma)$ one has the reproducing formula*

$$\varphi(z) = +\frac{12}{\pi} \iint_{\mathbb{H}} \frac{\eta^2 \varphi(\bar{\zeta})\, d\xi\, d\eta}{(\zeta - z)^4}, \tag{16}$$

where z is in $\mathbb{H}^*$ and $\zeta = \xi + i\eta$.

Proof. We showed in Lemma 7 that $\Phi(-2y^2\varphi(\bar{z})) = \varphi(z)$ for φ in $B(\Gamma)$ and $\|\varphi\|_B < 2$. The corollary follows by letting $\mu = -2y^2\varphi(\bar{z})$ in Theorem 5, dividing by t and letting $t \to 0$.

Formula (16) can be proved directly. In fact, on making the change of variable $\zeta \to \bar{\zeta}$, it can be rewritten as

$$\varphi(z) = \frac{12}{\pi} \iint_{\mathbb{H}^*} \frac{|\zeta - \bar{\zeta}|}{4} \cdot \frac{\varphi(\zeta)}{(\bar{\zeta} - z)^4}\, d\xi\, d\eta, \tag{17}$$

or, equivalently,

$$\varphi(z) = \frac{3}{\pi} \iint_{\mathbb{H}^*} \rho^{-2}(\zeta) K(z, \zeta) \varphi(\zeta)\, d\xi\, d\eta, \tag{18}$$

where ρ is the noneuclidean metric for $\mathbb{H}^*$ and $K(z, \zeta)$ is the kernel function we introduced in Section 4.4. Formula (18) was already verified in that section.

5.8. INFINITESIMALLY TRIVIAL BELTRAMI DIFFERENTIALS

We continue to assume $C = \hat{\mathbb{R}}$ so that the mapping $\bar{\Phi}$ from $T(\Gamma)$ in $B(\Gamma)$ is defined and one-to-one and, when restricted to a neighborhood of the origin, maps onto an open set. We have the following Banach spaces:

$$L_\infty(\Gamma) = \left\{ \mu : \begin{array}{l} \mu \text{ is a measurable function on } \mathbb{H}, \\ \mu(Az)\overline{A'(z)} = \mu(z)A'(z) \text{ and } \|\mu\|_\infty < \infty \text{ for all } A \text{ in } \Gamma \end{array} \right\}$$

$$A(\Gamma) = \left\{ \varphi : \begin{array}{l} \varphi \text{ is holomorphic in } \mathbb{H},\ \varphi(A(z))A'(z) = \varphi(z) \\ \text{for all } A \text{ in } \Gamma, \text{ and } \|\varphi\| = \iint_{\mathbb{H}/\Gamma} |\varphi|\, dx\, dy < \infty \end{array} \right\}$$

$$N(\Gamma) = \left\{ \mu : \begin{array}{l} \mu \in L_\infty(\Gamma) \text{ and } \iint_{\gamma/\Gamma} \mu\varphi\, dx\, dy = 0 \\ \text{for all } \varphi \text{ in } A(\Gamma) \end{array} \right\};$$

$M(\Gamma)$ is the open unit ball of $L_\infty(\Gamma)$ and $M_0(\Gamma)$ is the subset of Beltrami coefficients which are equivalent to 0. By Section 5.2, μ is in $M_0(\Gamma)$ if it is in $M(\Gamma)$ and if $w_\mu(x) = x$ for all x in $\hat{\mathbb{R}}$. Now $M_0(\Gamma)$ is called the space of *trivial* Beltrami coefficients and $N(\Gamma)$ is called the space of *infinitesimally trivial* Beltrami differentials.

The following theorem says that the infinitesimally trivial Beltrami differentials form the tangent space to $M_0(\Gamma)$ at the origin.

Theorem 6. *μ is an infinitesimally trivial Beltrami differential if, and only if, there exists a holomorphic curve σ_t of trivial Beltrami differentials for which $\sigma_t(z) = t\mu(z) + O(t^2)$ uniformly in z.*

Proof. Suppose such a curve exists. From Theorem 4 and Lemma 5 of Section 4.5, we know that

$$\left|\operatorname{Re} \iint\limits_{\mathbb{H}/\Gamma} \frac{\sigma_t \varphi}{1 - |\sigma_t|^2}\, dx\, dy\right| \leqslant \iint\limits_{\mathbb{H}/\Gamma} \frac{|\sigma_t|^2 |\varphi|}{1 - |\sigma_t|^2}\, dx\, dy. \tag{19}$$

Substituting in $\sigma_t = t\mu + O(t^2)$, we see the right-hand term vanishes of second order whereas the left-hand term is of first order. We obtain $\iint_{\mathbb{H}/\Gamma} \mu\varphi\, dx\, dy = 0$ for all φ in $A(\Gamma)$. Thus μ is in $N(\Gamma)$.

Conversely, suppose μ is in $N(\Gamma)$. Consider the quadratic differential $\varphi(\zeta) = \sum A'(\zeta)^2/(A(\zeta) - z)^4$, where $\zeta \in \mathbb{H}$, $z \in \mathbb{H}^*$, and the summation is over all A in Γ. (This is a Poincaré theta series.) Obviously,

$$\|\varphi(\zeta)\| \leqslant \iint\limits_{\mathbb{H}/\Gamma} \sum \left|\frac{A'(\zeta)^2}{(A(\zeta) - z)^4}\right| d\xi\, d\eta = \iint\limits_{\mathbb{H}} \frac{d\xi\, d\eta}{|\zeta - z|^4} = \frac{\pi}{4} y^{-2}.$$

By the hypothesis $\iint_{\mathbb{H}/\Gamma} \mu(\zeta)\varphi(\zeta)\, d\xi\, d\eta = 0$. Therefore,

$$\iint\limits_{\mathbb{H}} \frac{\mu(\zeta)}{(\zeta - z)^4}\, d\xi\, d\eta = 0 \qquad \text{for all } z \text{ in } \mathbb{H}^*$$

and we see that $\dot{\Phi}[\mu] = 0$. Now let $\varphi^t(z) = \Phi(t\mu)(z)$ and let $v_t(z) = -2y^2\varphi^t(\bar{z})$. By Lemma 7, $\Phi(v_t) = \Phi(t\mu)$ and, since Φ: $T(\Gamma) \to B(\Gamma)$ is one-to-one, this means $t\mu \sim v_t$. Moreover, by Theorem 5, $|\Phi(t\mu)(z)y^2| \leqslant Ct^2$ and so $\|v_t\|_\infty \leqslant Ct^2$.

Now form $w_{\sigma_t} = (w_{v_t})^{-1} \circ w_{t\mu}$. Note that $\sigma(t)$ is a trivial Beltrami differential because $t\mu \sim v_t$. Also, $\sigma(t) = (t\mu + \hat{v})/(1 + t\bar{\mu}\hat{v})$, where $\hat{v}$ is the Beltrami coefficient of the inverse mapping to w_{v_t} times a function of modulus constantly equal to one. We conclude that $\sigma_t = t\mu + O(t^2)$ and $\sigma_t \in M_0(\Gamma)$.

Remark. This theorem is the basis of a number of important existence theorems. We will use it to prove the existence part of Teichmüller's theorem and the Hamilton–Krushkal necessary condition for extremality.

5.9. TEICHMÜLLER SPACES OF FUCHSIAN GROUPS WITH BOUNDARY

In Section 5.4 through 5.8, we assumed $C = \hat{\mathbb{R}}$. This provided us with the very convenient fact that μ and v were Teichmüller equivalent if, and only if, $\{w^\mu, z\} = \{w^v, z\}$ in the lower half plane. We now treat the case where C is a Γ-invariant closed set containing the limit set of Γ and C is a proper subset of $\hat{\mathbb{R}}$. Our objective is to show that the Teichmüller space $T(\Gamma, C)$ is a real analytic manifold modeled on a Banach space and to obtain a theorem analogous to Theorem 6 for trivial curves. This objective naturally leads us to consideration of noneuclidean crystallographic groups which are Fuchsian groups to which anticonformal isometries have been adjoined.

Since C is Γ-invariant and closed, the complementary set $\Omega = \hat{\mathbb{C}} - C$ is Γ-invariant and open. Moreover, since C contains the limit set, Γ acts discontinuously on Ω. The Riemann surface Ω/Γ has an anticonformal involution induced by the conjugation $j\colon \Omega \to \Omega$ given by $j(z) = \bar{z}$.

Let Γ^+ be the group generated by Γ and j. Note that Γ^+ is a group of isometries of Ω in the noneuclidean metric for Ω and a fundamental domain for Γ acting on $\mathbb{H}$ is also a fundamental domain for Γ^+ acting on Ω.

Let μ be a Beltrami coefficient for Γ acting on $\mathbb{H}$. We extend μ to a Beltrami coefficient $\hat{\mu}$ for Γ^+ acting on Ω by the formula

$$\hat{\mu}(z) = \begin{cases} \mu(z) & \text{for } z \text{ in the upper half plane,} \\ \overline{\mu(\bar{z})} & \text{for } z \text{ in the lower half plane.} \end{cases}$$

By definition, a Beltrami coefficient for a group which contains some anticonformal as well as conformal transformations obeys the laws

$$\begin{aligned} &\text{(i)} \quad \mu(A(z))\frac{\overline{\partial A}}{\partial z} = \mu(z)\frac{\partial A}{\partial z} \qquad \text{when } A \text{ is conformal, and} \\ &\text{(ii)} \quad \mu(A(z))\left(\frac{\overline{\partial A}}{\partial \bar{z}}\right) = \overline{\mu(z)}\left(\frac{\partial A}{\partial \bar{z}}\right) \qquad \text{when } A \text{ is anticonformal.} \end{aligned} \tag{20}$$

We leave to the reader the verification that the mapping $\mu \to \hat{\mu}$ establishes an isometric isomorphism from the space $M(\Gamma)$ of Beltrami coefficients for Γ acting on $\mathbb{H}$ onto the space $M(\Gamma^+)$ of Beltrami coefficients for Γ^+ acting on Ω. Moreover, if w is a quasiconformal homeomorphism of $\mathbb{H}$ with Beltrami coefficient μ, then $\hat{w}$ given by w in $\mathbb{H}$ and by $\overline{w(\bar{z})}$ in the lower half plane is a quasiconformal homeomorphism of the extended complex plane with Beltrami coefficient $\hat{\mu}$.

Let $\pi\colon \mathbb{H} \to \Omega$ be the universal covering of Ω. Then the groups Γ and Γ^+ of isometries of Ω lift to groups G and G^+ of isometries of $\mathbb{H}$. In general, elements of

G^+ are either of the form

$$A(z) = \frac{az + b}{cz + d}$$

or of the form

$$A(z) = -\frac{a\bar{z} + b}{c\bar{z} + d},$$

where a, b, c, and d are real numbers with $ad - bc = 1$. The Beltrami coefficients for Γ^+ acting on Ω lift to Beltrami coefficients $M(G^+)$ for G^+ acting on $\mathbb{H}$ by the formula

$$\pi^*(\nu) = \nu(\pi(z))\overline{\pi'(z)}/\pi'(z). \tag{21}$$

It is obvious that the mapping $\mu \to \pi^*(\hat{\mu})$ gives an isomorphism of $M(\Gamma)$ onto $M(G^+)$.

We recall the definition of the set $M_0(\Gamma, C)$ of trivial Beltrami coefficients for the group Γ with respect to C. $M_0(\Gamma, C)$ consists of all Beltrami coefficients of quasiconformal self-mappings w of $\mathbb{H}$ for which

$$\begin{aligned} &\text{(i)} \quad w \circ A = A \circ w \qquad \text{for all } A \text{ in } \Gamma, \text{ and} \\ &\text{(ii)} \quad w(x) = x \qquad \text{for all } x \text{ in } C. \end{aligned} \tag{22}$$

Let $M_0(G^+)$ consist of Beltrami coefficients in $M(G^+)$ of quasiconformal self-mappings f of $\mathbb{H}$ for which $f(x) = x$ for all x in $\mathbb{R}$. A quasiconformal self-mapping f of $\mathbb{H}$ is called trivial for G^+ if its Beltrami coefficient is in $M_0(G^+)$.

Lemma 9. *The natural isometry $\mu \mapsto \pi^*(\hat{\mu})$ from $M(\Gamma)$ onto $M(G^+)$ takes $M_0(\Gamma, C)$ onto $M_0(G^+)$. In other words, trivial mappings w for Γ relative to C correspond uniquely to trivial mappings for G^+.*

Remark 1. The point of this lemma is that it replaces an equivalence relation based on two functions being equal on some Γ-invariant closed set C by an equivalence relation based on two functions being equal on all of $\mathbb{R}$.

Remark 2. The Fuchsian group G may or may not be of the first kind.

Proof. It is permissible to assume C contains ∞. The open set $\mathbb{R} - C$ is a countable union of open intervals I_k. Any one of these intervals, say I_1, is a noneuclidean goedesic in the domain Ω and, therefore, each component of $\pi^{-1}(I_1)$ is a hyperbolic line in $\mathbb{H}$.

Let $\hat{w}$ be a quasiconformal self-mapping of Ω whose Beltrami coefficient, when restricted to the upper half plane, is an element of $M_0(\Gamma, C)$. A lifting f of $\hat{w}$ satisfies $\pi \circ f = \hat{w} \circ \pi$. If $\hat{I}_1$ is one of the components of $\pi^{-1}(I_1)$, then $\pi(f(\hat{I}_1)) = \hat{w}(\pi(\hat{I}_1)) = \hat{w}(I_1) = I_1$ and, therefore, $f(\hat{I}_1)$ is also a component of

$\pi^{-1}(I_1)$. The covering group G_0 for the covering π permutes the different liftings of I_1 and thus there is an element A of G_0 for which $A \circ f(\hat{I}_1) = \hat{I}_1$. Since $A \circ f$ is also a lifting of π, by selecting the correct lifting, we may assume A is the identity and $f(\hat{I}_1) = \hat{I}_1$.

Pick a second open interval I_2 in $\mathbb{R} - C$ and a curve γ lying in the upper half plane which connects I_1 to I_2. This curve lifts to a curve $\hat{\gamma}$ which connects $\hat{I}_1$ to a hyperbolic line $\hat{I}_2$ in $\mathbb{H}$ for which $\pi(\hat{I}_2) = I_2$. By repeating this process for every interval I_k in $\mathbb{R} - C$, we obtain a family of hyperbolic lines in $\mathbb{H}$ which bound a noneuclidean convex set $\hat{\omega}$ and the covering mapping π restricted to $\hat{\omega}$ is a homeomorphism of $\hat{\omega}$ onto $\mathbb{H}$. Reflect $\hat{\omega}$ about $\hat{I}_1$ and let ω be the union of $\hat{\omega}$, its reflection about $\hat{I}_1$, and the common boundary line $\hat{I}_1$. Then ω is a fundamental domain for the covering group G_0 of the covering π. In ω the lifting f which preserves the hyperbolic line $\hat{I}_1$ is given by the formula

$$f = (\pi|\omega)^{-1} \circ \hat{w} \circ (\pi|\omega). \tag{23}$$

The fact that $A \circ f = f \circ A$ for every A in the covering group G_0 implies that formula (23) determines f everywhere in the upper half plane. Moreover, f must preserve any component of any one of the sets $\pi^{-1}(I_k)$, as well as the orientation of any of these components. Thus the unique quasiconformal extension of f to the real axis (which we also denote by f) fixes the endpoints of any of the liftings $\hat{I}_k$.

Any point p in $\mathbb{R}$ which is in the limit set for the covering group G_0 of the covering π is an accumulation point of endpoints of the hyperbolic lines $A(\hat{I}_k)$, where A is in G_0 and k varies. In this case $f(p) = p$ because the endpoints of the hyperbolic lines $A(\hat{I}_k)$ are held fixed by f. On the other hand, if p is not in the limit set, then there is an interval $\tilde{\alpha}$ in $\mathbb{R}$ containing p such that $\pi|\tilde{\alpha}$ is a homeomorphism from $\tilde{\alpha}$ onto an open interval α contained in C. In this case, $f(p) = p$ because of formula (23) and the fact that $\hat{w}(x) = x$ for all x in C.

Conversely, if $f(p) = p$ for all p in $\mathbb{R}$, then from formula (23) the same argument shows that $\hat{w}(x) = x$ for all x in C. The lemma follows.

The Teichmüller space $T(G^+)$ is defined to be the $M(G^+)$ factored by the equivalence relation induced by $M_0(G^+)$.

Theorem 7. *Assume Γ is of the second kind and C is a closed, proper, Γ-invariant subset of $\hat{\mathbb{R}}$ containing the limit set of Γ. Then the Teichmüller space $T(\Gamma, C)$ is naturally isomorphic to $T(G^+)$.*

Proof. We have already defined the natural isometric isomorphism from $M(\Gamma)$ onto $M(G^+)$. By Lemma 9 this mapping preserves the equivalence relations for the two Teichmüller spaces. Therefore, it induces an isomorphism.

To obtain the Bers embedding we form the space of quadratic differentials $B(G^+)$ for the group G^+. An element ψ of $B(G^+)$ is a holomorphic function in the upper half plane for which

$$\begin{aligned} &\text{(i)} \quad \psi(A(z))A'(z)^2 = \psi(z) \text{ for conformal elements } A \text{ of } G^+, \\ &\text{(ii)} \quad \psi(A(z))\left(\frac{\partial A}{\partial \bar z}\right)^2 = \overline{\psi(z)} \text{ for anticonformal elements } A \text{ of } G^+, \text{ and} \\ &\text{(iii)} \quad \|\psi\| = \sup_{z\in\mathbb{H}} |\psi(z)4y^2| \text{ is finite.} \end{aligned} \tag{24}$$

Given a Beltrami coefficient μ in $M(G^+)$, satisfying (20), we let f^μ be a quasiconformal homeomorphism of the sphere with Beltrami coefficient identically equal to zero in the lower half plane and equal to μ in the upper half plane. It follows from the Cayley identity (see Section 5.4) that the Schwarzian derivative

$$\varphi^\mu(z) = \{f^\mu, z\}$$

satisfies (i) of (24). It satisfies (iii) of (24) with $\|\varphi^\mu\| \leqslant 6$ because of the Nehari–Kraus lemma (Lemma 6 of Section 5.4). To prove property (ii) we introduce the notation $j(z) = \bar z$ and $S(f) = \{f, z\}$, the Schwarzian derivative of f. Corresponding to an anticonformal transformation A in G^+, there is an anticonformal transformation A^μ for which $f^\mu \circ A = A^\mu \circ f^\mu$. On applying the Schwarzian derivative to both sides of the equation $j \circ f^\mu \circ A = j \circ A^\mu \circ f^\mu$ and using the Cayley identity, we obtain

$$S(j \circ f^\mu \circ A) = S(f^\mu).$$

Now rewrite $j \circ f^\mu \circ A$ as $j \circ f^\mu \circ j \circ j \circ A$ and take the Schwarzian derivative and the complex conjugate. Since

$$\overline{S(j \circ f^\mu \circ j)}\left(\frac{\partial A}{\partial \bar z}\right)^2 = \overline{S(f^\mu)}$$

and since $j \circ S(j \circ f^\mu \circ j) \circ j \circ A = S(f) \circ A$, we obtain the required identity. It follows that φ^μ is an element of $B(G^+)$.

Since the equivalence relation which determines trivial elements of $M(G^+)$ involves the equality $f^\mu(x) = x$ for all x in $\mathbb{R}$ (see Lemma 3), the mapping

$$\Phi^+\colon M(G^+) \to B(G^+)$$

given by $\Phi^+(\mu) = \{f^\mu, z\}$ induces a one-to-one mapping $\bar\Phi^+$ from $T(G^+)$ into $B(G^+)$. It is not complex linear because the conditions (20, ii) and (24, ii) are invariant only under multiplication by real numbers. However, just as in the case of Theorem 1, the image of $\bar\Phi^+$ is contained in the ball of radius 6 and contains the ball of radius 2. The same bounds apply because Φ^+ is the restriction of the holomorphic mapping $\Phi\colon M(G) \to B(G)$.

Let $A_s(\Gamma, C)$ be all holomorphic quadratic differentials φ defined on $\mathbb{H}$ such that

(i) $\varphi(Bz)B'(z)^2 = \varphi(z)$ for all B in Γ;
(ii) $\varphi(x)$ is real if x is in $\hat{\mathbb{R}} - C$; and
(iii) $\|\varphi\| = \iint_{H/\Gamma} |\varphi|\, dx\, dy < \infty$.

The space of infinitesimally trivial Beltrami differentials for Γ relative to C is

$$N(\Gamma, C) = \left\{ \mu \text{ in } L_\infty(\Gamma) \,\middle|\, \iint_{\mathbb{H}/\Gamma} \mu\varphi = 0 \quad \text{for all } \varphi \text{ in } A_s(\Gamma, C) \right\}.$$

We obtain the following modification of Theorem 6.

Theorem 8. *Let C be a Γ-invariant closed set containing the limit set Λ of Γ and contained in $\hat{\mathbb{R}}$. A Beltrami differential μ is in $N(\Gamma, C)$ if, and only if, there exists a real analytic curve σ_t of trivial Beltrami coefficients in $M_0(\Gamma, C)$ for which*

$$\sigma_t(z) = t\mu(z) + O(t^2) \text{ uniformly in } z.$$

Proof. Because of Lemma 9 and Theorem 7 the proof is analogous to the proof of Theorem 6.

Notes. The major parts of this chapter are due to Teichmüller [T1, T2], Ahlfors [Ah4], and Bers [Ber2, Ber3]. Alternative interpretations of the Schwarzian derivative are given by Hawley and Schiffer [HS] and by Thurston [Th3]. The interpretation of the imaginary part of Schwarzian derivative as a measurement of the change of curvature appears as an exercise in Ahlfors's book [Ah6]. Exercise 12 of this chapter gives an approach used by Bers and Royden [BerR] to the lambda lemma of Mañé, Sad, and Sullivan [MañSS].

A great deal of work has been done by Wolpert in describing the geometric meaning of harmonic Beltrami differentials [Wo3, Wo4, Wo5]. One of his very significant results is that with regard to motion along a path in Teichmüller space tangent to a harmonic Beltrami differential, the geodesic length functions measured in the noneuclidean metric are convex [Wo5].

We have only briefly mentioned quasicircles and quasifuchsian groups in Lemma 4, Theorem 1, and the remarks at the end of Section 5.4. Quasicircles were characterized geometrically by Ahlfors [Ah4]. Gehring has written a monograph entitled *Characteristic Properties of Quasidisks* [Ge3], and quasicircles and their generalizations are an active subject of research ([Ge2], [GeO]).

We have omitted the important result of Ahlfors and Beurling [AhBeu] which characterizes the homeomorphisms of the real axis which can arise as boundary values of quasiconformal self-mappings of the upper half plane (see [LehtV]).

The infinitesimal theory worked out in Section 5.7 for Fuchsian groups also leads to an infinitesimal theory for Teichmüller spaces of Kleinian groups and to Eichler cohomology (see, for example [Kr1, Kr3, Kr4, Kr9, Kr10].

The result of Exercise 11 in this chapter appears in [Ga1].

A way to introduce a local complex structure for Teichmüller space arises from the variational techniques of Schiffer and Spencer [ScSp]. See also [Ga2] and [N2].

In Section 5.9 we considered discrete groups of conformal and anticonformal isometries of the noneuclidean plane. Such groups are called noneuclidean crystallographic groups and have been studied by Singerman [Si2], Bujalance [Bu], Macbeath [Mac2], and others.

The Ahlfors–Weill section was first presented in [AhW]. A generalization to quasi-disks is used by Ahlfors in [Ah4] to prove arbitrary Teichmüller spaces are bounded domains in certain Banach spaces. Earle and Nag [EN] use conformally natural reflections to show the existence of a section equivariant with respect of the action of a quasifuchsian group. The method of Earle and Nag is based on the conformal extension operator of Douady and Earle [DE]. There is also a recent paper of Velling [V] which gives a three-dimensional interpretation of the Ahlfors–Weill extension.

EXERCISES

1. Show that formula (4) in Section 5.3 gives a metric.
2. Let $R = \{z: 1 < |z| < \lambda\}$ and σ be the empty set. Show that $T(R, \sigma)$ is isometric to $\mathbb{R}$, with Euclidean metric.
3. Let $R = \mathbb{C}/L$, where L is the lattice generated by $z \to z + 1$ and $z \to z + i$. (Here σ must be the empty set since R has no boundary.) Show that $T(R)$ is isometric to the upper half plane with the noneuclidean metric.
4. Let f be a holomorphic function in a domain D and define
$$F(w, z) = \log \frac{f(w) - f(z)}{w - z}.$$
Show that $f(z)$ is univalent in the domain D if, and only if, $U(w, z) = (\partial^2/\partial z \partial w)F(w, z)$ is regular in the domain $D \times D$ [HS].
5. Show that for the function $U(z, w)$ defined in Exercise 4 that
$$-6 \lim_{w \to z} U(w, z) = \text{the Schwarzian derivative of } f \text{ at } z.$$
6. Let $f(z)$ be holomorphic in a neighborhood of the origin and suppose $f'(0)$ is not zero. Let L be the best approximating Möbius transformation to f at the origin. By this we mean that $L^{-1} \circ f(z)$ has a Taylor expansion of the form

$$L^{-1} \circ f(z) = z + B_3 z^3 + B_4 z^4 + \cdots.$$

Prove that B_3 is the Schwarzian derivative of f evaluated at $z = 0$.

7. (Ahlfors [Ah6]) Let $z = z(t)$ be a plane curve of class C^3. Show that the rate of change of the curvature of $z(t)$ is $|z'(t)|^{-1}$ multiplied by the imaginary part of the Schwarzian derivative of $z(t)$.

8. Let $f(z) = a_0(z_0) + a_1(z_0)(z - z_0) + a_2(z_0)(z - z_0)^2 + \cdots$ with $a_1 \neq 0$ and let

$$L(z) = \frac{(a_1^2 - a_2 a_0)(z - z_0) + a_0 a_1}{-a_2(z - z_0) + a_1}.$$

Show that $L(z)$ has the same 2-jet as f, that is,

$$L(z_0) = a_0, \qquad L'(z_0) = a_1, \quad \text{and} \quad L''(z_0) = 2a_2.$$

9. (Thurston). Let η_1 and η_2 be independent solutions to $\eta'' + \frac{1}{2}\varphi\eta = 0$ normalized so that $\eta_2\eta_1' - \eta_1\eta_2' \equiv 1$. Assume φ is holomorphic. We have seen that $\{f, z\} = \varphi$, where $f = \eta_1\eta_2^{-1}$. Show that the best approximating Möbius transformation L in Exercise 8 can be rewritten in matrix form as

$$L(z) = \frac{\eta_1'(z_0)(z - z_0) + \eta_1(z_0)}{\eta_2'(z_0)(z - z_0) + \eta_2(z_0)}.$$

[This means that to obtain the Ahlfors–Weill extension in formula (8) you evaluate the best approximating Möbius transformation at $\bar{z}_0$ and apply it to z_0.]

10. Show that the homeomorphisms of $\mathbb{R}$ which fix the points 0 and 1 and which extend to quasiconformal homeomorphisms of $\mathbb{H}$ form a group under composition.

11. Let w_μ be a quasiconformal extension of a homeomorphism of h in the group described in the previous exercise and let w^μ the normalized quasiconformal homeomorphism of the sphere $\hat{\mathbb{C}}$ with Beltrami coefficient μ in the upper half plane and zero in the lower half plane. Let $C^\mu = w^\mu(\mathbb{R})$.

(a) Show that the oriented curve C^μ determines the group element h.

(b) Show that the complex conjugate of the oriented curve C^μ is the oriented curve which determines h^{-1}.

12. [BeR] Let $f_k(\lambda)$ be a family of n holomorphic functions defined in $|\lambda| < 1$ and satisfying $f_k(\lambda)$ is not equal to 0 or 1 for any k with $1 \leqslant k \leqslant n$ and $f_k(\lambda) \neq f_j(\lambda)$ whenever $k \neq j$. The vector $\vec{f}(\lambda) = (0, 1, \infty, f_1(\lambda), f_2(\lambda), \ldots, f_n(\lambda))$ is called a holomorphic motion of $n + 3$ points in the extended complex plane $\hat{\mathbb{C}}$. The conditions on the f_k ensure that no two of the $n + 3$ coordinates of $\vec{f}(\lambda)$ coincide.

Let x_{n+1} be any point not equal to any of the coordinates of $f(0)$. Show that $\vec{f}$ extends to a holomorphic motion $\vec{g}(\lambda)$ of $n + 4$ points defined for $|\lambda| < \frac{1}{3}$. By this we mean the first $n + 3$ components of $\vec{g}$ are identical with the components of $\vec{f}$ and $g_{n+1}(0) = x_{n+1}$ and g_{n+1} has image in $\mathbb{C} - \{0, 1\}$ and $g_{n+1}(\lambda)$ is never equal to any $f_k(\lambda)$ for $1 \leqslant k \leqslant n$ and for $|\lambda| < \frac{1}{3}$.

Hint: Let T_n be the Teichmüller space of $\hat{\mathbb{C}} - \{0, 1, \infty, f_1(0), f_2(0), \ldots, f_n(0)\}$. Observe that $\vec{f}$ determines a holomorphic function from $|\lambda| < 1$ into T_n. Use the Ahlfors–Weill section.

Remark. It is not known whether a holomorphic motion of n points defined in $|\lambda|$ can be extended to a holomorphic motion of $n + 1$ points beyond the disk $|\lambda| < \frac{1}{3}$.

6

TEICHMÜLLER'S THEOREM

In this chapter we prove Teichmüller's theorem as well as the necessity and sufficiency of the Hamilton–Krushkal condition for a quasiconformal mapping to be extremal in its Teichmüller class. We deduce the uniqueness part of Teichmüller's theorem from the Reich–Strebel inequality (Theorem 3 of Section 2.4) and its consequence, Exercise 6 of Chapter 2. The Reich–Strebel inequality (and, consequently, the result of Exercise 6 of Chapter 2) is generalized to the setting of Fuchsian groups in Chapters 3 and 4.

We deduce the existence part of Teichmüller's theorem from the existence of coordinates for Teichmüller space and, in particular, the existence of a trivial curve tangent to a given infinitesimally trivial line (Theorems 6 and 8 of Chapter 5).

In Section 6.6 we give a variational formula for the extremal value and in Section 6.7 we prove that finite dimensional Teichmüller spaces are cells.

Finally, in the last section of the chapter, we give Strebel's frame mapping condition. The frame mapping condition is a sufficient condition for determining when a Teichmüller equivalence class can be represented in the form $k|\varphi|/\varphi$, where φ is a holomorphic quadratic differential of finite norm.

6.1. THE HAMILTON–KRUSHKAL CONDITION: NECESSITY

Let Γ be a Fuchsian group acting on the upper half plane $\mathbb{H}$ and let ω be a fundamental domain for Γ in $\mathbb{H}$. Let C be a Γ-invariant closed subset of $\hat{\mathbb{R}}$ with $\Lambda \subseteq C \subseteq \hat{\mathbb{R}}$.

Theorem 1. *Let $\mu \in M(\Gamma)$ and assume μ is extremal in its class, that is, $\|\mu\|_\infty \leqslant \|\nu\|_\infty$ for all ν in $M(\Gamma)$ for which $w_\mu(x) = w_\nu(x)$ for all x in C. Let $k_0 = \|\mu\|_\infty$. Then*

$$k_0 = \sup \left| \iint_\omega \mu\varphi \, dx \, dy \right|, \tag{1}$$

where the supremum is taken over all φ in $A_s(\Gamma, C)$ for which $\|\varphi\| = \iint_\omega |\varphi|\, dx\, dy = 1$.

Remark 1. Notice that this is an existence theorem because it implies the existence of a sequence φ_n in $A(\Gamma, \hat{\mathbb{R}})$ with $\|\varphi_n\| = 1$ such that $k_0 = \lim_{n\to\infty} \iint \mu\varphi_n\, dx\, dy$.

Remark 2. Using normal family arguments for quasiconformal mappings it is elementary to see that for any ν in $M(\Gamma)$, there is an extremal element in its class.

Proof. Start by assuming $k = \|\mu\|_\infty > k_0$, where k_0 is the supremum in (1). By the Hahn–Banach theorem and the Riesz representation theorem, there exists ν in $M(\Gamma)$ such that $\iint_\omega \mu\varphi\, dx\, dy = \iint_\omega \nu\varphi\, dx\, dy$ for all φ in $A_s(\Gamma, C)$ and such that $\|\nu\|_\infty = k_0$, the supremum in (1). Hence $\mu - \nu$ is infinitesmally trivial and, by Theorems 6 and 8 of Chapter 5, there exists a curve of trivial Beltrami differentials σ_t such that

$$\|\sigma_t - t(\mu - \nu)\|_\infty = O(t^2). \tag{2}$$

For brevity let $\sigma = \sigma_t$ and form

$$w_\tau = w_\mu \circ (w_\sigma)^{-1}. \tag{3}$$

Clearly, τ has the same Teichmüller class as μ. For sufficiently small $t > 0$, we will show that $\|\tau\|_\infty < \|\mu\|_\infty$ and this contradicts the assumption that μ is extremal in its class. An application of the result of Exercise 16 of Chapter 1 yields

$$\tau(w_\sigma(z)) = \frac{\mu - \sigma}{1 - \bar{\sigma}\mu} \cdot \frac{1}{\theta}, \tag{4}$$

where $\theta = \bar{p}/p$ and $p = (\partial/\partial z)w_\sigma$. Clearly, (4) implies

$$|\tau \circ w_\sigma|^2 = \frac{|\mu|^2 - 2\operatorname{Re}\mu\bar{\sigma} + |\sigma|^2}{1 - 2\operatorname{Re}\mu\bar{\sigma} + |\sigma\mu|^2},$$

which gives

$$|\tau \circ w_\sigma| = |\mu| - \frac{1 - |\mu|^2}{|\mu|}\operatorname{Re}\mu\bar{\sigma} + O(t^2). \tag{5}$$

Combining (5) with (2), we find that

$$|\tau \circ w_\sigma| = |\mu| - t\frac{1 - |\mu|^2}{|\mu|}\operatorname{Re}(|\mu|^2 - \mu\bar{\nu}) + O(t^2). \tag{6}$$

Recall that $k_0 = \|\nu\|_\infty < k = \|\mu\|_\infty$. Let

$$S_1 = \{z \in \mathbb{H} \mid |\mu(z)| \leqslant (k + k_0)/2\}, \qquad \text{and}$$
$$S_2 = \{z \in \mathbb{H} \mid (k + k_0)/2 < |\mu(z)| \leqslant k\}.$$

Clearly, $S_1 \cup S_2 = \mathbb{H}$ and (4) implies there exists $\delta_1 > 0$ and $c_1 > 0$ such that for $0 < t < \delta_1$,

$$|\tau \circ w_\sigma(z)| \leqslant k - c_1 t \qquad \text{for } z \text{ in } S_1. \tag{7}$$

For z in S_2 the coefficient of t in (6) is bounded below by

$$\frac{1 - k^2}{k} \cdot \left[\left(\frac{k + k_0}{2}\right)^2 - k_0 k\right] = \frac{1 - k^2}{k} \cdot \left(\frac{k - k_0}{2}\right)^2 > 0.$$

Therefore, (6) implies there exists $\delta_2 > 0$ and $c_2 > 0$ such that for $0 < t < \delta_2$,

$$|\tau \circ w_\sigma(z)| \leqslant k - c_2 t \qquad \text{for } z \text{ in } S_2. \tag{8}$$

Putting (7) and (8) together, we find that $\|\tau\|_\infty < k$ for sufficiently small $t > 0$, and this proves the theorem.

6.2. TEICHMÜLLER'S THEOREM: EXISTENCE

We will prove a more general existence theorem in Section 6.8. In this section we assume that $T(\Gamma, C)$ is finite dimensional. Then $A_s(\Gamma, C)$ is finite dimensional and there must be an element φ_0 in $A(\Gamma, C)$ with $\|\varphi_0\| = 1$ for which the supremum in (1) is achieved. This means that if μ is extremal in its class, then

$$\|\mu\|_\infty = \iint_\omega \mu\varphi_0 \, dx\, dy \leqslant \|\mu\|_\infty \iint_\omega |\varphi_0| \, dx\, dy = \|\mu\|_\infty. \tag{9}$$

It is elementary to see that the only way (9) can hold with $\|\varphi_0\| = 1$ is for

$$\mu(z) = \|\mu\|_\infty |\varphi_0(z)|/\varphi_0(z) \text{ almost everywhere.} \tag{10}$$

Theorem 2. *Suppose $T(\Gamma, C)$ is a finite dimensional Teichmüller space. Then any ν in $M(\Gamma)$ is equivalent to a Beltrami differential of form $k|\varphi|/\varphi$, where $0 \leqslant k \leqslant \|\nu\|_\infty < 1$ and φ is an element of $A(\Gamma, C)$ with $\|\varphi\| = 1$.*

Proof. Given a ν in $M(\Gamma)$, we construct a μ in the same class as ν and which has minimal norm. This is done by taking a sequence of quasiconformal mappings w_{μ_n} with $\mu_n \in M(\Gamma)$ such that $w_{\mu_n}(x) = w_\nu(x)$ for all x in $\hat{\mathbb{R}}$ and such that $\|\mu_n\|$ approaches the extremal value in the Teichmüller class of ν. Then by a normal families argument, w_{μ_n} has a quasiconformal limit w_μ in the same class as w_ν and $k \leqslant \|\mu\|_\infty \leqslant \|\mu_n\|_\infty$ for all n, so $\|\mu\|_\infty = k$. We have already shown that such an extremal element must be of the form (10) and, obviously, $k \leqslant \|\nu\|_\infty$.

6.3. TEICHMÜLLER'S THEOREM: UNIQUENESS

We now bring in the Reich–Strebel inequality (Theorem 3 of Section 2.4), which is generalized to the setting of Fuchsian groups in Chapters 3 and 4. Suppose μ and ν are equivalent Beltrami differentials in $M(\Gamma)$, that is, $w_\mu(x) = w_\nu(x)$ for all x in a Γ-invariant closed subset $C \subseteq \hat{\mathbb{R}}$. Then $w_\nu^{-1} \circ w_\mu$ is trivial, and, if we apply the Reich–Strebel inequality to $w_\tau = w_\nu^{-1} \circ w_\mu$, we get (using the result of Exercise 6 of Chapter 2)

$$1 \leqslant \iint\limits_{\omega} |\varphi(z)| \frac{|1 - \mu\varphi/|\varphi||^2}{1 - |\mu|^2} \cdot \frac{|1 + \nu\theta\varphi/|\varphi||^2}{1 - |\nu|^2}\, dx\, dy, \tag{11}$$

where $\theta = (1 - \overline{\mu\varphi/|\varphi|})(1 - \mu\varphi/|\varphi|)^{-1}$ and $\|\varphi\| = 1$.

Suppose w_μ is a Teichmüller mapping. By this we mean $\mu = k|\varphi|/\varphi$ for some φ in $A_s(\Gamma, C)$ with $\|\varphi\| = 1$ and $0 < k < 1$. Then $\theta = 1$ and $|1 - \mu\varphi/|\varphi||^2/1 - |\mu|^2 = K^{-1}$, where $K = (1 + k)/(1 - k)$. Therefore, (11) becomes

$$K \leqslant \iint\limits_{\omega} |\varphi| \frac{|1 + \nu\varphi/|\varphi||^2}{1 - |\nu|^2}\, dx\, dy \tag{12}$$

for every ν equivalent to $k|\varphi|/\varphi$, where $\|\varphi\| = 1$. From (12) we see that $K \leqslant (1 + \|\nu\|_\infty)/(1 - \|\nu\|_\infty)$ which implies $k \leqslant \|\nu\|_\infty$ and, therefore, $k|\varphi|/\varphi$ has minimal norm among all equivalent Beltrami differentials ν. Moreover, if $k = \|\nu\|_\infty$, then (12) yields

$$K \leqslant \iint\limits_{\omega} |\varphi| \frac{|1 + \nu\varphi/|\varphi||^2}{1 - |\nu|^2}\, dx\, dy \leqslant K, \tag{13}$$

and so (13) is an equality and this obviously implies that $\nu = k|\varphi|/\varphi$ almost everywhere. We have proved the following theorem.

Theorem 3 (Teichmüller's Uniqueness Theorem). *Let Γ be a Fuchsian group and C an invariant closed subset with $\Lambda \subseteq C \subseteq \hat{\mathbb{R}}$. Suppose $\mu = k|\varphi|/\varphi$, where* $0 < k < 1$ and φ is an element of norm 1 in $A_s(\Gamma, C)$. *Then any ν in $M(\Gamma)$ for which $w_\nu(x) = w_\mu(x)$ for all x in C satisfies $\|\nu\|_\infty \geqslant k$. Moreover, if $\|\nu\|_\infty = k$, then $\nu = k|\varphi|/\varphi$ almost everywhere.*

Corollary. *Suppose* φ_1 *and* φ_2 *are nonzero elements of* $A(\Gamma, C)$ *and* $0 < k_1 < 1$. *Suppose also that* $k_1|\varphi_1|/\varphi_1$ *is equivalent to* $k_2|\varphi_2|/\varphi_2$. *Then* $k_1 = k_2$ *and* φ_1 *is a positive multiple of* φ_2.

Proof. We apply Theorem 3 to both Beltrami differentials $k_1|\varphi_1|/\varphi_1$ and $k_2|\varphi_2|/\varphi_2$. We get $k_1 \leqslant k_2$ and $k_2 \leqslant k_1$. The equation $|\varphi_1|/\varphi_1 = |\varphi_2|/\varphi_2$ for holomorphic functions φ_1 and φ_2 implies φ_1 is a positive multiple of φ_2.

6.4. INEQUALITIES FOR FUNCTIONALS OF BELTRAMI COEFFICIENTS

Following Reich and Strebel [ReiS2], we introduce three functionals:

$$I[\mu] = \sup\left|\mathrm{Re}\iint \frac{\mu\varphi}{1-|\mu|^2}\,dx\,dy\right| \tag{14}$$

$$H[\mu] = \sup\left|\mathrm{Re}\iint \mu\varphi\,dx\,dy\right| \tag{15}$$

$$J[\mu] = \sup\iint \frac{|\mu|^2|\varphi|}{1-|\mu|^2}\,dx\,dy, \tag{16}$$

where in all three integrals the domain of integration is ω, a fundamental domain for Γ, and the supremum is over all φ in $A_s(\Gamma, C)$ for which $\iint_\omega |\varphi|\,dx\,dy = 1$.

If we let ν be extremal in its class and equivalent to μ, inequality (11) yields

$$\frac{1}{K_0} \leqslant \iint_\omega |\varphi|\frac{|1-\mu\varphi/|\varphi||^2}{1-|\mu|^2}\,dx\,dy, \tag{17}$$

where $K_0 = (1+k_0)/(1-k_0)$ and k_0 is the extremal value of $\|\nu\|_\infty$. If we expand the numerator of the integrand in (17) and simplify, we find that

$$\frac{1}{K_0} - 1 \leqslant -2\,\mathrm{Re}\iint \frac{\mu\varphi}{1-|\mu|^2}\,dx\,dy + 2\iint \frac{|\mu|^2|\varphi|}{1-|\mu|^2}\,dx\,dy,$$

which leads to

$$\mathrm{Re}\iint \frac{\mu\varphi}{1-|\mu|^2}\,dx\,dy \leqslant \frac{k_0}{1+k_0} + \iint \frac{|\mu|^2|\varphi|}{1-|\mu|^2}\,dx\,dy$$

and consequently

$$I[\mu] \leqslant \frac{k_0}{1 + k_0} + J[\mu]. \tag{18}$$

To obtain an inequality of opposite type, let us temporarily assume that $T(\Gamma, C)$ is a finite dimensional Teichmüller space. Thus, by Teichmüller's existence theorem, within the equivalence class of ν there is a Beltrami differential of the form $\mu = k|\varphi|/\varphi$, where $0 \leqslant k < 1$. Substituting this value of μ in (11), we find that

$$K \leqslant \iint_{\omega} |\varphi| \frac{|1 + \nu\varphi/|\varphi||^2}{1 - |\mu|^2} \, dx\, dy, \tag{19}$$

To extend this inequality to the case where Γ is finitely generated and $T(\Gamma, C)$ is infinite dimensional, we take a sequence of closed Γ-invariant subsets C_n of C such that $\bigcup C_n$ is dense in C and such that $T(\Gamma, C_n)$ is finite dimensional. Let ν be an element of $M(\Gamma)$ and assume $k_0 = \|\mu\|_\infty$ is extremal among all Beltrami coefficients in $M(\Gamma)$ for which $w_\mu(x) = w_\nu(x)$ for all x in C. Let μ_n be extremal among all Beltrami coefficients in $M(\Gamma)$ for which $w_{\mu_n}(x) = w_\nu(x)$ for all x in C_n and let $k_n = \|\mu_n\|$. Clearly $k_n \leqslant k_0$ and k_n increases to k_0 because w_{μ_n} approaches a quasiconformal mapping which agrees with w_ν on C and which has dilatation $\leqslant \lim k_n$.

Let $K_0 = (1 + k_0)/(1 - k_0)$ and $K_n = (1 + k_n)/(1 - k_n)$. Formula (19) applied to $T(\Gamma, C_n)$ implies

$$K_n \leqslant \iint_{\omega} |\varphi_n| \frac{|1 + \nu\varphi_n/|\varphi_n||^2}{1 - |\nu|^2} \, dx\, dy, \tag{20}$$

where $\|\varphi_n\| = 1$, φ_n is in $A(\Gamma, C_n)$, and k_n is unique with the property that $k_n|\varphi_n|/\varphi_n \sim \nu$ in $T(\Gamma, C_n)$. Since $A(\Gamma, C_n) \subset A(\Gamma, C)$, we see that

$$K_0 \leqslant \sup \iint_{\omega} |\varphi| \frac{|1 + \nu\varphi/|\varphi||^2}{1 - |\nu|^2} \, dx\, dy, \tag{21}$$

where the supremum is over φ in $A(\Gamma, C)$ such that $\|\varphi\| = 1$.

Expanding out the numerator in (21) and simplifying, we get

$$K_0 - 1 \leqslant 2(I[\nu] + J[\nu])$$

or

$$\frac{k_0}{1 - k_0} \leqslant I[\nu] + J[\nu]. \tag{22}$$

This inequality can easily be extended to the case of infinitely generated Fuchsian groups by the same method we used for the proof of Theorem 4 of Chapter 4. From inequalities (18) and (22), the following theorem is an easy consequence.

Theorem 4. *Let* Γ *be a Fuchsian group and* C *a closed* Γ*-invariant set with* $\Lambda \subseteq C \subseteq \hat{\mathbb{R}}$. *Let* v *and* μ *be Beltrami coefficients in* $M(\Gamma)$ *and* k_0 *the minimum value of* $\|v\|_\infty$ *for which* $w_v(x) = w_\mu(x)$ *for all* x *in* C. *Then*

$$I[\mu] - \delta[\mu] \leqslant \frac{k_0}{1 - k_0^2} \leqslant I[\mu] + \delta[\mu], \tag{23}$$

where $\delta[\mu] = J[\mu] - k_0^2/(1 - k_0^2)$.

6.5. THE HAMILTON–KRUSHKAL CONDITION: SUFFICIENCY

A theorem proved by Reich and Strebel states that (1) is also a sufficient condition for extremality. Before proving this, we give an analogous theorem for the functional $I[\mu]$.

Theorem 5. *A necessary and sufficient condition for* μ *in* $M(\Gamma)$ *to be extremal in its class relative to the closed* Γ*-invariant set* C *is that*

$$I[\mu] = \frac{k}{1 - k^2}, \tag{24}$$

where $\|\mu\|_\infty = k$ *and where the supremum for* $I[\mu]$ *in* (14) *is over all* φ *in* $A_s(\Gamma, C)$ *with* $\|\varphi\| = 1$.

Proof. Suppose (24) holds. Then the left side of (23) gives

$$\frac{k}{1 - k^2} \leqslant \frac{k_0}{1 - k_0^2} + \frac{k^2}{1 - k^2} - \frac{k_0^2}{1 - k_0^2},$$

which implies $k/(1 + k) \leqslant k_0/(1 + k_0)$ and hence $k \leqslant k_0$. Conversely, suppose μ is extremal. Then $\|\mu\|_\infty = k_0$ and the right side of (23) gives $k_0/(1 - k_0^2) \leqslant I[\mu]$. Since the opposite inequality is obvious, we see that $I[\mu] = k_0/(1 - k_0^2)$.

To prove the sufficiency of Hamilton's condition, we must relate $I[\mu]$ to $H[\mu]$. We need the following lemma.

Lemma 1. *Suppose* φ_n *is a sequence for which* $\|\varphi_n\| = 1$ *and*

$$\iint_\omega \varphi_n \mu \, dx \, dy \to k$$

as $n \to \infty$, *where* $k = \|\mu\|_\infty$. *Then* φ_n *converges uniformly to zero on any set* $E' = \{z \in \omega : |\mu(z)| \leqslant k' < k\}$.

Proof.

$$\iint_\omega \varphi_n \mu = \iint_{\omega - E'} \varphi_n \mu + \iint_{E'} \varphi_n \mu.$$

The second integral on the right side is bounded above by $k' \|\varphi_n\|_{E'}$. Thus

$$\left| \iint_\omega \varphi_n \mu \right| \leqslant k' \|\varphi_n\|_{E'} + k \|\varphi_n\|_{\omega - E'} = (k' - k)\|\varphi_n\|_{E'} + k.$$

Since the left side of this inequality approaches k as $n \to \infty$ and since $k' < k$, we see that $\|\varphi_n\|_{E'} \to 0$.

Lemma 2. *Suppose* φ_n *is a sequence for which* $\|\varphi_n\| = 1$ *and*

$$\iint_\omega \varphi_n \mu \, dx \, dy \to k$$

as $n \to \infty$, *where* $k = \|\mu\|_\infty$. *Then*

$$\iint_\omega \frac{\varphi_n \mu}{1 - |\mu|^2} \, dx \, dy \to \frac{k}{1 - k^2}.$$

Proof. Let $k' < k$ and $E' = \{z \in \omega \,|\, |\mu| \leqslant k'\}$. We know that $\|\varphi_n\|_{E'} \to 0$. Thus

$$\iint_{\omega - E'} \varphi_n \mu \to k$$

and

$$\iint_{\omega - E'} \frac{\varphi_n \mu}{1 - k'^2} \to \frac{k}{1 - k'^2}.$$

Since $|\mu| \geqslant k'$ on $\omega - E'$, this implies $\iint_{\omega - E'} \varphi_n / 1 - |\mu|^2$ is a sequence whose cluster points are between $k/1 - k'^2$ and $k/1 - k^2$. Since $\|\varphi_n\|_{E'} \to 0$, this implies that

$$\iint_\omega \frac{\varphi_n \mu}{1 - |\mu|^2}$$

is a sequence whose cluster points are also between $k/(1-k'^2)$ and $k/(1-k^2)$. Since k' is an arbitrary number with $0 < k' < k$, we see that

$$\lim_{n\to\infty} \iint_{\omega} \frac{\varphi_n \mu}{1-|\mu|^2}\, dx\, dy = \frac{k}{1-k^2}.$$

Theorem 6 (Hamilton's Condition: Sufficiency). *Suppose μ is a Beltrami coefficient in $M(\Gamma)$ and $H[\mu] = k$. Then μ is extremal. That is, any other ν in $M(\Gamma)$ for which $w_\gamma(x) = w_\mu(x)$ for all x in C satisfies $\|\nu\|_\infty \geqslant \|\mu\|_\infty$.*

Proof. Lemma 2 implies that $I[\mu] = k/1 - k^2$. Thus Theorem 5 gives the result.

Remark. We have already proved that this condition is necessary in Section 6.1.

6.6. VARIATION OF THE EXTREMAL VALUE

Inequality (23) in Theorem 4 is fundamental in consideration of the infinitesimal form of Teichmüller's metric.

Theorem 7. *Let μ be a Beltrami differential for Γ (so $t\mu$ is in $M(\Gamma)$ for $t < 1/\|\mu\|_\infty$). Let $k_0(t)$ be the minimum value of $\|\nu\|_\infty$, where ν is a Beltrami coefficient equivalent to $t\mu$. Then for $t > 0$ and t approaching 0,*

$$k_0(t) = t \sup \left| \operatorname{Re} \iint_{\omega} \mu\varphi\, dx\, dy \right| + O(t^2) \tag{25}$$

where the supremum is over all φ in $A_s(\Gamma, C)$ for which $\|\varphi\| = 1$.

Proof. We replace μ by $t\mu$ in inequality (23). Clearly $I[t\mu]$ differs from $H[t\mu]$ by a term of order t^2. The same statement applies to the difference between $k_0(t)$ and $k_0(t)/1 - k_0(t)^2$. Finally, the term $\delta[t\mu]$ is of order t^2 and (25) follows.

6.7. FINITE DIMENSIONAL TEICHMÜLLER SPACES ARE CELLS

We call a topological space a cell if it is homeomorphic to $\mathbb{R}^n$ for some positive integer n.

Let $T(\Gamma, C)$ be a Teichmüller space for which the space of holomorphic quadratic differentials $A_s(\Gamma, C)$ is finite dimensional. Let B be the interior of the unit ball in $A_s(\Gamma, C)$ and let $M(\Gamma)$ be the Beltrami coefficients for Γ (with L_∞-norm less than one). Consider the sequence of mappings

$$B \xrightarrow{\Psi} M(\Gamma) \xrightarrow{\Phi} T(\Gamma, C), \tag{26}$$

where Φ is the mapping which assigns a Beltrami coefficient to its Teichmüller class and Ψ is given by $\Psi(\varphi) = \|\varphi\| \, |\varphi(z)|/\varphi(z)$ when φ is not identically zero and $\Psi(0) = 0$.

Theorem 8. *The mapping $\Phi \circ \Psi$ is a homeomorphism and, therefore, any finite dimensional Teichmüller space is a cell of dimension equal to the dimension of $A_s(\Gamma, C)$.*

Proof. The existence part of Teichmüller's theorem tells us that $\Phi \circ \Psi$ is a surjection. The uniqueness part tells us that $\Phi \circ \Psi$ is one-to-one. To show that $\Phi \circ \Psi$ is continuous assume that φ_n is in B and $\varphi_n \to \varphi_0$. If φ_0 is identically zero, the continuity is obvious and, if φ_0 is not zero, then the Beltrami coefficients $\mu_n(z) = \|\varphi_n\| \, |\varphi_n(z)|/\varphi_n(z)$ converge in the bounded pointwise sense to $\mu_0(z) = \|\varphi_0\| \, |\varphi_0(z)|/\varphi_0(z)$. Then from the convergence principle (Lemma 5 of Section 1.8) the quasiconformal mappings w_{μ_n} converge uniformly on compact sets to w_{μ_0}. The mapping w_{μ_0} gives a translation isometry of the Teichmüller space $T(\Gamma, C)$ onto $T(\Gamma_{\mu_0}, C_{\mu_0})$, where $\Gamma_{\mu_0} = w_{\mu_0} \circ \Gamma \circ w_{\mu_0}^{-1}$ and $C_{\mu_0} = w_{\mu_0}(C)$. This isometry is obtained by composition on the right with $(w_{\mu_0})^{-1}$ and it maps the point $[\mu_0]$ in $T(\Gamma, C)$ to the point $[0]$ in $T(\Gamma_{\mu_0}, C_{\mu_0})$. Note that $w_{\mu_n} \circ w_{\mu_0}^{-1}$ converges uniformly on compact sets to the identity. Because of finite dimensionality and the Ahlfors–Weill coordinates (Lemma 7 of Section 5.4) this implies the Teichmüller class of $w_{\mu_n} \circ w_{\mu_0}^{-1}$ converges to zero in $T(\Gamma_{\mu_0}, C_{\mu_0})$. Hence the class of μ_n converges to the class of μ_0 in $T(\Gamma, C)$.

Thus we have a continuous one-to-one mapping from the unit ball B onto a finite dimensional manifold. By invariance of domain the mapping must be a homeomorphism.

6.8. STREBEL'S FRAME MAPPING CONDITION

In this section, for the sake of simplicity, we treat only the case where Γ is a torsion free Fuchsian group action on $\mathbb{H}$ and R is the Riemann surface $\mathbb{H}/\Gamma$. We will consider the Teichmüller space $T(R, \sigma)$ of the Riemann surface R where σ is the entire border of R.

In order to state Strebel's frame mapping condition, we need to define the boundary dilatation of the mapping f from R into $f(R)$, which we will denote by $H(f)$. For every compact set S contained in R we let $K(R - S, f)$ be the maximal dilatation of f on R minus S. If R is compact, then S can equal R and, in that case, we let $K(R - S, f) = 1$. We also let $K_0(R - S, f)$ be the infimum of all the numbers $K(R - S, f_1)$, where f_1 is any quasiconformal mapping in the same equivalence class as f. Obviously, the numbers $K_0(R - S, f)$ decrease as the set S increases. The boundary dilatation $H(f)$ of f is defined to be the direct limit as S increases to R of the numbers $K_0(R - S, f)$. Note that if R is compact or compact except for a finite number of punctures then $H(w) = 1$ for every quasiconformal mapping f. The following theorem is due to Strebel; he states the theorem for the unit disk and applies it to plane domains.

Theorem 9 (The Frame Mapping Condition [St2]). *Suppose f is a quasiconformal mapping from a Riemann surface R to a Riemann surface $f(R)$. Suppose $H(f) < K_0(f)$. Then the Teichmüller equivalence class of f is represented by a quasiconformal mapping f_0 whose Beltrami coefficient has the form*

$$k\frac{|\varphi|}{\varphi},$$

where k is a number between 0 and 1 and φ is an integrable, holomorphic, quadratic differential on R. Moreover, k and φ (up to multiplication by a positive number) are uniquely determined by the Teichmüller equivalence class of f, and $k|\varphi|/\varphi$ is the Beltrami coefficient of the unique extremal representative of the Teichmüller equivalence class of f.

Proof. We let f_0 and f_1 be equivalent mappings defined on R with Beltrami coefficients μ_0 and μ_1. Assume that $\|\varphi\| = \iint_R |\varphi(z)|\, dx\, dy = 1$. Then it is convenient to rewrite inequality (11) in the following form.

$$1 \leqslant \iint_R |\varphi(z)| \frac{|1 - \mu_0(z)\varphi(z)/|\varphi(z)||^2}{(1 - |\mu_0(z)|^2)} \times \frac{|1 + (\nu_1/\nu_0)\mu_0\alpha\varphi(z)/|\varphi(z)||^2}{(1 - |\mu_1(z)|^2)}\, dx\, dy. \tag{27}$$

where

$$\alpha = \frac{(1 - \overline{\mu_0\varphi/|\varphi|})}{(1 - \mu_0\varphi/|\varphi|)}$$

and ν_1 and ν_0 are, respectively, the Beltrami coefficients of the mappings $(f_1)^{-1}$ and $(f_0)^{-1}$. In this inequality, it is important to note that $\|\nu_1\|_\infty = \|\mu_1\|_\infty$ and that μ_0/ν_0 and α are measurable functions whose absolute value is almost everywhere equal to 1.

Assume f_0 is an extremal quasiconformal mapping in the class of f with Beltrami coefficient μ_0 such that $K_0(f) = (1 + \|\mu_0\|_\infty)/(1 - \|\mu_0\|_\infty)$.

The first step in the proof is to show that the hypothesis $H(f) < K_0(f)$ implies the Beltrami coefficient μ_0 of an extremal representative of the class of f

cannot have a degenerating Hamilton sequence. A sequence φ_n is a Hamilton sequence for μ_0 if $\|\varphi_n\| = 1$ and if

$$\operatorname{Re} \iint_{\Omega} \varphi_n(z)\mu_0(z)\, dx\, dy \text{ converges to } \|\mu_0\|_\infty. \tag{28}$$

The sequence φ_n is degenerating if, in addition, it converges to 0 uniformly on compact subsets of R.

First notice that if R is compact, there cannot be any degenerating Hamilton sequence, because normal convergence of the sequence φ_n would imply uniform convergence. Thus, the assertion that μ_0 can have no degenerating Hamilton sequence is interesting only when R is noncompact.

If $H(f_0) < K_0(f_0) = K_0$, then there exists a quasiconformal mapping f_1 equivalent to f_0 which has dilatation H_1 off of a compact subset S of R, where $H_1 < K_0$ (f_1 is called a frame mapping for f_0). Moreover, the maximal dilatation of f_1 on the whole surface R is bounded by some constant K_1 (K_1 is possibly much larger than K_0). Then inequality (27) yields

$$1 \leqslant K_1 K_0 \iint_S |\varphi_n|\, dx\, dy + H_1 \iint_{R-S} \frac{|1 - \mu_0\varphi_n/|\varphi_n||^2}{1 - |\mu_0|^2}\, dx\, dy. \tag{29}$$

If we assume φ_n is a degenerating Hamilton sequence, then the first term on the right side of (29) approaches zero as n approaches infinity. Let $k_0 = \|\mu_0\|_\infty$ and $K_0 = (1 + k_0)/(1 - k_0)$. Using Lemma 1 and Lemma 2, we see that

$$\lim_{n\to\infty} \iint_R |\varphi_n| \frac{|\mu_0|^2}{1 - |\mu_0|^2} = \frac{k_0^2}{1 - k_0^2} \tag{30}$$

and

$$\lim_{n\to\infty} \iint_R \frac{\varphi_n \mu_0}{1 - |\mu_0|^2}\, dx\, dy = \frac{k_0}{1 - k_0^2}. \tag{31}$$

On the other hand, from (28), (30) and (31), one can show that the second integral on the right side of (29) approaches K_0^{-1}. Putting this result together with the fact that the first integral on the right side of (29) approaches zero, we see that $1 \leqslant H_1 K_0^{-1}$, which contradicts the hypothesis that $H_1 < K_0$.

The second step in the proof is to show that a Hamilton sequence φ_n for μ_0 must converge in norm to a quadratic differential φ with norm equal to 1. Obviously, the sequence φ_n with $\|\varphi_n\| = 1$, which in local coordinates is a normal family of holomorphic functions, must have a subsequence which converges to a holomorphic quadratic differential φ with $\|\varphi\| \leqslant 1$. Without changing notation, we denote the convergent subsequence by the same symbol φ_n. We have just shown that the norm of φ cannot be zero because μ_0 cannot have a degenerating Hamilton sequence. If $\|\varphi\|$ were strictly less than 1, then we could construct another Hamilton sequence $\tilde{\varphi}_n = (\varphi_n - \varphi)/\|(\varphi_n - \varphi)\|$ which would be degenerating. Since μ_0 cannot have a degenerating Hamilton sequence, we conclude that $\|\varphi\| = 1$. On the other hand, there is the obvious inequality

$$\iint_R |\varphi_n - \varphi| \leqslant \|\varphi_n\| - \|\varphi\| + 2 \iint_{R-S} |\varphi| + 2 \iint_S |\varphi_n - \varphi|. \tag{32}$$

Since the norms of φ_n and φ are both equal to 1, this inequality implies that φ_n converges to φ in norm.

From Theorem 1 we know that a Hamilton sequence exists. By the preceding arguments, such a sequence must converge in norm to a φ with $\|\varphi\| = 1$. Therefore, on passing to the limit in (28), we obtain

$$\iint_\Omega \varphi\mu_0 \, dx\, dy = \|\mu_0\|_\infty = k_0 \qquad \text{with} \qquad \iint_\Omega |\varphi| \, dx\, dy = 1. \tag{33}$$

This is possible only if $\mu_0 = k_0|\varphi|/\varphi$. Now, let μ_1 be any other extremal Beltrami coefficient whose Teichmüller equivalence class is the same as that of μ_0. Substituting $\mu_0 = k_0|\varphi|/\varphi$ into inequality (27), we obtain

$$K_0 \leqslant \iint |\varphi| \frac{\left|1 + \dfrac{v_1}{v_0} k_0\right|^2}{1 - |\mu_1|^2} \, dx\, dy. \tag{34}$$

Since μ_1 is also extremal, we have $\|\mu_1\|_\infty = \|v_1\|_\infty = k_0$. Furthermore, since v_0 is the Beltrami coefficient of the mapping inverse to f_0, the absolute value of v_0 is constantly equal to k_0. Thus the right-hand side of (34) must be less than or equal to K_0. The only way this can happen is for v_1 to be identically equal to v_0, which implies that μ_1 is identically equal to μ_0, and completes the proof of the theorem.

Notes. Much of this chapter follows closely arguments given by Reich and Strebel in [ReiS1] and [ReiS2]. The proof of the necessity of the Hamilton–Krushkal condition is due to Hamilton [H] and, independently, to Krushkal [Kru1]. Bers adapted the same argument to the setting of Kleinian groups [Ber7]. Teichmüller's existence theorem can also be proved by first obtaining global coordinates for Teichmüller space and then applying the uniqueness theorem and invariance of domain. This is done by Bers in [Ber2].

The sufficiency of the Hamilton–Krushkal condition was first proved by Reich and Strebel [ReiS2]. In the same paper they gave the infinitesimal formula in Theorem 7.

There are many papers by Reich and Strebel on the subject of whether and when a given Teichmüller class contains nonuniquely extremal representatives. We refer to only one of them [Rei3].

Section 6.7 contains the important result that finite dimensional Teichmüller spaces are contractible (in fact, homeomorphic to $\mathbb{R}^n$). The proof we give is quite similar to the one given by Bers in [Ber2]. However, it differs from Bers's proof in an essential way because it does not depend on the introduction of other coordinates for Teichmüller space. Bers uses coordinates coming from generators for the Fuchsian group, the so-called Fenchel–Nielsen coordinates. Our proof avoids the necessity for these coordinates because we already know Teichmüller's existence theorem. Bers's method of obtaining Teichmüller's existence theorem is via the Fenchel–Nielsen coordinates [FenN] combined with invariance of domain and the Teichmüller uniqueness theorem.

The fact that infinite dimensional Teichmüller spaces are contractible has been proved by Tukia [Tu2]. It is also a consequence of results in a paper by Douady and Earle [DE]. Putting Douady and Earle [DE] together with results in a paper by Earle and Eells [EE2], we know that the space of trivial Beltrami coefficients $M_0(\Gamma)$ is contractible and, in particular, connected. Thus, the equivalence relation of homotopy which we have used to define equivalent Beltrami differentials may be replaced by isotopy through quasiconformal mappings. This fact is also a consequence of results in a paper by Reich [Rei1] and a paper of Earle and McMullen [EM]. The ideas developed in [EE2] lead to results about the diffeomorphism group of a Riemann surface (see [EE1]).

There is a notion of extremal mappings whose Beltrami coefficients have support on a fixed subset and also a theorem analagous to the necessity and sufficiency of the Hamilton–Krushkal condition. This question has been studied by Gardiner [Ga4], Fehlmann and Sakan [Feh, Sa, FehS].

The main ideas for Section 6.8 on Strebel's frame mapping condition come from [St2]. Here the condition is generalized to the setting of Riemann surfaces. The version of the main inequality given in formula (27) of Section 6.8 first appears in a paper by Reich [Rei2].

A formula for the variation of Teichmüller's metric can be derived from various forms of the main inequality (17), (21) and (27) (see [Ga3]).

EXERCISES

1. Let Ω be a plane domain bounded by smooth curves and suppose the Euclidean area of Ω is finite. Let f_K be the affine map: $f_K(z) = Kx + iy$. We call f_K extremal if all other quasiconformal mappings from Ω to $f_K(\Omega)$ which agree with f_K on $\partial\Omega$ and which are homotopic to f_K by a homotopy with the same boundary values as f_K have dilatation $\geqslant K$. Show that f_K is uniquely extremal.

2. Let Ω be a domain of the same type as in the previous exercise but not necessarily of finite area. Let

$$H(\Omega) = \sup \left| \iint_{\Omega} \varphi(z)\, dx\, dy \right|,$$

where the supremum is taken over all functions φ which are holomorphic in Ω and for which $\iint_{\Omega} |\varphi(z)|\, dx\, dy = 1$. Show that f_K is extremal if and only if $H(\Omega) = 1$.

3. [Reich and Strebel] In the same setting as the previous exercise, let K^* be the minimal dilation of a mapping in the same Teichmüller class as f_K. Show that $H(\Omega) \leqslant 1 - \delta$ implies $K^* \leqslant K - (\delta/2)(K - 1/K)$ and $H(\Omega) \geqslant 1 - \delta$ implies

$$K^* \geqslant \frac{K}{1 + \dfrac{\delta}{2}(K^2 - 1)}.$$

Conclude that

$$H(\Omega) = \lim_{K \to 1} \frac{K^* - 1}{K - 1}.$$

4. Let Ω be the complex plane with n points removed, n finite, and $\geqslant 3$. Show that f_K with $K > 1$ cannot be extremal for the domain Ω.

5. Show that any finite dimensional normed space is homeomorphic to the interior of its unit ball.

7

TEICHMÜLLER'S AND KOBAYASHI'S METRICS

We recall the definition of Teichmüller's metric given in Section 5.3. For two elements $[\mu]$ and $[\nu]$ of $T(\Gamma, C)$,

$$d([\mu], [\nu]) = \inf \tfrac{1}{2} \log K(w_\mu \circ w_\nu^{-1}) \tag{1}$$

where the infimum is over all μ and ν in the equivalence classes $[\mu]$ and $[\nu]$, respectively. Normally we will write $d(\mu, \nu)$ instead of the more cumbersome $d([\mu], [\nu])$, if no confusion is possible. In particular,

$$d(0, \mu) = \frac{1}{2} \log \frac{1 + k_0}{1 - k_0},$$

where k_0 is the minimal value of $\|\mu\|_\infty$, where μ ranges over a given Teichmüller class.

Let $k_0 = k_0(t)$ be a differentiable function of t with $k_0(0) = 0$. Then clearly $(d/dt)(k(t))|_{t=0} = (d/dt)\frac{1}{2}\log(1 + k_0(t))/(1 - k_0(t))|_{t=0}$. Therefore, Theorem 7 of the previous chapter tells us that $d(0, t\mu)$ has a derivative from the right at $t = 0$ and, for small $t \geqslant 0$,

$$d(0, t\mu) = t \sup \left| \operatorname{Re} \iint_\omega \varphi\mu \, dx \, dy \right| + O(t^2), \tag{2}$$

where the supremum is over φ in $A_s(\Gamma, C)$ for which $\|\varphi\| = 1$.

Formula (2) is the infinitesimal form of Teichmüller's metric at the origin. It can be used to find the infinitesimal form at every point in Teichmüller space. In Section 7.1 we give the formula for the infinitesimal form and show that it is a continuous function on the tangent bundle to Teichmüller space. In Section 7.2, we show that Teichmüller's metric is the integral of its infinitesimal form. This fact was first proved by O'Byrne [O] for infinite dimensional spaces by use of a general theorem in Finsler geometry and by appealing to a result of Earle and Eells [EE2, EE3, EE4], on fibrations over Teichmüller spaces.

Section 7.3 gives the definition of Kobayashi's metric and the obvious statement that it is less than or equal to Teichmüller's metric. Section 7.4 completes the proof of Royden's theorem that Teichmüller's metric equals Kobayashi's metric for finite dimensional spaces [Ko]. The argument uses the infinitesimal form of the metric and a curvature argument based on a lemma of Ahlfors [Ah6].

Section 7.5 extends Royden's theorem to infinite dimensional spaces.

7.1. THE INFINITESIMAL METRIC ON THE TANGENT BUNDLE TO $T(\Gamma, C)$

Let Γ be a Fuchsian group acting on $\mathbb{H}$ and C be a closed invariant set such that $\Lambda \subseteq C \subseteq \hat{\mathbb{R}}$. Let w be a quasiconformal mapping of $\mathbb{C}$ to $\mathbb{C}$ with Beltrami coefficient μ such that $\overline{w(z)} = w(\bar{z})$ and $w \circ A \circ w^{-1}$ is a Möbius transformation for each A in Γ. Let $\Gamma_\mu = w \circ \Gamma \circ w^{-1}$ and let $C_\mu = w(C)$. We have seen in Section 5.5 that w induces an isometric mapping between Teichmüller spaces. Here, we use the notation

$$\alpha\colon T(\Gamma, C) \to T(\Gamma_\mu, C_\mu), \tag{3}$$

where by definition, $\alpha([\tau])$ is the Teichmüller equivalence class of $w_\tau \circ w_\mu^{-1}$.

Notice that $\alpha([\mu]) = [0]$ and so

$$d([\mu], [\nu]) = d([0], \alpha([\nu])). \tag{4}$$

We now will use (4) to calculate the infinitesimal length $F([\mu], \nu)$ of a tangent vector ν at an arbitrary point $[\mu]$. By definition F is the derivative from the right of the function $d([\mu], [\mu + t\nu])$ with respect to t at $t = 0$. On replacing ν by $\mu + t\nu$ in (4) and using (2), we get

$$\begin{aligned} d(\mu, \mu + t\nu) &= d(0, \alpha(\mu + t\nu)) \\ &= \sup \left| \operatorname{Re} \iint_{\omega_\mu} t\varphi S(\nu)\, du\, dv \right| + O(t^2), \end{aligned}$$

where S is the derivative at μ of α and $u + iv = w = w_\mu$, and the supremum is over all φ in $A(\Gamma_\mu, C_\mu)$ with $\|\varphi\| = 1$. Also, ω_μ is a fundamental domain for Γ_μ.

By the result of Exercise 16 of Chapter 1, one finds that

$$\alpha(\tau) = \left[\frac{\tau - \mu}{1 - \bar{\mu}\tau} \cdot \frac{1}{\theta}\right] \circ w_\mu^{-1},$$

where $\theta = \bar{p}/p$ and $p = (\partial/\partial z)w_\mu$. Letting $\tau = \mu + tv$, we find $S(v) = v/((1 - |\mu|^2)\theta)$ and so

$$F([\mu], v) = \sup \left| \operatorname{Re} \iint \varphi(w) \left[\frac{v}{1 - |\mu|^2} \cdot \frac{1}{\theta}\right] du\, dv \right|, \tag{5}$$

where the supremum is over all φ in $A(\Gamma_\mu, C_\mu)$ with $\|\varphi\| = 1$ and the integral is over $\omega_\mu = w(\omega)$.

Lemma 1. *The function F in (5) from the tangent bundle of $T(\Gamma, C)$ to $\mathbb{R}$ is continuous.*

Proof. Since the translation mapping α is an isometry, it suffices to show F is continuous at $\mu = 0$. In other words, we must show $|F(\mu, v_1) - F(0, v)| < \varepsilon$ when $\|\mu\|_\infty < \delta$ and $\|v - v_1\|_\infty < \delta$. Since F is a seminorm in the variable v, it is enough to prove the same inequality with v_1 replaced by v and for $\|v\|_\infty < 1$. In fact, it suffices to prove that for every $\varepsilon > 0$ there is a $\delta > 0$ for which

$$F(0, v) < F(\mu, v) + \varepsilon \tag{6}$$

whenever $\|\mu\|_\infty < \delta$ and $\|v\|_\infty < 1$. For, if we show this, then by applying the translation mapping α and letting α'_μ denote the derivative of α at μ, we find that

$$F(\mu_1, \alpha'_0(v)) < F(0, \alpha'_\mu(v)) + \varepsilon,$$

where μ_1 is the Beltrami coefficient of the inverse mapping to w_μ. Obviously $\|\mu\|_\infty = \|\mu_1\|_\infty$. Moreover, by differentiating the formula for $\alpha(\tau)$, we obtain

$$\alpha'_\mu(v) = \left[\frac{v}{1 - |\mu|^2} \cdot \frac{1}{\theta}\right] \circ w_\mu^{-1}$$

$$\alpha'_0(v) = \left[v(1 - |\mu|^2)\frac{1}{\theta}\right] \circ w_\mu^{-1}.$$

It is clear that the L_∞-norm of the difference between the two quantities $\alpha'_\mu(v)$ and $\alpha'_0(v)$ is bounded by a constant times $\|\mu\|_\infty^2 \|v\|_\infty$ for small $\|\mu\|_\infty$. Thus, for $\|\mu\|_\infty < \delta$, from (6) we obtain the inequality $F(\mu_1, v) < F(0, v) + 2\varepsilon$.

Hence, we have reduced the lemma to proving (6). To accomplish this, we must show that for every φ in $A_s(\Gamma, C)$ with $\|\varphi\| = 1$, there is ψ in $A_s(\Gamma_\mu, C_\mu)$

with $\|\psi\| \leqslant 1$ such that

$$\operatorname{Re} \iint_{\omega} \varphi \nu \, dx \, dy \leqslant \operatorname{Re} \iint_{\omega_\mu} \psi \frac{\nu}{1 - |\mu|^2} \cdot \frac{1}{\theta} \, du \, dv + \varepsilon$$

for all μ with $\|\mu\|_\infty$ sufficiently small and all ν with $\|\nu\|_\infty \leqslant 1$.

We construct the quadratic differential ψ by using theta series. Let Θ_μ be the theta series operator for the group Γ_μ. Thus, $\Theta_0 = \Theta$ is the theta series operator for Γ. By the surjectivity of the theta series operator (Theorem 3 in Section 4.3) there is a function G in $A_s(C)$ such that $\Theta G = \varphi$. Note that

$$\iint_{\omega} \varphi \nu \, dx \, dy = \iint_{\mathbb{H}} G \nu \, dx \, dy. \tag{7}$$

Our candidate for ψ is $\psi = \Theta_\mu G / \|\Theta_\mu G\|$. Note that $\|\psi\| \leqslant 1$ and

$$\iint_{\omega_\mu} (\Theta_\mu G) \left[\frac{\nu}{1 - |\mu|^2} \cdot \frac{1}{\theta} \right] \circ w_\mu^{-1} \, du \, dv = \iint_{\mathbb{H}} G(w(z)) w_z^2(z) \nu(z) \, dx \, dy. \tag{8}$$

It suffices to show that for $\|\mu\|_\infty$ sufficiently small and for $\|\nu\|_\infty \leqslant 1$ that the difference between (7) and (8) is arbitrarily small and that $\|\Theta_\mu G\|$ converges to $\|\Theta G\|$. Moreover, it suffices to show the inequalities for any particular representative of the Teichmüller class of μ. Since we may take $\|\mu\|_\infty < \frac{1}{3}$, we are permitted to choose μ of the particular form which comes from the Ahlfors–Weill section (Lemma 7 of Section 5.4). Hence we may assume that w_z converges uniformly to 1 on compact subsets of $\mathbb{H}$ and $w(z)$ converges uniformly to z on compact subsets of $\mathbb{H}$. Let $\hat{G}(z) = G(w(z)) w_z^2(z)$, where $w = w_\mu$. It is easy to see that $\hat{G}(z)$ converges uniformly on compact subsets of $\mathbb{H}$ to G and $\|\hat{G}\| \to \|G\|$, where the norm is the L_1-integral of the absolute value over the upper half plane. For any compact subset D of $\mathbb{H}$ we have the elementary inequality

$$\iint_{\mathbb{H}} |\hat{G} - G| \leqslant (\|\hat{G}\| - \|G\|) + 2 \iint_{\mathbb{H} - D} |G| + 2 \iint_{D} |\hat{G} - G|. \tag{9}$$

Putting these observations together we see that (7) and (8) are arbitrarily close for sufficiently small $\|\mu\|_\infty$.

Clearly, to complete the proof of the lemma, it suffices to prove the following lemma.

Lemma 2. *Let G be in $A(C)$. Then $\iint_{\omega_\mu} |\Theta_\mu G|$ approaches $\iint_{\omega} |\Theta G|$ as $\|\mu\|_\infty$ approaches zero.*

Proof. Let A_μ^n be an enumeration of Γ_μ and $\Theta_{\mu n}$ be the truncation of Θ_μ to the first n elements of Γ_μ. Let Θ_n be the corresponding finite sum for Γ. Clearly $\iint_{\omega_\mu} |\Theta_{\mu_n} G|$ approaches $\iint_\omega |\Theta_n G|$ as $\|\mu\|_\infty \to 0$. In order to pass to the sum over all elements of Γ_μ and Γ, we will show that for any $\varepsilon > 0$, there exists n_0 and $\delta > 0$ such that, for $\|\mu\|_\infty < \delta$,

$$\iint_{\omega_\mu} \Sigma_{n > n_0} |G(A_\mu^n z) A_\mu^{n'}(z)^2| < \varepsilon. \tag{10}$$

To simplify notation, let $\mathbb{H}$ be replaced by the unit disk Δ and assume the groups Γ and Γ_μ act on Δ. Pick $r < 1$ so that $\iint_{r<|z|<1} |G| < \varepsilon$ and pick n_0 so that $D = \bigcup_{n=1}^{n_0} A^n(\omega)$ contains the disk of radius $(r+1)/2$. Now $\omega_\mu = w_\mu(\omega)$ and, hence, $w_\mu(D) = \bigcup_{n=1}^{n_0} A_\mu^n(\omega_\mu)$. Since w_μ converges uniformly in Δ to the identity as $\|\mu\|_\infty$ converges to zero, we see that for sufficiently small $\|\mu\|_\infty$, $w_\mu(D) \supseteq \{z: |z| < r\}$. This implies (10) is bounded by $\iint_{r<|z|<1} |G|$, which is less than ε. This completes the proofs of Lemmas 1 and 2.

7.2. INTEGRATION OF THE INFINITESIMAL METRIC

Let $\bar{d}$ be the integrated form of (7). This means $\bar{d}(p, q) = \inf L(\gamma)$, where γ is a piecewise smooth path joining p to q and $L(\gamma) = \int_0^{t_0} F(\gamma(t), \gamma'(t))\, dt$ and $\gamma(0) = p$, $\gamma(t_0) = q$. This integral is well defined because F is a continuous function on the tangent bundle. It is an elementary fact that if F is the differentiated form of d and if F is continuous, then $d \leqslant \bar{d}$.

Theorem 1. *For any Teichmüller space $T(\Gamma, C)$, $d = \bar{d}$, that is, Teichmüller's metric is the integral of its differentiated form.*

By the remarks preceding the theorem, what remains to be shown is that $\bar{d} \leqslant d$. Assume $\|\mu\|_\infty = 1$ and $k\mu$ is extremal so $d(0, [k\mu]) = \frac{1}{2}\log[(1+k)/(1-k)]$. Let $\gamma(t) = [t\mu]$, $0 \leqslant t \leqslant k$. We will show that $L(\gamma) = d(0, [k\mu])$ and it will follow that $\bar{d} \leqslant d$. Since $\gamma'(t) = \mu$, we must calculate $F([t\mu], \mu)$. We know that $w_{t\mu}$ and $(w_{t\mu})^{-1}$ are both extremal. The Beltrami coefficient of $(w_{t\mu})^{-1}$ is $-t\mu\theta^{-1}$, where $\theta = \bar{p}/p$ and $p = (\partial/\partial z) w_{t\mu}$. Since $-t\mu\theta^{-1}$ is extremal, Theorem 5 of Section 6.5 tells us that

$$\sup \operatorname{Re} \iint_{\omega_\mu} \varphi \frac{t\mu}{1 - t^2|\mu|^2} \cdot \frac{1}{\theta}\, du\, dv = \frac{t}{1 - t^2}, \tag{11}$$

where the supremum is over all φ in $A(\Gamma_\mu, C_\mu)$ with $\iint_{\omega_\mu} |\varphi| = 1$. From (11) and

(5), one sees that $F([t\mu], \mu) = 1/(1 - t^2)$, and, therefore,

$$\int_0^k F(\gamma(t), \gamma'(t))\, dt = \int_0^k \frac{dt}{1 - t^2} = \frac{1}{2} \log \frac{1 + k}{1 - k}.$$

7.3. THE KOBAYASHI METRIC

Let M be any complex manifold modeled on a Banach space. The Kobayashi metric d_K for M is defined in a way which depends on the family $\mathscr{F}$ of holomorphic functions from the open unit disk Δ into M. Let P and Q be points in M and let $d_1(P, Q) = \frac{1}{2}\log[(1 + r)/(1 - r)]$, where r is the infimum of the nonnegative numbers s for which there exists a holomorphic function f in $\mathscr{F}$ with $f(0) = P$ and $f(s) = Q$. Obviously $d_1(P, Q) = d_1(Q, P)$ since for any f in $\mathscr{F}$ we can consider $f \circ h$ in $\mathscr{F}$, where $h(z) = -(z - s)/(1 - sz)$.

Let

$$d_n(P, Q) = \inf \sum_{i=1}^{n} d_1(P_{i-1}, P_i) \tag{12}$$

where the infimum is taken over all points $P_0, \ldots, P_n$ in M for which $P_0 = P$ and $P_n = Q$. Obviously, $d_{n+1} \leqslant d_n$ for all n.

Definition. *The Kobayashi (pseudo)metric* $d_K(P, Q) = \lim_{n \to \infty} d_n(P, Q)$.

As long as M is connected, d_K is finite valued. Obviously d_K is symmetric since d_1 is. Now d_K satisfies the triangle inequality since $d_{2n}(P, R) \leqslant d_n(P, Q) + d_n(R, Q)$. It is not always the case that d_K is nondegenerate; for example, it is obvious that if $M = \mathbb{C}$, then $d_K \equiv 0$.

Observe that from (12), if d_1 satisfies the triangle inequality, then $d_1 = d_n$ for all n and so $d_1 = d_K$. Ultimately, we will show that for all Teichmüller spaces with complex structure which are modeled on a Fuchsian group, $d_1 = d_K = d_T$, where d_T is Teichmüller's metric.

Lemma 3. *Let $T(\Gamma, C)$ be a Teichmüller space which has complex structure. Let d_K and d_T be Kobayashi's and Teichmüller's metrics, respectively. Then $d_K \leqslant d_T$.*

Proof. The assumption that $T(\Gamma, C)$ has complex structure is tantamount to saying that $C = \hat{\mathbb{R}}$. This hypothesis is required because there is no Kobayashi metric unless the space has complex structure. Let $[\mu] \in T(\Gamma, \sigma)$ and let μ be extremal in its class and $\|\mu\|_\infty = k$. We have already observed that every class possesses at least one extremal representative. Then by definition (1), $d_T(0, \mu) = \frac{1}{2}\log[(1 + k)/(1 - k)]$. For such a μ, let $f(z) = [z\mu/k]$. Clearly, f is a holomorphic function from Δ into Teichmüller space, $f(0) = 0$ and $f(k) = [\mu]$. Hence, $d_K(0, \mu) \leqslant d_1(0, \mu) \leqslant d_T(0, \mu)$. Now the translation mapping $\alpha([w_\tau]) = [w_\tau \circ w_\mu^{-1}]$ is holomorphic, so obviously it is an isometry in

Kobayashi's metric. We already know that it is an isometry in Teichmüller's metric. Therefore, the inequality $d_K(p,q) \leqslant d_1(p,q) \leqslant d_T(p,q)$ holds for an arbitrary pair of points p and q in Teichmüller space.

The remainder of this chapter is devoted to showing the reverse inequality, $d_T \leqslant d_K$, for arbitrary Teichmüller spaces with complex structure.

7.4. THE FINITE DIMENSIONAL CASE

The infinitesimal metric F defined by (5) is obviously dual to the infinitesimal cometric G defined on the cotangent bundle to Teichmüller space, where G is given by the formula

$$G([\mu], \varphi) = \iint_{\omega_\mu} \frac{|\varphi(w)|\, du\, dv}{1 - |\mu|^2}, \tag{13}$$

where μ is in $M(\Gamma)$, φ is in $A(\Gamma_\mu, C_\mu)$, and $w = u + iv = w_\mu$. We have seen in Lemma 1 that the infinitesimal metric F is continuous. We have a much stronger result for the finite dimensional case, namely, F is of class C^1 on the tangent bundle except at the zero section. This statement is a consequence of the following lemma.

Lemma 4. *Assume $A(\Gamma, C)$ is a finite dimensional vector space. Then the cometric G defined in* (13) *is C^1 except where $\varphi \equiv 0$.*

We postpone the proof of this lemma until Section 9.3. We note that since the dual of a strictly convex C^1 cometric is a C^1 metric (a fact which is also proved in Chapter 9), this lemma implies F is C^1.

For the remainder of this section we assume the Teichmüller space T is a finite dimensional complex manifold. Therefore, by Chapter 5, it has local coordinates in $\mathbb{C}^n$ and, following Royden [Ro], we use coordinate notation. An element of the tangent bundle to T is a pair $(x; \xi)$, where x is in the base space and ξ is a tangent vector. We will write $x = (x_1, x_i)$, where $2 \leqslant i \leqslant n$, and, similarly, $\xi = (\xi_1, \xi_j)$, where $2 \leqslant j \leqslant n$. Always, i and j will be integers ranging from 2 to n.

Our objective is to write the Taylor series for F up to first order and then adjust the coordinates $(x; \xi)$ so that F has a special form. First of all, we pick the first coordinate x_1 so that

$$F(x_1, 0; 1, 0) = \frac{1}{1 - |x_1|^2}. \tag{14}$$

This is achieved by letting x_1 correspond to a Beltrami differential of the form $x_1|\varphi|/\varphi$, where φ is a holomorphic quadratic differential of norm 1. Then

$(\partial/\partial x_1)(x_1|\varphi|/\varphi) = |\varphi|/\varphi$ and to find $F(x_1, 0; 1, 0)$ we substitute $\mu = x_1|\varphi|/\varphi$ and $v = |\varphi|/\varphi$ in formula (5). An elementary calculation [using formula (4) of Chapter 6] for $p = (\partial/\partial z)w_\mu(z)$ yields

$$-\frac{p}{\bar{p}}\frac{|\varphi(z)|}{\varphi(z)} = \frac{|\psi_0(w)|}{\psi_0(w)},$$

where ψ_0 is the quadratic differential which corresponds to the extremal map in the class of $(w_\mu)^{-1}$. Hence, formula (5) simplifies to

$$F(x_1, 0; 1, 0) = \frac{1}{1 - |x_1|^2} \sup \operatorname{Re} \iint \psi(w) \frac{|\psi_0(w)|}{\psi_0(w)}\, dx\, dy,$$

where the integral is over a fundamental domain for $\Gamma_\mu = w_\mu \circ \Gamma \circ w_\mu^{-1}$ and the supremum is over all holomorphic quadratic differentials ψ for Γ_μ with $\|\psi\| = 1$. This supremum is 1 and we obtain (14).

The Taylor expansion for $F(x_1, x_i; 1, \xi_j)$ is

$$F(x_1, 0; 1, \xi_j) + \operatorname{Re} \sum C_i(x_1, \xi_j)x_i + o(|x_i|). \tag{15}$$

Expanding $F(x_1, 0; 1, \xi_j)$ relative to the variable ξ_j, we get

$$F(x_1, 0; 1, \xi_j) = F(x_1, 0; 1, 0) + \operatorname{Re} \sum B_j(x_1)\xi_j + o(|\xi_j|). \tag{16}$$

Here B_j and C_i are continuous functions and the error estimates $o(|x_i|)$ and $o(|\xi_j|)$ hold uniformly in a neighborhood of the origin in the variables x_1 and ξ_j for formula (15) and in the variable x_1 for formula (16). Now, make the change of coordinates

$$\begin{aligned} x_1 &= y_1 - \operatorname{Re} \sum B_i(0)\, y_i, \\ x_i &= y_i, \end{aligned} \tag{17}$$

which on the fibers of the tangent bundle yields

$$\begin{aligned} \xi_1 &= \eta_1 - \operatorname{Re} \sum B_j(0)\eta_j, \\ \xi_j &= \eta_j. \end{aligned} \tag{18}$$

Notice that (14) is preserved for the (y, η) coordinates and $\partial F/\partial\eta_j(x_1, 0; \xi_1, \xi_j) = (\partial F/\partial\xi_1)(-\operatorname{Re} B_j(0)) + \partial F/\partial\xi_j$. But $\partial F/\partial\xi_j(x_1, 0; 1, 0) = \operatorname{Re} B_j(x_1)$ and $\partial F/\partial\xi_1(x_1, 0; \xi_1, 0) = 1/1 - |x_1|^2$. Hence, $\partial F/\partial\eta_j(0, 0; 1, 0) = 0$. This means that for the coordinates (y, η) the functions $B_j(y_1)$ satisfy $B_j(0) = 0$.

Thus, we can assume (16) holds for coordinates (x, ξ) with $B_j(0) = 0$. We now make a second change of coordinates which does not alter these properties and for which $C_i(0,0)=0$. Let

$$\begin{aligned} x_1 &= y_1 - \operatorname{Re}\sum C_i(0,0)\, y_i y_1, \\ x_i &= y_i, \end{aligned} \tag{19}$$

which on the fibers of the tangent bundle yields

$$\begin{aligned} \xi_1 &= \eta_1 - \operatorname{Re}\sum C_j(0,0)\, y_j \eta_1 - \operatorname{Re}\sum C_j(0,0)\eta_j y_1, \\ \xi_j &= \eta_j. \end{aligned} \tag{20}$$

Again, notice that when $x_i = 0$ and $\xi_j = 0$, we have $x_1 = y_1$ and $\xi_1 = \eta_1$, so (14) remains true in the new coordinates. In the (y, η) coordinates we have

$$\frac{\partial F}{\partial \eta_j}(y_1, 0;\, 1, \eta_j) = \frac{\partial F}{\partial \xi_1}\operatorname{Re}(-C_j(0,0)\, y_1) + \frac{\partial F}{\partial \xi_j}. \tag{21}$$

When we evaluate (21) at $(y, \eta) = (0, 0; 1, 0)$, we see that, because in the first change of coordinates we achieved $\partial F/\partial \xi_j(0, 0; 1, 0) = 0$, we now have $\partial F/\partial \eta_j(0, 0; 1, 0) = 0$. Furthermore,

$$\frac{\partial F}{\partial y_i} = \frac{\partial F}{\partial x_1}\operatorname{Re}(-C_i(0,0)\, y_1) + \frac{\partial F}{\partial x_i} + \frac{\partial F}{\partial \xi_1}\operatorname{Re}(-C_i(0,0)\eta_1). \tag{22}$$

Evaluating (22) at $(y, \eta) = (0, 0; 1, 0)$, the first term on the right is zero, the second is $+\operatorname{Re} C_i(0, 0)$, and the third is $-\operatorname{Re} C_i(0, 0)$. We have proved the following lemma.

Lemma 5. *Given any unit tangent vector $|\varphi|/\varphi$ to the finite dimensional complex Teichmüller space T, there is a choice of coordinates $(x_1, x_i; \xi_1, \xi_j)$ for the tangent bundle such that*

(a) $(x_1, 0) = [x_1|\varphi|/\varphi]$ *for* $|x_1| < 1$;
(b) $F(x_1, 0; 1, 0) = 1/(1 - |x_1|^2)$ *for* $|x_1| < 1$;
(c) $B_j(0) = 0$ *for* $2 \leqslant j \leqslant n$ *in (16); and*
(d) $C_i(0, 0) = 0$ *for* $2 \leqslant i \leqslant n$ *in (15).*

Lemma 6. *For the coordinates of Lemma 5, $B_i(x_1) = o(|x_1|)$ as $x_1 \to 0$.*

Proof. Let $g(t)$ be a smooth function from $[0, 1]$ into $\mathbb{C}^n$ with $g(0) = g(1) = 0$ and for which the first coordinate of g is identically zero. Let $\gamma(t) = t\zeta e_1$, where $e_1 = (1, 0, 0, \ldots, 0)$ and $0 \leqslant t \leqslant 1$ and $|\zeta| < 1$. By part (a) of Lemma 5, γ is a geodesic path and the distance from 0 to ζe_1 is less than or equal to the length of the path $\gamma + \varepsilon g$. Thus if ε is positive and small enough so that the path $\gamma + \varepsilon g$ is in the same coordinate neighborhood, one has

$$0 \leqslant \lim_{\varepsilon \to 0} + \frac{1}{\varepsilon} \int_0^1 [F(t\zeta e_1 + \varepsilon g(t), \zeta e_1 + \varepsilon g'(t)) - F(t\zeta e_1, \zeta e_1)]\, dt. \tag{23}$$

By applying (15) and (16) and using the fact that F is positive homogeneous in ξ, one finds that

$$0 = \int_0^1 \operatorname{Re}(B_j(t\zeta)g_j'(t) + \zeta C_j(t\zeta, 0) g_j(t))\, dt. \tag{24}$$

Since g can be multiplied by an arbitrary complex number of absolute value 1, we can drop the real part symbol in (24). Thus, for all C^1 functions g_j with $g_j(0) = g_j(1) = 0$, we have $\int_0^1 h(t) g_j'(t)\, dt = 0$, where $h(t) = B_j(t\zeta) - \zeta \int_0^t C_j(s\zeta, 0)\, ds$. Thus $h(t)$ is constant. It is equal to zero because $h(0) = 0$ and so

$$B_j(\zeta) = \zeta \int_0^1 C_j(s\zeta, 0)\, ds. \tag{25}$$

But (25) is $o(|\zeta|)$ since C_j is continuous.

Lemma 7. $F(x_1, 0; 1, \xi_j) \geqslant F(x_1, 0; 1, 0) + \operatorname{Re} \sum B_i(x_1)\xi_i$.

Proof. Let $h(t) = F(x_1, 0; 1, t\xi_j)$. It is a convex C^1 function of t. Thus, the lemma follows from (16).

Remark. This inequality is the starting point for constructing a supporting metric referred to in Lemma 8, below. It is a different inequality for different choices of local coordinate. By taking coordinates satisfying Lemma 5, we get the required inequality.

Lemma 8. *Let f be a holomorphic function from the unit disk into T and suppose $f(0) = 0$. Let $\lambda(\zeta) = F(f(\zeta); f'(\zeta))$. Then there exists a smooth function $\lambda_0(\zeta)$ defined for ζ in a neighborhood of 0 such that $\lambda_0(0) = \lambda(0) = 1$, $\lambda_0(\zeta) \leqslant \lambda(\zeta) + o(\zeta^2)$ and $-(\Delta \log \lambda_0)/\lambda_0^2 \leqslant -4$.*

Proof. The expression $-(\Delta \log \lambda)/\lambda^2$, where Δ is the Laplacian, is called the curvature of $\lambda(\zeta)|d\zeta|$. In Section 1.3 we saw that the curvature is invariant under holomorphic changes of coordinates. Now suppose $f(z)$ from the unit disk into T is holomorphic and $f(0) = 0$, $f'(0) = k|\varphi|/\varphi$, where φ is an integrable holomorphic quadratic differential. Pick local coordinates satisfying the conditions of Lemma 5 for the unit tangent vector $|\varphi|/\varphi$ and let $f = (f_1, f_2, \ldots, f_n)$ in these coordinates. Clearly, from (a) of Lemma 5, $f_1'(0) \neq 0$ and $f_i'(0) = 0$. Now alter the local parameter ζ at the origin of the unit disk so that $f_1(\zeta) = \zeta$ and

$f_i(\zeta) = a_i\zeta^2 + O(\zeta^3)$. Then $\lambda(\zeta) = F(\zeta, f_i(\zeta); 1, f_i'(\zeta))$ and from Lemma 7 and (15), we have

$$\lambda(\zeta) \geqslant F(\zeta, 0; 1, 0) + o(|\zeta|^2).$$

Since $F(\zeta, 0; 1, 0) = 1/(1 - |\zeta|^2)$ and this has curvature -4, the lemma is proved.

The following lemma is called Ahlfors's version of Schwarz's lemma. We present a modified version due to Earle. If $\lambda|dz|$ is a conformal metric on the disk, λ_0 is called a supporting metric for λ at p if $\lambda_0(p) = \lambda(p)$ and $\lambda_0(\zeta) \leqslant \lambda(\zeta) + o(|\zeta|^2)$ for a local parameter ζ centered at p.

Lemma 9. *Let $\lambda|dz|$ be a conformal metric on the unit disk Δ such that at each point p in Δ there is a smooth supporting metric for λ_0 which has curvature at most -4. Then $\lambda(z) \leqslant (1 - |z|^2)^{-1}$, that is, λ is bounded by the noneuclidean metric on the unit disk.*

The proof of this lemma can be found in Ahlfors's book [Ah6]. For the convenience of the reader we include a proof in the appendix to this chapter.

Lemma 10. *Let f be a holomorphic function from Δ into T. Then $F(f(\zeta), f'(\zeta)) \leqslant 1/(1 - |\zeta|^2)$.*

Proof. Let $\lambda(\zeta) = F(f(\zeta); f'(\zeta))$. Let $p = f(\zeta_0)$. Then let the Teichmüller space T be constructed from the equivalence classes of Beltrami differentials on the Riemann surface represented by the point p. With this basepoint the zero Beltrami differential corresponds to the point p. Changing the local coordinate in the unit disk via the mapping $\zeta \mapsto (\zeta - \zeta_0)/(1 - \bar{\zeta}_0\zeta)$ does not change the curvature of λ. Hence, by Lemma 8, there is a smooth supporting metric for λ at ζ_0 with curvature $\leqslant 4$. Hence, the result follows from Lemma 9.

Theorem 2 (Royden [Ro]). *For finite dimensional Teichmüller spaces which have complex structure, $d_K(p, q) = d_1(p, q) = d_T(p, q)$, that is, Kobayashi's and Teichmüller's metrics coincide.*

Proof. From Lemma 4, it suffices to show that $d_T \leqslant d_1$. Then it follows that d_1 is itself a metric, and so $d_1 = d_n$ for all n, which in turn implies $d_1 = d_K$.

Let f be a holomorphic function from Δ into T with $f(\zeta_0) = p$ and $f(\zeta_1) = q$. From Lemma 10, we know that $F(f(\zeta); f'(\zeta)) \leqslant 1/(1 - |\zeta|^2)$. Let α be the noneuclidean geodesic in Δ joining ζ_0 to ζ_1. From this inequality we see that the Teichmüller length of $f(\alpha)$ is less than or equal to the noneuclidean distance from ζ_0 to ζ_1. Thus, from Theorem 1, the Teichmüller distance from $f(\zeta_0)$ to $f(\zeta_1)$ is less than or equal to the noneuclidean distance from ζ_0 to ζ_1. This shows $d_T(p, q) \leqslant d_1(p, q)$ and proves the theorem.

7.5. THE INFINITE DIMENSIONAL CASE

Our objective now is to show that whenever T has complex structure, even when T is infinite dimensional, Teichmüller's and Kobayashi's metrics coincide. In view of Lemma 3, what we must show is that whenever $f: \Delta \to T$ is holomorphic and $f(0) = 0$ and $f(r) = [\mu]$, where μ is extremal and $\|\mu\|_\infty = k$ and $r > 0$, then $r \geqslant k$. From Theorem 2, we know this inequality holds whenever T is finite dimensional.

All Teichmüller spaces with complex structure (with one exception) can be viewed as coming from Beltrami differentials in the upper half plane $\mathbb{H}$. The one exception is the case of a torus, but in that case the Teichmüller space is isomorphic to $\mathbb{H}$ and it is easy to see that for the torus d_K and d_T are both identical with the noneuclidean metric.

Since we are assuming $T = T(\Gamma, C)$ and $C = \hat{\mathbb{R}}$, μ and ν are equivalent ($\mu \sim \nu$) if $w_\mu(x) = w_\nu(x)$ for all x in $\hat{\mathbb{R}}$. It is clear that there exists a sequence of finitely generated subgroups Γ_n of Γ and subsets C_n of $\hat{\mathbb{R}}$ with the following properties:

(i) $\Gamma_n \subseteq \Gamma_{n+1}$ and $\bigcup \Gamma_n = \Gamma$;
(ii) each Γ_n contains elements with fixed points in the intervals $I_{kn} = ((k-1)/n, k/n)$ for $-n^2 \leqslant k \leqslant n^2$ whenever $I_{kn} \cap \Lambda \neq \emptyset$;
(iii) C_n is invariant under Γ_n, $C_n \supseteq \Lambda_n$, and $(C_n - \Lambda_n)/\Gamma_n$ is a finite set; and
(iv) $C_n \subseteq C_{n+1}$ and the closure of $\bigcup C_n = \hat{\mathbb{R}}$.

Now, let $\Omega_n = \mathbb{C} \cup \{\infty\} - C_n$. We introduce a new set of Beltrami coefficients. It consists of complex valued measurable functions μ with support in Ω_n for which $\|\mu\|_\infty < 1$ and for which

$$\mu(Az)\overline{A'(z)} = \mu(z)A'(z)$$

for all A in Γ_n. The set of such Beltrami coefficients is denoted by $M(\Gamma_n, \Omega_n)$.

Let $w = w^\mu$ be the unique homeomorphic self-mapping of $\mathbb{C} \cup \{\infty\}$ satisfying

$$\frac{\partial w}{\partial \bar{z}} = \mu \frac{\partial}{\partial z} w$$

and normalized to fix 0, 1, and ∞. Define μ to be strongly equivalent to ν and write $\mu \equiv_n \nu$ if $w^\mu(x) = w^\nu(x)$ for all x in C_n and if w^μ is homotopic to w^ν in Ω_n. Define $\tilde{T}(\Gamma_n, \Omega_n)$ to be $M(\Gamma_n, \Omega_n)$ factored by the strong equivalence relation.

Let $\pi: M(\Gamma) \to M(\Gamma_n, \Omega_n)$ be defined by $\pi(\mu) = \mu(z)$ for z in $\mathbb{H}$ and $\pi(\mu) = 0$ for z in L.

Lemma 11. *If $\mu \sim \nu$, then $\pi(\mu) \equiv_n \pi(\nu)$.*

Proof. The hypothesis tells us that $w_\mu(x) = w_\nu(x)$ for all x in $\mathbb{R}$, since in the case under consideration $C = \hat{\mathbb{R}}$. This implies $w^{\pi(\mu)}(x) = w^{\pi(\nu)}(x)$ for all x in $\mathbb{R}$

and hence for all x in C_n. For the proof see Lemma 3 in Section 5.4. Furthermore, there is a homotopy $h_t\colon \mathbb{H} \to w^{\pi(\mu)}(\mathbb{H})$ for which $h_0(z) = w^{\pi(\mu)}(z)$ for z in $\mathbb{H}$ and $h_1(z) = w^{\pi(\nu)}(z)$ for z in $\mathbb{H}$ and $h_t(x) = w^{\pi(\mu)}(x) = w^{\pi(\nu)}(x)$ for x in $\mathbb{R}$ and $0 \leqslant t \leqslant 1$. This homotopy extends to a homotopy h_t from Ω_n to $w^{\pi(\mu)}(\Omega_n)$ by setting $h_t(z) = w^{\pi(\mu)}(z) = w^{\pi(\nu)}(z)$ for z in the lower half plane. It follows that $\pi(\mu)$ and $\pi(\nu)$ are strongly equivalent.

Lemma 11 implies that the mapping π induces a mapping from $T(\Gamma)$ to $\tilde{T}(\Gamma_n, \Omega_n)$. We denote this new mapping by the same letter π. Since the complex structures on $T(\Gamma)$ and on $\tilde{T}(\Gamma_n, \Omega_n)$ are inherited from $M(\Gamma)$ and $M(\Gamma_n, \Omega_n)$, this new mapping is holomorphic.

Now suppose μ is extremal in its class in $M(\Gamma)$. By this we mean that $k = \|\mu\|_\infty \leqslant \|\nu\|_\infty$ for all ν in $M(\Gamma)$ for which $\nu \sim \mu$. Let

$$k_n|\eta_n|/\eta_n \tag{26}$$

be extremal in the class of $\pi(\mu)$ in $M(\Gamma_n, \Omega_n)$ under the equivalence relation $\equiv_n$. Since $\tilde{T}(\Gamma_n, \Omega_n)$ is isomorphic to a finite dimensional Teichmüller space, from Teichmüller's theorem it follows that the class of $\pi(\mu)$ in $M(\Gamma_n, \Omega_n)$ possesses a unique extremal element of the form (26) where $0 < k_n < 1$ and η_n is a holomorphic quadratic differential form on Ω_n except for at most simple poles at points of $C_n - \Lambda_n$ and η_n is integrable on Ω_n/Γ_n.

Obviously, $k_n \leqslant k_{n+1} \leqslant k$ for all n.

Lemma 12. *The sequence k_n monotonically increases to k.*

Proof. Consider the mappings w^{ν_n}, where $\nu_n = k_n|\eta_n|/|\eta_n|$. By hypothesis $w^{\nu_n}(x) = w^{\pi(\mu)}(x)$ for all x in C_n. Let w^ν be a normalized limit of some subsequence of w^{ν_n}. Such a limit exists because $\|\nu_n\|_\infty = k_n \leqslant k < 1$ for all n. Moreover, the Beltrami coefficient ν of the limit satisfies $\nu(Az)\overline{A'(z)} = \nu(z)A'(z)$ for all A in Γ. From the fact that the closure of $\bigcup_n C_n$ is $\hat{\mathbb{R}}$, it follows that $w^\nu(x) = w^{\pi(\mu)}(x)$ for all x in $\hat{\mathbb{R}}$. Thus by Lemma 3 in Section 5.4, ν restricted to the upper half plane is equivalent to μ. [Actually, ν restricted to the lower half plane $\mathbb{H}^*$ is trivial in $M(\Gamma, \mathbb{H}^*)$, but ν might not be identically zero in $\mathbb{H}^*$.] By the fact that μ is extremal in its class, the L_∞-norm of ν restricted to $\mathbb{H}$ is bigger than or equal to $\|\mu\|_\infty = k$. But if $k_n \leqslant k - \varepsilon$ for all n and some positive ε, one would have $\|\nu\|_\infty \leqslant k - \varepsilon$, a contradiction. Hence, the lemma follows.

Theorem 3. *For any complex Teichmüller space of a Fuchsian group, the Kobayashi and Teichmüller metrics coincide.*

Proof. From Lemma 4, it suffices to show that given a holomorphic mapping from Δ into $T(\Gamma)$ with $f(0) = 0$ and $f(r) = [\mu]$, where μ is extremal and $0 < r < 1$, then $r \geqslant \|\mu\|_\infty$. The mapping $\pi \circ f$ from Δ into $\tilde{T}(\Gamma_n, \Omega_n)$ is holomorphic and takes 0 into 0 and maps into a finite dimensional Teichmüller space. Therefore, by Theorem 2, $r \geqslant k_n$, where k_n is defined in (26). From Lemma 12 this implies $r \geqslant k$ and this concludes the proof of the theorem.

APPENDIX: A Lemma of Ahlfors

Definition. *Let $\lambda(z)$ be a nonnegative function defined in the unit disk and let $\lambda(z)|dz|$ be a metric. Then $\lambda_0(z)|dz|$ is a supporting metric for $\lambda|dz|$ at a point z_0 if there is a neighborhood V of z_0 such that*

(i) *$\lambda_0(z)$ is of class C^2 for z in V;*

(ii) *$\lambda_0(z) \leqslant \lambda(z) + o(|z - z_0|^2)$; and*

(iii) *$\lambda_0(z_0) = \lambda(z_0)$.*

Ahlfors's Lemma: *Suppose $\lambda(z)|dz|$ is a metric in the unit disk for which $\lambda(z) \geqslant 0$* and for which

(a) *$\lambda(z)$ is upper semicontinuous; and*

(b) *at every point z_0 in the unit disk with $\lambda(z_0) > 0$ there is a supporting metric λ_0 at z_0 with curvature $\leqslant -4$ at z_0* (i.e., $-\lambda_0^{-2}\,\Delta \log \lambda_0$ *evaluated at z_0 is* $\leqslant -4$).

Then $\lambda(z) \leqslant (1 - |z|^2)^{-1}$ for z in the unit disk.

Proof. We first prove the lemma in the case where λ is positive and of class C^2 in an open set containing the closed unit disk and the curvature of λ is $\leqslant -4$ at every point. Then for $\rho(z) = (1 - |z|^2)^{-1}$ we have $\Delta \log \rho = +4\rho^2$ and $\Delta \log \lambda \geqslant 4\lambda^2$. Thus, $\Delta \log \rho - \Delta \log \lambda \leqslant 4(\rho^2 - \lambda^2)$. The function $\log \rho - \log \lambda$ tends to $+\infty$ when $|z| \to 1$ and, therefore, has a minimum in the unit disk. At the point of minimum, $\Delta(\log \rho - \log \lambda) \geqslant 0$ and hence $\rho^2(z) \geqslant \lambda^2(z)$ for all z in the unit disk.

To extend to the case where λ is of class C^2 only in the interior of the unit disk, we replace $\lambda(z)$ by $r\lambda(rz)$ for $r < 1$. Then this new metric is of class C^2 in an open set containing the closed unit disk and has curvature $\leqslant -4$. The above result shows $\rho(z) \geqslant r\lambda(rz)$ and $\rho(z) \geqslant \lambda(z)$ follows by continuity.

Now assume $\lambda(z)$ is not of class C^2 but satisfies only the weaker hypotheses of the lemma. By the same device of replacing $\lambda(z)$ by $r\lambda(rz)$ and using upper semicontinuity, we may assume $\lambda(z)$ satisfies the hypothesis of the lemma in an open set containing the closed unit disk. Thus, we have $\log \rho - \log \lambda$ is lower semicontinuous in the closed unit disk and this function tends to $+\infty$ as $|z| \to 1$. Thus, the existence of a minimum is still assured. The minimum cannot occur at a point where λ is zero because then ρ would take on the value $-\infty$. Assume the minimum occurs at a point z_0 where $\lambda(z_0) > 0$. Then let λ_0 be a supporting metric to λ at z_0 in a neighborhood V. Then $\log \rho - \log \lambda \leqslant \log \rho - \log \lambda_0 + o(|z - z_0|^2)$ with equality when $z = z_0$. Thus $\log \rho - \log \lambda_0 + o(|z - z_0|^2)$ has a minimum in V when $z = z_0$ and so its Laplacian is zero at z_0. Obviously, the Laplacian of the term $o(|z - z_0|^2)$ is zero at z_0. Since λ_0 is C^2 and has curvature $\leqslant -4$, we can apply the earlier result and we obtain $\rho(z_0) - \lambda_0(z_0) = \rho(z_0) - \lambda(z_0) \geqslant 0$. Thus the minimum of $\log \rho - \log \lambda$ occurs at z_0 and is nonnegative there. We conclude that $\lambda(z) \leqslant \rho(z)$ everywhere in the unit disk.

Notes. The deepest part of this chapter is the theorem that Teichmüller's metric equals Kobayashi's metric in finite dimensional spaces, and this theorem is due to Royden [Ro]. The statement that Teichmüller's metric is the integral of its infinitesimal form is due to O'Byrne [O]. O'Byrne's proof depends on a general theorem in Finsler geometry and a theorem on fibrations over Teichmüller space due to Earle and Eells [EE2]. The proof given in Section 7.2 relies on Hamilton's condition for extremality and is due to Gardiner [Ga5].

Many of the details of Royden's proof were clarified in unpublished notes of Earle. Ahlfors's lemma can be found in Ahlfors [Ah6]. The version presented in the appendix to this chapter has a slightly modified definition of supporting metric. This modification and formulation is due to Earle. The extension of Royden's theorem to infinite dimensional Teichmüller spaces was first proved by Gardiner in [Ga5].

The question of whether or not Carathéodory's natural metric for a complex manifold is equal to Teichmüller's metric is partially answered by Kra in [Kr8].

Another natural but distinct metric for Teichmüller space is the Weil–Petersson metric (see the article by Wolpert [Wo6]).

In [Th2] Thurston has introduced yet another important metric on Teichmüller space which is nonsymmetric and which is defined in a way which depends on the hyperbolic structures.

EXERCISES

1. Let M_1 and M_2 be complex manifolds with Kobayashi metrics d_1 and d_2 and suppose f is a holomorphic mapping from M_1 to M_2. Show that $d_2(f(x), f(y)) \leqslant d_1(x, y)$.
2. Define the Carathéodory (pseudo)metric d_C on a complex manifold M as follows:

$$d_C(x, y) = \sup d_\Delta(f(x), f(y))$$

where the supremum is taken over all holomorphic mappings f from M into the unit disk and d_Δ is the noneuclidean metric for the unit disk. Show that d_C is a pseudometric, (it can be degenerate). Show that d_C has contraction property; if f is a holomorphic function from a complex manifold M_1 to a complex manifold M_2 and d_1 and d_2 are the respective Carathéodory metrics, then $d_2(f(x), f(y)) \leqslant d_1(f(x), f(y))$.
3. Show that for any complex manifold the Carathéodory metric is less than or equal to the Kobayashi metric.

4. (Open Problem) Let d be Teichmüller's metric on a finite dimensional Teichmüller space T. Let $S_R = \{P \in T: d(0, P) = R\}$. Determine whether the sphere S_R is strictly convex with respect to Teichmüller geodesics.

8

DISCONTINUITY OF THE MODULAR GROUP

The modular group of a surface is the group of homotopy classes of orientation-preserving homeomorphisms of the surface. This group is also called the mapping class group. It has a natural action on Teichmüller space. In fact, it acts as a group of biholomorphic mappings and a group of isometries in the Teichmüller metric. Except in a few low dimensional cases, the action is faithful. Moreover, if the surface is of finite analytic type, then the modular group acts discontinuously on Teichmüller space. The discontinuity of the action of the modular group is the primary result of this chapter.

In Section 8.1, we define the modular group of a surface and the modular group of a Fuchsian group and show that under appropriate assumptions the two modular groups are isomorphic. In Section 8.2, we define the action of the modular group, describe its isotropy groups, and show that the action is faithful except in a few low dimensional cases. In Section 8.3, we show that knowledge of the absolute traces of a sufficiently large finite, ordered set of words in the generators determines a Fuchsian group up to conjugacy. In Section 8.4, we show that the length spectrum of a finitely generated Fuchsian group is a discrete set. In Section 8.5 we prove the modular group acts discontinuously, and in the last section we show that isotropy subgroups of the modular type are finite.

8.1. DEFINITION OF THE MODULAR GROUP OF A SURFACE

Let Γ be a Fuchsian group acting on the upper half plane $\mathbb{H}$. In this section, we make the following assumptions on Γ:

(i) Γ is of the first kind, that is, its limit set is the whole real axis;
(ii) Γ is finitely generated; and
(iii) Γ is torsion free.

These assumptions assure that the surface $S = \mathbb{H}/\Gamma$ is of type (g, n), where g is the genus and n is he number of punctures of S and $2g - 2 + n > 0$.

Definition. *The Teichmüller modular group* Mod S *of the surface* S *is the group of sense-preserving quasiconformal homeomorphisms of* S *onto itself modulo the normal subgroup of homeomorphisms homotopic to the identity. The extended modular group* $\widetilde{\text{M}}$od S *is the same group except that one allows the homeomorphisms which are orientation-reversing as well as orientation-preserving.*

We now define Mod Γ and $\widetilde{\text{M}}$od Γ in a parallel manner. Let $QC(\Gamma)$ be the group of quasiconformal (orientation-preserving) homeomorphisms w of $\mathbb{H}$ onto itself which satisfy $w \circ \Gamma \circ w^{-1} = \Gamma$. Let $QC_0(\Gamma)$ be the normal subgroup of $QC(\Gamma)$ whose extensions to the real axis satisfy $w(x) = x$ for all x in $\mathbb{R}$. From Proposition 3 section 3.2 we know when Γ is torsion free that a mapping w in $QC_0(\Gamma)$ is homotopic to the identity in $\mathbb{H}$ through a homotopy which is compatible with Γ. The quotient group $QC(\Gamma)/QC_0(\Gamma)$ naturally contains Γ as a normal subgroup, since, for w in $QC(\Gamma)$, $w \circ \Gamma \circ w^{-1} = \Gamma$.

Definition. *The Teichmüller modular group of the Fuchsian group* Γ *is* $[QC(\Gamma)/QC_0(\Gamma)]/\Gamma$. *The extended modular group* $\widetilde{\text{M}}$od Γ *is obtained in the same way except one replaces* $QC(\Gamma)$ *by* $\widetilde{Q}C(\Gamma)$, *the group of all quasiconformal homeomorphisms* w *of* $\mathbb{H}$ *for which* $w \circ \Gamma \circ w^{-1} = \Gamma$, *but where* w *is not necessarily orientation-preserving.*

By using the methods of Sections 3.2 and 5.2, one obtains the following theorem.

Theorem 1. *Let* Γ *be finitely generated, of the first kind and without elliptic elements. Let* $S = \mathbb{H}/\Gamma$. *Then* Mod $\Gamma \cong$ Mod S *and* $\widetilde{\text{M}}$od $\Gamma \cong \widetilde{\text{M}}$od S.

We leave the proof of this theorem as an exercise.

8.2. THE ACTION OF THE MODULAR GROUP ON TEICHMÜLLER SPACE

Even when Γ contains elliptic elements, Mod Γ is defined in the same way as in Section 8.1. The results of Theorem 2 in this section apply whether or not Γ contains elliptic elements.

Let a Beltrami coefficient μ in M(Γ) and an element h in $QC(\Gamma)$ be given. Since $h \circ \Gamma \circ h^{-1} = \Gamma$, it is clear that the Beltrami coefficient σ of $w_\mu \circ h$ is an element of $M(\Gamma)$, where w_μ is the normalized quasiconformal self-mapping of $\mathbb{H}$ with Beltrami coefficient μ. Since $\sigma(z) = (w_\mu \circ h)_{\bar{z}}/(w_\mu \circ h)_z$, the formula for σ is

$$\sigma(z) = \frac{v(z) + \mu(h(z))\theta(z)}{1 + \bar{v}(z)\mu(h(z))\theta(z)}, \tag{1}$$

where $v(z) = h_{\bar{z}}/h_z$ and $\theta(z) = \bar{p}/p$, where $p = \partial h/\partial z$. (See Exercise 16 of Chapter 1.) This gives an action of $QC(\Gamma)$ on $M(\Gamma)$ as a group of holomorphic mappings because, for fixed h, formula (1) depends holomorphically on μ. Using Sections 5.1 and 5.2, one sees that this action induces an action of Mod Γ on $T(\Gamma)$ as a group of biholomorphic mappings.

The sense-reversing elements h of $\tilde{Q}C(\Gamma)$ induce an action on $M(\Gamma)$ by letting $\overline{h^*(\mu)} = (w_\mu \circ h)_z/(w_\mu \circ h)_{\bar{z}}$. On working out the formula analogous to (1), one sees that the dependence of $h^*(\mu)$ on μ is antiholomorphic.

Recall that Teichmüller's metric is defined by

$$d(\mu_1, \mu_2) = \tfrac{1}{2} \inf \log K(w_1 \circ w_2^{-1}), \tag{2}$$

where $K(f)$ is the maximal dilatation of a quasiconformal mapping f. Here w_i has Beltrami coefficient μ_i for $i = 1$ and 2 and the infinum is taken over all Beltrami coefficients μ_1 in a given Teichmüller class and μ_2 in another given Teichmüller class. Since the action of h translates w_1 to $w_1 \circ h$ and w_2 to $w_2 \circ h$, it is clear from (2) that h induces an isometry.

Notice that if an element of Mod Γ is represented by h in QC, then $[h]([0]) = [\sigma]$, where $\sigma = h_{\bar{z}}/h_z$. Let w_σ be the unique quasiconformal self-mapping of $\mathbb{H}$ with Beltrami coefficient σ normalized to fix the points 0, 1, and ∞. Then there exists a Möbius transformation B with $h = B \circ w_\sigma$ and $w_\sigma \circ \Gamma \circ w_\sigma^{-1} = \Gamma_\sigma$, a Fuchsian group, where $B \circ \Gamma_\sigma \circ B^{-1} = \Gamma$. If $[h]([0]) = [0]$, then $[\sigma] = [0]$, which implies $w_\sigma(x) = x$ for all x in $\mathbb{R}$. Thus $w_\sigma \circ A \circ w_\sigma^{-1} = A$ for all A in Γ and, therefore, $h \circ A \circ h^{-1} = B \circ A \circ B^{-1}$ for A in Γ. Thus the automorphism of Γ induced by h is the same as the automorphism induced by B, where B is a Möbius transformation in the normalizer $N(\Gamma)$ of Γ. If B is in Γ, then from the definition of Mod Γ, h induces the identity in Mod Γ. If B is not in Γ, then the element $[h]$ of Mod Γ is induced by a nontrivial conformal self-mapping of $\mathbb{H}/\Gamma$. If Γ contains an elliptic element E with a fixed point at p in $\mathbb{H}$, then $B \circ E \circ B^{-1}$ is elliptic with fixed point at $B(p)$. Since p and $B(p)$ project to the same point on the quotient surface $S = \mathbb{H}/\Gamma$, the conformal self-mapping of S induced by B fixes elliptic points of S. Since an arbitrary point in $T(\Gamma)$ can be translated from the origin in an equivalent Teichmüller space by an isometry, the preceding argument gives the following theorem.

Theorem 2. *The isotropy subgroup of* Mod Γ *which fixes the origin in* $T(\Gamma)$ *is isomorphic to* $N(\Gamma)/\Gamma$, *which is the group of holomorphic, sense-preserving homeomorphisms of* $\mathbb{H}/\Gamma$ *which fix the elliptic points of* $\mathbb{H}/\Gamma$. *The isotropy subgroup of* Mod Γ *which fixes an arbitrary point* $[\mu]$ *in* $T(\Gamma)$ *is isomorphic to the group of holomorphic, sense-preserving homeomorphisms of the surface corresponding to* $[\mu]$ *which fix elliptic points of the deformed surface* $\mathbb{H}/\Gamma_\mu$, *where* $\Gamma_\mu = w_\mu \circ \Gamma \circ w_\mu^{-1}$.

Now assume Γ is torsion free and $\mathbb{H}/\Gamma$ has genus g with n punctures. The fact that $\mathbb{H}/\Gamma$ has hyperbolic structure (positive area in the Poincaré metric) forces

$2g - 2 + n \geqslant 1$. It is an algebraic fact that groups corresponding to surfaces of genus two with no punctures must all be hyperelliptic. As such, they all have a conformal involution. This conformal involution does not change Teichmüller class, and hence every point in Teichmüller space for a surface of genus 2 with no punctures has a nontrivial isotropy group in Mod Γ. Therefore, the action of the modular group is not faithful, by which we mean that distinct elements of Mod Γ do not necessarily induce distinct mappings of $T(\Gamma)$. The same remark is true for any group Γ for which the surface $\mathbb{H}/\Gamma$ is obtained from a hyperelliptic surface of genus two by pinching simple curves to punctures. We call such a group a group of *exceptional type*, by which we mean (g, n) is either $(2, 0)$, $(1, 2)$, $(1, 1)$, $(0, 4)$, or $(0, 3)$.

On the other hand, assume Γ is not of exceptional type and that $h([\mu]) = [\mu]$ for all $[\mu]$ in $T(\Gamma)$ and some element h of the modular group. In particular, $h([0]) = [0]$, and we have just seen that this implies the action of h is equal to the action of a Möbius transformation B in the normalizer of Γ. This means we may take $v = 0$ and $h = B$ in formula (1), and the hypothesis therefore tells us that $\mu^* = \mu(B)\bar{B}'/B'$ is equivalent to μ for every μ. Thus, for $|t| < 1$, $t\mu - t\mu^*$ is tangent to a trivial curve and we conclude that $\mu - \mu^*$ is infinitesimally trivial in the sense that

$$\iint_{\mathbb{H}/\Gamma} \mu(z)\varphi(z)\, dx\, dy = \iint_{\mathbb{H}/\Gamma} \mu(B(z))\overline{B'(z)}B'(z)^{-1}\varphi(z)\, dx\, dy \tag{3}$$

for every holomorphic quadratic differential on $\mathbb{H}/\Gamma$ and every Beltrami coefficient μ. By changing the variable of integration from z to $B(z)$ in the left side of (3), we find that $\varphi(Bz)B'(z)^2 = \varphi(z)$ for every holomorphic quadratic differential φ on $\mathbb{H}/\Gamma$. Whenever there are sufficiently many quadratic differentials, this is impossible unless B is in Γ. Using the Riemann–Roch theorem, one finds that the only cases in which such a B not in Γ can exist for all integrable holomorphic quadratic differential on $\mathbb{H}/\Gamma$ are the cases where Γ is of exceptional type. For the part of the proof depending on the Riemann–Roch theorem we refer to the book of Farkas and Kra [FarK].

These facts are summarized in the following theorem.

Theorem 3. *Let Γ be a finitely generated Fuchsian group of the first kind. Then* Mod Γ *acts as a group of holomorphic, invertible isometries of* $T(\Gamma)$. $\widetilde{\text{Mod}}\ \Gamma$ *acts as a group of invertible isometries, each element acting either holomorphically or antiholomorphically. When Γ is torsion free and not of exceptional type, then* Mod Γ *acts faithfully.*

Remark. From a theorem of Bers and Greenberg, Theorem 1 of the next chapter, it will follow that Mod Γ acts faithfully on $T(\Gamma)$ even when Γ contains elliptic elements provided that the covered surface with elliptic points removed is not of exceptional type.

8.3. MODULI SETS

In Section 1.5 we saw that a hyperbolic Möbius transformation A has two fixed points. If a_0 is its attracting fixed point and b_0 its repelling fixed point, A is determined by the equation

$$\frac{A(z) - b_0}{A(z) - a_0} = \lambda \frac{z - b_0}{z - a_0}, \qquad \text{where} \quad \lambda > 1. \tag{4}$$

If $A(\mathbb{H}) = \mathbb{H}$, then the fixed points a_0 and b_0 must lie on the real axis and the semicircle with endpoints at a_0 and b_0 which meets the real axis orthogonally is the axis of A. This axis is a geodesic in the noneuclidean metric for the upper half plane.

At this point, it is convenient (and standard) to normalize the noneuclidean metric differently from the way we normalized it in Chapter 1. We now let it be $|dz|/y$ instead of $|dz|/2y$. Then for a point p on the axis of A the distance from p to $A(p)$ is $\log \lambda$. See Exercise 7 of Chapter 1. It is called the translation length of A and we denote it by $\ell(A)$. If A is an element of a Fuchsian group Γ, then the geodesic segment which joins p to $A(p)$ along the axis of A projects via π: $\mathbb{H} \to \mathbb{H}/\Gamma$ onto a closed curve. If Γ is torsion free, this curve has minimum length among all curves freely homotopic to it on the surface $\mathbb{H}/\Gamma$.

Since there is a simple formula relating the trace of A to the multiplier of A,

$$(\text{trace } A)^2 = \lambda + \lambda^{-1} + 2,$$

there is also a simple formula relating the absolute value of the trace (hereinafter called the absolute trace) to the translation length of A:

$$\cosh\left(\frac{\ell(A)}{2}\right) = \tfrac{1}{2}|\text{trace}(A)|. \tag{5}$$

We need to know to what extent lengths of closed curves on $\mathbb{H}/\Gamma$ determine the group Γ. The following theorem gives a partial answer to this question if the lengths are associated with specific generators of Γ. The theorem says that if you know the absolute traces of a large enough but finite ordered set of elements of a finitely generated group Γ, then Γ is determined up to conjugation by a Möbius transformation.

Theorem 4. *Let Γ be a Fuchsian group which is generated by $A_1, \ldots, A_n$. Assume A_1 and A_2 are hyperbolic and the axis of A_2 intersects the axis of A_1 from left to right. Then there exists a Möbius transformation D such that the group elements DA_jD^{-1} for $j = 1, 2, \ldots, n$ are real analytic functions of the absolute traces of finitely many words in $A_1, \ldots, A_n$. If Γ is normalized so that A_1 has attracting fixed point at ∞ and repelling fixed point at 0 and A_2 has attracting fixed point 1, then D is the identity.*

Remark 1. This theorem is an extension of a result of Teichmüller [T2].

Remark 2. We call the finite set S_0 of words in the generators, whose existence is assured by the theorem, a moduli set.

Remark 3. If all that is known is the set of absolute traces, without knowing with which words they are associated, the conjugacy class of Γ cannot in general be determined. However, for generic groups it can be determined. This has been shown by Wolpert [Wo1; Wo2].

The following corollary is an obvious consequence of the theorem.

Corollary. *Suppose α is an isomorphism from a Fuchsian group Γ to a Fuchsian group Γ' and that α preserves the intersection of the axes of A_1 and A_2, where A_j, $1 \leqslant j \leqslant n$, are a set of generators as in the theorem. Suppose further that* $|\text{trace } A| = |\text{trace } \alpha(A)|$ *for every A in the moduli set S_0. Then there exists a Möbius transformation D such that $\alpha(A) = DAD^{-1}$ for every A in Γ.*

Proof of Theorem 4. We may assume that the repelling and attracting fixed points of A_1 are 0 and ∞, those of A_2 are a negative number x_1 and 1. This normalization can be achieved by a unique conjugation by a Möbius transformation D. After this conjugation, A_1 is conjugated into a Möbius transformation represented by a matrix

$$A = \begin{pmatrix} m & 0 \\ 0 & m^{-1} \end{pmatrix}, \qquad m > 1. \tag{6}$$

On the other hand, A_2 will be conjugated into a Möbius transformation represented by a matrix B satisfying the relation

$$\frac{B(z) - x_1}{B(z) - 1} = r\,\frac{z - x_1}{z - 1}, \qquad r > 1. \tag{7}$$

A matrix

$$\begin{pmatrix} \alpha & \beta \\ \gamma & \delta \end{pmatrix}$$

for B can be selected so that $\alpha > 0$, $\beta > 0$, $\gamma > 0$, $\delta > 0$, as one can see by solving (7) for α, β, γ, and δ. Since $B(1) = 1$, it is obvious that $\alpha + \beta = \gamma + \delta$.

The number m obviously can be computed from

$$|\text{trace } A| = m + m^{-1}$$

and the numbers α and δ can then be calculated from

$$\alpha + \delta = |\text{trace } B| \quad \text{and} \quad m\alpha + m^{-1}\delta = |\text{trace } AB|.$$

From the fact that 1 is the positive root of the quadratic equation $B(z) = z$, one sees that $2\gamma = \alpha - \delta + \sqrt{(\alpha + \delta)^2 - 4}$. This determines γ and thus $\beta = \gamma + \delta - \alpha$ is also determined. Thus, we see the matrices A and B are determined from the absolute traces of A, B, and AB.

Next, let C be a matrix representing the element that any one of the generators A_j, $j \geqslant 3$, becomes after the conjugation. We may choose C in $SL(2, \mathbb{R})$ so that

$$C = \begin{pmatrix} a & b \\ c & d \end{pmatrix} \qquad \text{with } a + d \geqslant 0, \qquad ad - bc = 1.$$

Now we need to use the identity

$$(\text{trace } P)(\text{trace } Q) = \text{trace } PQ + \text{trace } P^{-1}Q, \tag{8}$$

which is valid for any 2×2 matrices P and Q in $SL(2, \mathbb{C})$. Suppose you know the left side of (8) is positive. If you know the absolute traces on the right side of (8), then you know the term with larger or equal absolute value on the right side of (8) must be positive. From this knowledge, the only remaining unknown trace in equation (8) is also determined.

If a transformation is not elliptic of order 2, we know its trace is nonzero. Therefore, from (8), it is possible to find the traces of AC, BC, and ABC from the absolute traces of AC, $A^{-1}C$, BC, $B^{-1}C$, ABC, and $A^{-1}BC$. To see this, you merely substitute into (8) first $P = A$, $Q = C$, second $P = B$, $Q = C$, and finally $P = A$ and $Q = BC$. Finally, the entries in the matrix C are obtained from the relations

$$\begin{gathered} a + d = |\text{trace } C|, \qquad \lambda a + \lambda^{-1} d = \text{trace } AC \\ \alpha a + \beta c + \gamma b + \delta d = \text{trace } BC \\ \lambda\alpha a + \lambda\beta c + \lambda^{-1}\gamma b + \lambda^{-1}\delta d = \text{trace } ABC. \end{gathered} \tag{9}$$

Thus we can calculate the matrix C, where C is any of the group elements A_j conjugated by D, from absolute traces of certain words in the elements $A_1, \dots, A_n$.

If trace $C = 0$ but trace$(AC) \neq 0$ [or trace$(BC) \neq 0$], we compute as before AC or BC and obtain C.

Finally, if trace $C =$ trace$(AC) =$ trace$(BC) = 0$, then an easy calculation shows that

$$C = \pm\begin{pmatrix} 0 & \sqrt{\beta/\gamma} \\ -\sqrt{\gamma/\beta} & 0 \end{pmatrix}. \tag{10}$$

In all cases, even when trace $C = 0$, there is a solution for the entries of the matrix C in terms of the absolute traces. This completes the proof.

8.4. THE LENGTH SPECTRUM

Definition. *The length spectrum of a Fuchsian group* Γ *is the set* $LS(\Gamma)$ *of all positive numbers* $\log \lambda$, *where* λ *is the multiplier of a hyperbolic element of* Γ [*as defined in formula* (4)].

From the discussion at the beginning of the previous section we know that $LS(\Gamma)$ is the set of all possible noneuclidean lengths of closed geodesics on the surface $\mathbb{H}/\Gamma$.

Theorem 5. *Let* Γ *be any finitely generated Fuchsian group. Then* $LS(\Gamma)$ *is a discrete subset of* $\mathbb{R}$.

Remark 1. We will prove a slightly stronger result. For A in Γ, let $[A]$ denote the conjugacy class of A. We will show that if M is an arbitrary positive number, then the set of $[A]$ for which $\ell(A) \leqslant M$ is a finite set.

Remark 2. We prove the result for groups of both the first and second kind since the proof easily covers both cases.

Proof. The Nielsen kernel of a Fuchsian group Γ is a subset of $\mathbb{H}$. It is obtained by cutting away from $\mathbb{H}$ every half disk whose diameter is contained in the part of the real axis where Γ acts discontinuously. If an open interval in $\mathbb{R}$ with an endpoint at ∞ or $-\infty$ is in the set of discontinuity, then one cuts away a half plane. We introduce the *modified Nielsen kernel* K by cutting away, in addition, every horocycle tangent to a parabolic fixed point which has area one on $\mathbb{H}/\Gamma$. This horocycle is most easily described by applying a conjugation to Γ which takes the given parabolic transformation into the element $z \to z + 1$. Then the horocycle one cuts away is $\{z\colon \operatorname{Im} z \geqslant 1\}$.

From the result on cusp neighborhoods which we will prove in Lemma 1, no two points in $D = \{z\colon \operatorname{Im} z \geqslant 1, 0 \leqslant x \leqslant 1\}$ are identified by an element Γ as long as the transformation $z \mapsto z + 1$ is primitive. Therefore, the image of D in $\mathbb{H}/\Gamma$ is a punctured disk whose area is the area of D in $\mathbb{H}$. This area is 1 because $\int_1^\infty \int_0^1 y^{-2}\, dx\, dy = 1$.

Let ω be a fundamental domain for Γ. Since Γ is finitely generated, ω can be selected so it has a finite number of sides which are geodesics in the hyperbolic metric. Clearly, $\omega \cap K$ is a compact subset of $\mathbb{H}$. Let A be a hyperbolic element of Γ. The axis of A cannot be contained in any parabolic horocycle because then A itself would have to be parabolic. Thus the axis of A meets K. Since ω is a fundamental domain, for some element B in Γ, BAB^{-1} has an axis which meets $\omega \cap K$.

Let A_n represent a sequence of distinct hyperbolic conjugacy classes and select B_n so that the axis of $B_n A_n B_n^{-1}$ has nonempty intersection with $\omega \cap K$. Let p_n be a point in this intersection. Since the sequence p_n is in a compact set, it has a convergent subsequence. Without introducing a new index, we see there is a sequence of distinct hyperbolic conjugacy classes represented by $B_n A_n B_n^{-1}$ for which p_n converges to a point p_0 in the closure of $\omega \cap K$.

If the distances $\log \lambda(A_n)$ are bounded by M, then the points $q_n = B_n A_n B_n^{-1}(p_n)$ lie in the compact subset of points whose noneuclidean distance from $\omega \cap K$ is less than or equal to M. By taking a subsequence, we may assume q_n converges to a point q_0. Since Γ acts discontinuously on $\mathbb{H}$, $q_0 \neq p_0$. Hence the sequence $B_n A_n B_n^{-1}$ would converge to the hyperbolic transformation whose axis passes through p_0 and q_0 and which translates p_0 to q_0. This is a contradiction because it shows Γ is not discrete. Q.E.D.

The following lemma is due to Shimizu [Sh] (see also Leutbecher [Leu]). It can be viewed as a special case of Jørgensen's inequality [Jo] (see Exercise 4 of this chapter).

Lemma 1. *Let Γ be a Fuchsian group containing a parabolic element $A(z) = z + 1$. If*

$$B = \begin{pmatrix} a & b \\ c & d \end{pmatrix}$$

represents element of Γ with $ad - bc = 1$ and with $c \neq 0$, then $|c| \geqslant 1$. Moreover, if A is not equal to a power of any element of Γ, then no two points of the domain $D = \{z\colon y \geqslant 1 \text{ and } 0 \leqslant x < 1\}$ are identified by any element of Γ.

Proof. Assume that B is an element of Γ with $|c| < 1$. Form, inductively, a sequence of parabolic transformations as follows:

$$\begin{aligned} A_1 &= BAB^{-1} \\ A_2 &= A_1 A A_1^{-1} \\ &\vdots \\ A_{n+1} &= A_n A A_n^{-1}. \end{aligned}$$

The sequence A_n is a sequence of distinct elements and we will show that $A_n \to A$, thus contradicting the discreteness of the group Γ. Write

$$A_n(z) = \frac{a_n z + b_n}{c_n z + d_n}$$

where a_n, b_n, c_n, and d_n are in $\mathbb{R}$ and $a_n d_n - b_n c_n = 1$. A computation gives

$$\begin{aligned} A_{n+1} &= \begin{pmatrix} a_n & b_n \\ c_n & d_n \end{pmatrix} \begin{pmatrix} 1 & 1 \\ 0 & 1 \end{pmatrix} \begin{pmatrix} d_n & -b_n \\ -c_n & a_n \end{pmatrix} \\ &= \begin{pmatrix} 1 - a_n c_n & a_n^2 \\ -c_n^2 & 1 + a_n c_n \end{pmatrix} = \begin{pmatrix} a_{n+1} & b_{n+1} \\ c_{n+1} & d_{n+1} \end{pmatrix}. \end{aligned}$$

Thus $c_n = -c^{2^n}$ and $\lim_{n\to\infty} c_n = 0$ since $|c| < 1$. Let M be bigger than the maximum of $(1 - |c|)^{-1}$ and $|a|$. Then $|a_0| = |a| < M$ and, if $|a_n| < M$, then $|a_{n+1}| = |1 - a_n c_n| \leqslant 1 + |a_n||c_n| < 1 + M|c|^{2^n} < 1 + |c|M < M$. Thus, by induction, $|a_n| < M$ for all n and so a subsequence of a_n will converge. From the equation $a_{n+1} = 1 - a_n c_n$, this implies $\lim_{n\to\infty} a_n = 1$. From $b_{n+1} = a_n^2$, it follows that $\lim_{n\to\infty} b_n = 1$. Finally, from $d^{n+1} = 1 + a_n c_n$, we see that $\lim_{n\to\infty} d_n = 1$.

Our next assertion is that Γ cannot contain any element with a fixed point at ∞ except for elements of the form $A^n(z) = z + n$, where n is an integer. We leave this verification to the reader; it depends on the hypothesis that A is not a power of any element of Γ.

We now know that any element

$$B = \begin{pmatrix} a & b \\ c & d \end{pmatrix}$$

of Γ not in the cyclic group generated by A has $|c| \geqslant 1$. Thus B applied to the part of the upper half plane lying above the horizontal line $y = 1$ is the interior of a circle of diameter equal to $c^{-2} \leqslant 1$ which is tangent to the real axis at the point $-d/c$. We conclude that no two points of the strip domain $D = \{z: y \geqslant 1$ and $0 \leqslant x < 1\}$ are identified by any element of Γ.

8.5. THE DISCONTINUITY OF THE MODULAR GROUP

Our aim is to show the modular group acts discontinuously on Teichmüller space. By this we mean that if K is any compact subset of $T(\Gamma)$, then the set of elements h in $\operatorname{Mod}\Gamma$ for which $h(K) \cap K$ is empty is a finite set. From Theorem 3 of Section 8.2, the modular group acts as a group of isometries. Therefore, to show the modular group acts discontinuously depends on showing

(1) isotropy groups are finite; and
(2) orbits are discrete.

Here, we are applying Lemma 2 of Section 1.4. By Theorem 2 of this chapter we know that to show isotropy groups are finite we must show that the group of conformal self-mappings of $\mathbb{H}/\Gamma$ is finite. We will prove this result in the next section if $\mathbb{H}/\Gamma$ has genus g and n punctures and $2g - 2 + n \geqslant 1$.

To show that orbits are discrete we use three ingredients:

(a) the discreteness of the length spectrum as expressed in Theorem 5;
(b) the existence of a finite moduli set as expressed in Theorem 4 and its corollary; and
(c) an inequality for the distortion of the length of a hyperbolic geodesic under the action of a quasiconformal mapping.

The third ingredient is provided by the following lemma of Teichmüller:

Lemma 2. *Suppose A_1 and A_2 are hyperbolic Möbius transformations each with two fixed points and multipliers λ_1 and λ_2, where $\lambda_1 > 1$ and $\lambda_2 > 1$. Suppose w is a quasiconformal homeomorphism of the complex plane with dilatation K and that $w \circ A_1 = A_2 \circ w$. Then*

$$K^{-1} \log \lambda_2 \leqslant \log \lambda_1 \leqslant K \log \lambda_2 \tag{11}$$

Proof. Let Γ_i be the cyclic group generated by A_i. By conjugation, we can assume A_1 and A_2 both have attracting fixed points at ∞ and repelling fixed points at 0. The conjugation will not affect the dilatation of w. The curve family F_1 which joins 0 to ∞ and is invariant under multiplication by λ_1 is taken into the curve family F_2 joining 0 to ∞ and invariant under multiplication by λ_2. Clearly, w transforms the family F_1 into the family F_2. Define the extremal length of F_1 relative to Γ_1 to be the extremal length of the curve family which is the image of F_1 in the torus $(\mathbb{C} - \{0\})/\Gamma_1$. It is easy to show that the extremal length of F_1 is $\log \lambda_1$. Of course, the extremal length of F_2 is $\log \lambda_2$ and, since quasiconformal mappings do not distort extremal length by an amount greater than their dilatation, inequality (11) follows.

We are now ready to show that orbits of points in $T(\Gamma)$ under the action of $\operatorname{Mod}\Gamma$ are discrete. Let S_0 be a moduli set for Γ, whose existence is guaranteed by Theorem 4. For each $\log \lambda_i$ in $LS(S_0)$, let $\log \lambda_i^+$ be the least element of $LS(\Gamma)$ which is larger than $\log \lambda_i$ and let $\log \lambda_i^-$ be the largest element of $LS(\Gamma)$ which is smaller than $\log \lambda_i$. If $\log \lambda_i$ is the smallest element of $LS(\Gamma)$, let $\log \lambda_i^- = 0+$.

Let

$$K_i = \min \left\{ \frac{\log \lambda_i^+}{\log \lambda_i}, \frac{\log \lambda_i}{\log \lambda_i^-} \right\} \tag{12}$$

and $K_0 =$ the minimum of K_i, where $\log \lambda_i$ is in the finite set $LS(S_0)$.

Let h represent an element of $\operatorname{Mod}\Gamma$. Then h determines an automorphism of Γ and, therefore, a permutation of $LS(\Gamma)$. From (11) we know that the dilatation $K(h)$ of the quasiconformal mapping representing h satisfies

$$K(h) \geqslant \max \left\{ \frac{\log \lambda(A)}{\log \lambda(h(A))}, \frac{\log \lambda(h(A))}{\log \lambda(A)} \right\}$$

for all A in Γ. But if $K(h) < K_0$, then since $\log \lambda(h(A))$ is in $LS(\Gamma)$ for all A, we see that $\log \lambda(h(A_i)) = \log \lambda(A_i)$ for each A_i in the moduli set S_0. By Theorem 4, this implies there exists a Möbius transformation D such that h is the automorphism $A \to DAD^{-1}$, and h fixes the origin in $T(\Gamma)$.

We have shown that the origin in $T(\Gamma)$ is not equivalent under $\operatorname{Mod}\Gamma$ to any other point in $T(\Gamma)$ whose Teichmüller distance from the origin is less than $\frac{1}{2} \log K_0$.

The argument clearly translates to any point in $T(\Gamma)$ and, thus, shows that the orbits are discrete sets. Except for showing that isotropy groups are finite, we have shown the following theorem.

Theorem 6. *Let* Γ *be a finitely generated Fuchsian group. Then* $\operatorname{Mod}\Gamma$ *acts discontinuously on* $T(\Gamma)$.

8.6. AUTOMORPHISM GROUPS

To finish the proof of discontinuity as outlined in Section 8.5 we need to show that isotropy groups of the modular group are finite. We know from Theorem 2 that this comes down to showing that, for the Fuchsian group Γ, and its normalizer $N(\Gamma)$ in $\mathrm{PSL}(2, \mathbb{R})$, the quotient group $N(\Gamma)/\Gamma$ is finite. This is true when Γ is finitely generated and contains noncommuting elements. The result is classical and appears in many textbooks; we include the proof here for the benefit of the reader. We first need an elementary lemma.

Lemma 3. *If a Fuchsian group* Γ *is not cyclic, then* $N(\Gamma)$ *is also Fuchsian.*

Proof. Γ must contain two elements A and B which do not have common fixed points, for if they had common fixed points, the discreteness of Γ would imply that A and B are in the same cyclic group. Assume the normalizer $N(\Gamma)$ is not discrete. Then we can find a sequence of distinct elements C_n in $N(\Gamma)$ such that C_n converges to the identity. Consider the sequence $A_n = C_n \circ A \circ C_n^{-1}$ and $B_n = C_n \circ B \circ C_n^{-1}$. We know that A_n and B_n are in Γ and $A_n \to A$ and $B_n \to B$. Thus, since Γ is discrete, there is an integer n_0 such that $n \geqslant n_0$ implies $A_n = A$ and $B_n = B$. Thus C_{n_0} commutes with A and with B. This implies that A, B, and C_n all have common fixed points, and this contradiction proves the lemma.

Theorem 7. *Suppose* Γ *is a finitely generated Fuchsian group whose limit set has more than two points. Then* $N(\Gamma)/\Gamma$ *is a finite group.*

Remark. Although it is unnecessary in our proof, it is a fact that if the limit set contains at least three points, then it automatically contains infinitely many points.

Proof. First assume Γ is of the first kind. Clearly $N(\Gamma)$ is also of the first kind and, by Theorem 2 of Section 1.4, $\mathbb{H}/N(\Gamma)$ has a Riemann surface structure. Thus, there is a ramified covering mapping

$$\pi\colon \mathbb{H}/\Gamma \to \mathbb{H}/N(\Gamma).$$

Since $\mathbb{H}/\Gamma$ and $\mathbb{H}/N(\Gamma)$ are compact surfaces except for at most a finite number of punctures, the analytic mapping π has finite degree, and the degree is the order of $N(\Gamma)/\Gamma$.

Next assume Γ is of the second kind. If we let $\Omega = \mathbb{C} - \{\text{limit set of } \Gamma\}$, then we can apply the same argument to the surfaces Ω/Γ and $\Omega/N(\Gamma)$.

One can find an explicit bound for $N(\Gamma)/\Gamma$ by comparing the Poincaré area of $\mathbb{H}/\Gamma$ to that of $\mathbb{H}/N(\Gamma)$. Obviously,

$$\operatorname{order}(N(\Gamma)/\Gamma) = \operatorname{area}(\mathbb{H}/\Gamma)/\operatorname{area}(\mathbb{H}/N(\Gamma)).$$

For example, if Γ is a fixed point free group covering a surface of genus $g \geqslant 2$, then

$$\operatorname{area}(\mathbb{H}/\Gamma) = 2\pi(2g - 2).$$

Now, if $N(\Gamma)$ is not equal to Γ, it will have elliptic fixed points, but from the Gauss–Bonnet theorem,

$$\operatorname{area}(\mathbb{H}/N(\Gamma)) = 2\pi\left(2\tilde{g} - 2 + \sum_p \left(1 - \frac{1}{\nu(p)}\right)\right) \tag{13}$$

where $\nu(p)$ is the order of an elliptic fixed point p, or $\nu(p) = \infty$ if p is parabolic, and the summation runs over all ramified points p in $\mathbb{H}/N(\Gamma)$. Also $\tilde{g}$ is the genus of $\mathbb{H}/N(\Gamma)$.

By elimination of cases, one finds that the smallest possible value of (13) is $\pi/21$, and hence

$$\operatorname{order}(N(\Gamma)/\Gamma) \leqslant 84(g - 1). \tag{14}$$

For more details see the books by Beardon [Bea] and Farkas and Kra [FarK].

Notes. The first proof of the discontinuity of the action of the modular group appears in Kravetz [Krav]. Bers also gives a proof in [Ber8] and Abikoff gives another proof in [Ab]. The proof given in this chapter is different and gives a way to estimate the Teichmüller distance from a point to the nearest distinct point in its orbit under the action of the modular group. I am indebted to L. Keen for pointing out that the "exceptional type" Riemann surfaces all come from surfaces of genus two or degenerations of such surfaces. A summary of the types of Fuchsian groups for which the modular group does not act faithfully is given by Earle and Kra [EK2]. The result of Theorem 4 is essentially due to Teichmüller. A similar result also appears in the book by Fricke and Klein [FrK]. The formulation given in Theorem 4 is due to Bers [BerGa].

Using a finite ordered set of lengths of closed geodesics to determine a marked Fuchsian group is a classical method used by Fricke and Klein in [FrK] and by Keen in [Ke2] and [Ke3].

There is a famous problem called the Hurwitz–Nielsen realization problem. It concerns whether or not a finite subgroup of the modular group is always a subgroup of the isotropy group of some point in Teichmüller space. This problem was believed to have been solved by Kravetz [Krav]. Later, an essential mistake in Kravetz's proof was found by Masur [Masu1]. The problem was solved in the affirmative by Kerckhoff [Ker] by use of a theorem of Thurston on earthquakes. Wolpert has obtained an independent proof involving the use of geodesics in the Weil–Petersson metric [Wo5].

The problem of deciding to what extent a compact Riemann surface can be determined by its length spectrum is studied by Wolpert [Wo1, Wo2]. This problem is naturally related to questions about Selberg's trace formula for Fuchsian groups. A good introduction to these problems is given by McKean [Mc]. Compactness conditions for spaces of Fuchsian groups are discussed by Matelski in [Mat] and Harvey in [Ha1].

Shimizu's lemma (Lemma 1 of section 8.4) [Sh, Leu] can be regarded as a special case of Jørgensen's inequality (see Exercise 4). Jørgensen's inequality applies to a pair of transformations A and B in PSL(2, $\mathbb{C}$) which generate a discrete group and for which BAB^{-1} is not in the cyclic group generated by A. The inequality says

$$|\mathrm{trace}(A)^2 - 4| + |\mathrm{trace}(ABA^{-1}B^{-1}) - 2| \geqslant 1.$$

For an exposition of this result see Beardon [Bea, chap. V, p. 105]. Brooks and Matelski have generalized Jørgensen's inequality to a whole sequence of inequalities in the traces which are necessary conditions for the group generated by A and B to be nonelementary and discrete [BM]. Important work on the geometric meaning of Jørgensen's inequality for Fuchsian groups has recently been done by Gilman in [Gi1, Gi2].

EXERCISES

1. Let R be a Riemann surface of infinite genus which has a conformal homeomorphism L such that $R/\langle L\rangle$ is a compact surface with genus >1, where $\langle L\rangle$ is the cyclic group generated by L. Here R could be the biinfinite periodic chain shown in Figure 8.1.
   (a) Show that the length spectrum of R is a discrete set.
   (b) Show that R is a point of discontinuity in $T(R)$ with respect to the modular group, in the sense that there exists $\varepsilon > 0$ such that $d(R, h(R)) \geqslant \varepsilon$ for all h in the modular group of R except the identity.
2. Let $R = \mathbb{C} - \{n \pm 1/2^n\colon n \text{ a positive integer}\}$. Show that the origin of the Teichmüller space $T(R)$ is not a point of discontinuity.
3. Let Γ be a finitely generated Fuchsian group and for each A in Γ let $\lambda(A) \geqslant 1$ be the multiplier of A. Show that $\Sigma\lambda(A)^{-2} < \infty$, where the summation is over a single representative of every conjugacy class in Γ.

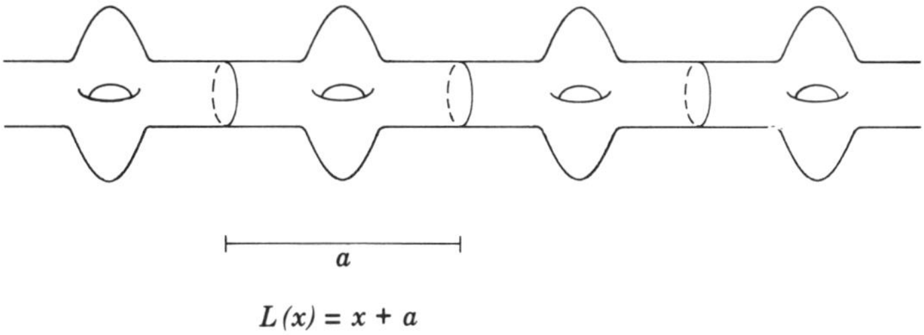

Figure 8.1

4. Let $[AB] = ABA^{-1}B^{-1}$, the commutator of A and B. Assume A and B are Möbius transformations represented by 2×2 matrices with determinant one. Notice that whereas trace (A) is defined only up to plus or minus sign trace, ($[AB]$) is well defined. Jørgensen shows that if A and B generate a discrete subgroup of PSL(2, ℂ), then

$$|\mathrm{trace}(A)^2 - 4| + |\mathrm{trace}([AB]) - 2| \geqslant 1$$

except when BAB^{-1} is in the cyclic group generated by A (see [Jø], [JøK]). From Jørgensen's inequality prove Lemma 1 of Section 8.4.

9

HOLOMORPHIC SELF-MAPPINGS OF TEICHMÜLLER SPACE

A theorem of Royden says that for compact surfaces of genus larger than two the Teichmüller modular group coincides with the full group of biholomorphic self-mappings of Teichmüller space. The case when the genus is two is exceptional because every such surface has a conformal hyperelliptic involution. This prevents the modular group from acting faithfully; however, even when the genus is two it is still true that every biholomorphic mapping is realized by an element of the modular group. Except for a few low dimensional cases occurring when the genus is less than or equal to two, the same theorem is true when the surface is of finite analytic type. The extension of the theorem to surfaces of finite analytic type is due to Earle and Kra [EK1].

To prove the result we use another theorem of Royden, namely, Theorem 2 of Chapter 7, which tells us that for Teichmüller spaces which have complex structure the Teichmüller and Kobayashi metrics coincide. From that theorem, we know that any biholomorphic mapping must induce isometries of the fibers of the tangent and cotangent bundles with respect to the infinitesimal forms of Teichmüller's metric and cometric [see formulas (5) and (13) of Chapter 7]. With this as a starting point, Royden's theorem is obtained by giving an analysis of the differentiability properties of the cometric.

We prove that the cometric is of class C^1 but not, in general, C^2. To show this, we end up considering, for quadratic differentials φ and ψ, real valued functions $f(t) = \|\varphi + t\psi\|$, where the norm is the L^1-norm. The function f is of class $C^{1+\varepsilon}$ but ε becomes smaller and smaller as the order of the highest zero of φ increases. Similarly, a pole of ψ tends to make ε even smaller. A linear isometry of the space of holomorphic quadratic differentials must preserve the Hölder exponent ε, and Royden's theorem comes from exploiting these observations.

This chapter contains several auxiliary results. First, there is a theorem of Bers and Greenberg [BerGr] (also proved independently by Marden [Mar]) which says that Teichmüller space of a finitely generated Fuchsian group of the first kind depends only on the type of the group and not on its signature. Second there is a theorem of D. B. Patterson which tells when two Teichmüller spaces can be biholomorphically equivalent [P]. In fact, if g is the genus and n is the number of punctures, then distinct ordered pairs (g, n) determine nonisomorphic Teichmüller spaces except in certain cases when $g \leqslant 2$ and $n \leqslant 6$. Earle and Kra [EK1] were the first to observe that Royden's technique could be used to prove Patterson's theorem.

9.1. SIGNATURE AND TYPE OF FUCHSIAN GROUPS: A THEOREM OF BERS AND GREENBERG

Let Γ be a finitely generated Fuchsian group of the first kind and let $\mathbb{H}_\Gamma$ be the set of z in $\mathbb{H}$ which are not fixed by any elliptic element of Γ. Then $\mathbb{H}_\Gamma/\Gamma$ is a compact surface of genus g with n punctures. The pair (g, n) is called the *type* of Γ or of $\mathbb{H}_\Gamma/\Gamma$. Let $p_1, \ldots, p_n$ be the punctures of $\mathbb{H}_\Gamma/\Gamma$. The ramification number v_i of p_i is the order of the subgroup of Γ fixing any point of $\mathbb{H}$ that lies over p_i if such a point exists. If no such point exists, p_i is called parabolic and $v_i = \infty$. We have seen in Chapter 1 that when $v_i = \infty$, p_i is represented by a parabolic cusp. (See also Lemma 1, Section 8.4.). The vector

$$(g, n; v_1, \ldots, v_n) \tag{1}$$

is the *signature* of Γ.

It is a known theorem [Leh] that there is a Fuchsian group Γ with signature (1) if, and only if,

$$2\pi\left[2g - 2 + \sum_{i=1}^{n} (1 - v_i^{-1})\right] > 0. \tag{2}$$

The quantity in (2) is the area in the noneuclidean metric of a fundamental domain for Γ and inequality (2) expresses the fact that $\mathbb{H}/\Gamma$ must have positive area.

We always assume that (2) is satisfied and consequently that $2g - 2 + n > 0$. The quantity $2g - 2 + n$ is the negative Euler characteristic of $\mathbb{H}_\Gamma/\Gamma$. If $\mathbb{H}_\Gamma/\Gamma$ is cut along a maximal system of simple closed curves, none of which is homotopic to a puncture or homotopically trivial, then the surface breaks up into $2g - 2 + n$ components each of which has genus zero and three boundary contours. We prove this elementary topological fact in the next chapter, Section 10.1. Notice that for $g \geqslant 2$ or $n \geqslant 5$ inequality (2) is always satisfied.

The next theorem is true for any Fuchsian group and tells us that the structure of Teichmüller space depends only on the type of Fuchsian group and not its signature.

Theorem 1 (Bers and Greenberg [BerGr]). *Let Γ be a Fuchsian group of the first kind acting on $\mathbb{H}$ and let $R = \mathbb{H}_\Gamma/\Gamma$, that is, the Riemann surface obtained by deleting all of the elliptic fixed points from $\mathbb{H}/\Gamma$. Then there is a natural map from $T(R)$ onto $T(\Gamma)$ which is a holomorphic isometry.*

Our proof also works for groups Γ of the second kind in the following general setting. Let C be a closed subset of $\hat{\mathbb{R}}$ invariant under the action of Γ and containing the limit set. Let $T(\Gamma, C)$ be the Teichmüller space of Γ relative to the closed set C, as defined in Chapter 5, and let $T(R, \sigma)$ be the Teichmüller space of the bordered surface R relative to the closed subset σ of its border.

Theorem 1′. *If $R = \mathbb{H}_\Gamma/\Gamma$ and $\sigma = (C - \Lambda)/\Gamma$, then the natural mapping from $T(R, \sigma)$ onto $T(\Gamma, C)$ is an isometry. When $C = \hat{\mathbb{R}}$, this mapping is holomorphic.*

Proofs of Theorems 1 and 1′. We construct the natural mapping from $T(R, \sigma)$ onto $T(\Gamma, C)$ just as we did in Lemma 2 of Section 5.2. The fact that a homotopy of a self-mapping of R relative to the subset σ of its border lifts to a homotopy of H_Γ relative to C, and conjugating Γ to itself is proved in Proposition 2 of Section 3.2. Thus we have a surjective mapping $\tilde{\Phi}$: $M(R) \to M(\Gamma)$ which induces a well-defined, surjective mapping Φ from $T(R, \sigma)$ to $T(\Gamma, C)$.

Our strategy is to use the necessity and sufficiency of the Hamilton–Krushkal condition (Chapter 6) to show that Φ is an isometry. First we observe that Φ is functorial with respect to translation mappings between Teichmüller spaces. Specifically, let Γ_1 and Γ_2 be Fuchsian groups acting on $\mathbb{H}$ and let h be a quasiconformal homeomorphism of $\mathbb{H}$ which conjugates Γ_1 into Γ_2, that is $h \circ \Gamma_1 \circ h^{-1} = \Gamma_2$. Moreover assume C_1 and C_2 are closed invariant sets of the extended real axis containing the limit sets Λ_1 and Λ_2 of Γ_1 and Γ_2, respectively, with $h(C_1) = C_2$.

Let $\sigma_j = (C_j - \Lambda_j)/\Gamma_j$ for $j = 1$ and 2. The sets σ_j are closed subsets of the borders of the surfaces $R_j = \mathbb{H}/\Gamma_j$.

The mapping h induces an isometry $\tilde{S}$ of $T(\Gamma_2, C_2)$ onto $T(\Gamma_1, C_1)$ by associating to a Beltrami coefficient ν in $M(\Gamma_2)$ the Beltrami coefficient of $w_\nu \circ h$ in $M(\Gamma_1)$. It also induces an isometry S from $T(R_2, \sigma_2)$ onto $T(R_1, \sigma_1)$ in a similar manner. The diagram in Figure 9.1 is obviously commutative where Φ_1 and Φ_2 are the mappings defined analogously to Φ.

Let d_1 and d_2 be the Teichmüller metrics for $T(R_1, \sigma_1)$ and $T(R_2, \sigma_2)$, respectively, and let d_{Γ_1} and d_{Γ_2} be the Teichmüller metrics for $T(\Gamma_1, C_1)$ and $T(\Gamma_2, C_2)$. Note that if $w \circ h(z) = z$ and w has Beltrami coefficient μ, then

$$\begin{array}{ccc} T(\Gamma_2, C_2) & \xrightarrow{\tilde{S}} & T(\Gamma_1, C_1) \\ \uparrow \Phi_2 & & \uparrow \Phi_1 \\ T(R_2, \sigma_2) & \xrightarrow{S} & T(R_1, \sigma_1). \end{array}$$

Figure 9.1

$S[\mu] = [0]$. Therefore,

$$d_2(\mu, v) = d_1(0, S(v)).$$

If we know that $d_1(0, S(v)) = d_{\Gamma_1}(0, \Phi_1 \circ S(v))$, then by the commutativity of the above diagram we obtain

$$d_1(0, S(v)) = d_{\Gamma_2}(0, \tilde{S} \circ \Phi_2(v)) = d_{\Gamma_2}(\tilde{S}^{-1}(0), \Phi_2(v))$$
$$= d_{\Gamma_2}(\Phi_2(\mu), \Phi_2(v)).$$

We conclude that $d_2(\mu, v) = d_{\Gamma_2}(\Phi(\mu), \Phi(v))$. Thus, in order to show that Φ is an isometry, it suffices to show that

$$d_\Gamma(0, \Phi(\mu)) = d(0, \mu)$$

for every μ in R.

Assume that μ is extremal for its class in $T(R, \sigma)$. Then $\|\mu\|_\infty = k$ and $d(0, \mu) = \frac{1}{2}\log((1 + k)/(1 - k))$. From the necessity of Hamilton's condition, this implies

$$k = \sup \operatorname{Re} \iint_R \mu\varphi \, dx \, dy,$$

where the supremum is over all holomorphic quadratic differentials φ on R satisfying

$$\|\varphi\| = \iint_R |\varphi(z)| \, dx \, dy = 1$$

and φ is real with respect to real boundary uniformizers along any part of the border of R in the complement of σ. However, from Theorem 1 of Section 3.1, this supremum is the same as

$$\sup \operatorname{Re} \iint_{\mathbb{H}/\Gamma} \pi^*(\mu) q \, dx \, dy,$$

where $\pi^*(\mu)$ is the lift of the Beltrami coefficient μ to $\mathbb{H}$ and where the supremum is over all holomorphic quadratic differential forms q for which

$$\|q\| = 1 \quad \text{and} \quad q \text{ is real valued on } \hat{\mathbb{R}} - C.$$

From the sufficiency of Hamilton's condition, this implies that $\pi^*(\mu)$ is extremal and, therefore,

$$d_\Gamma(0, \Phi([\mu])) = \frac{1}{2}\log\frac{1+k}{1-k}.$$

Q.E.D.

9.2. ROYDEN'S THEOREM ON ISOMETRIES

From the previous section, when $\mathbb{H}_\Gamma/\Gamma$ is of finite analytic type, the group of isometries of $T(\Gamma)$ depends only on the type of Γ and not on its signature. It is fairly easy to see that if Γ and Γ' have the same type and if Γ has no elliptic elements, then $\operatorname{Mod}\Gamma'$ is contained in $\operatorname{Mod}\Gamma$ in a natural way. However, $\operatorname{Mod}\Gamma'$ will, in general, be a proper subgroup of $\operatorname{Mod}\Gamma$ because elements of $\operatorname{Mod}\Gamma'$ can permute punctures on H/Γ' only if those punctures correspond to elliptic elements of the same order.

We observed in Theorem 2 of Chapter 8 that every element of $\operatorname{Mod}\Gamma$ gives a biholomorphic isometry of either of the Teichmüller spaces $T(\Gamma)$ or $T(\Gamma')$. Royden's theorem gives a converse to this statement. Let aut $T(\Gamma)$ denote the full group of biholomorphic automorphisms of $T(\Gamma)$. When Γ contains no elliptic elements and has type (g, n), often we will write $\operatorname{Mod}(g, n)$ in place of $\operatorname{Mod}\Gamma$.

Theorem 2. *Suppose Γ is a finitely generated Fuchsian group of the first kind with no elliptic elements. Suppose further that Γ is not one of the exceptional types* (0, 3), (0, 4), (1, 1), (1, 2), or (2, 0). *Then* aut $T(\Gamma) = \operatorname{Mod}\Gamma$. *In the exceptional cases we have:*

for (0, 3), aut $T(\Gamma) = \{id\}$;
for (0, 4) *and* (1, 1), aut $T(\Gamma) \cong \mathrm{PSL}(2, \mathbb{R})$;
for (1, 2), aut $T(\Gamma) = \operatorname{Mod}(0, 5)$; *and*
for (2, 0), aut $T(\Gamma) = \operatorname{Mod}(0, 6) \cong \operatorname{Mod}(2, 0)/\mathbb{Z}_2$.

We will not complete the proof of this theorem until the end of Section 9.6.

9.3. THE SMOOTHNESS OF TEICHMÜLLER'S METRIC

Let μ be in $M(\Gamma)$ and ν in $L_\infty(\Gamma)$. Form the quasifuchsian group $\Gamma^\mu = w^\mu \circ \Gamma \circ (w^\mu)^{-1}$, where w^μ has Beltrami coefficient μ in the upper half plane and zero in the lower half plane. The infinitesimal form F of Teichmüller's metric can be written in a way which depends on the holomorphic quadratic differential forms for the quasifuchsian group Γ^μ:

$$F([\mu], \nu) = \sup\left|\iint \varphi(w)\left[\frac{\nu}{1-|\mu|^2}\cdot\frac{1}{\theta}\right] du\, dv\right|, \tag{3}$$

where the supremum is over all φ in $A(\Gamma^{\mu})$ with $\|\varphi\| = 1$ and the integral is over $w^{\mu}(\mathbb{H})/\Gamma^{\mu}$. Here $\theta = \bar{p}/p$, where $p = \partial w^{\mu}/\partial z$ and $w = w^{\mu} = u + iv$.

We assume Γ is finitely generated and of the first kind, which implies the dimension of $A(\Gamma^{\mu})$ is $3g - 3 + n$, where g is the genus and n is the total number of punctures (either elliptic or parabolic) on $\mathbb{H}_{\Gamma}/\Gamma$. From Section 9.1 we know that if Γ and Γ' have the same type, then $T(\Gamma)$ is isometric to $T(\Gamma')$. Therefore, in order to prove smoothness properties of the metric, we may assume the punctures are elliptic or parabolic, whichever assumption works to our advantage. In this section it is helpful to assume all punctures are elliptic and hence that the group Γ has a compact fundamental domain.

The metric F in (3) is dual to the cometric G given by

$$G([\mu], \varphi) = \iint_{\mathbb{H}/\Gamma^{\mu}} \frac{|\varphi|\, du\, dv}{1 - |\mu|^2} = \iint_{\mathbb{H}/\Gamma} |\varphi(w(z)) w_z^2(z)|\, dx\, dy, \tag{4}$$

where φ is in $A(\Gamma^{\mu})$ and $w = w^{\mu} = u + iv$. Since Teichmüller space is a manifold, the cotangent bundle to Teichmüller space lying over a neighborhood U of the origin can be trivialized in the form $U \times A(\Gamma)$. This trivialization induces a holomorphic family of complex linear isomorphisms $L_{\mu}: A(\Gamma) \to A(\Gamma^{\mu})$, where L_{μ} depends only on the Teichmüller class of μ and L_0 is the identity.

If we let $\tilde{G}$ be the cometric G expressed in terms of this trivialization, then for $[\mu]$ in U and for φ in $A(\Gamma)$ we have

$$\tilde{G}([\mu], \varphi) = \iint_{H/\Gamma} |L_{\mu}(\varphi)(w(z)) w_z^2(z)|\, dx\, dy.$$

In order to show that the cometric is of class C^1 except at points on the zero section (where $\varphi = 0$), we develop formulas for the partial derivatives

$$\frac{d}{dt} \tilde{G}([\mu + t\Delta\mu], \varphi)|_{t=0} \tag{5}$$

and

$$\frac{d}{dt} \tilde{G}([\mu], \varphi + t\Delta\varphi)|_{t=0} \tag{5'}$$

and show that they are continuous functions of $[\mu]$ and φ, provided that $[\mu]$ is in a sufficiently small neighborhood of the origin and φ is not identically zero. We need two lemmas.

Lemma 1. *Suppose φ and ψ are integrable quadratic differentials on a Riemann surface R and suppose $\varphi(z) \neq 0$ almost everywhere. Let*

$$f(t) = \iint_R |\varphi + t\psi|\, dx\, dy$$

Then f has a derivative at $t = 0$ and

$$f'(0) = \operatorname{Re} \iint_R \psi \frac{|\varphi|}{\varphi} \, dx \, dy.$$

Proof. We need to find $\lim_{t \to 0} (f(t) - f(0))/t$. Note that

$$\left| \frac{|\varphi + t\psi| - |\varphi|}{t} \right| \leqslant |\psi|$$

and so by dominated convergence the limit for the derivative $f'(0)$ can be taken under the integral. Also

$$\frac{|\varphi + t\psi| - |\varphi|}{t} = \frac{|\varphi + t\psi|^2 - |\varphi|^2}{t(|\varphi + t\psi| + |\psi|)} = \frac{2 \operatorname{Re} \bar{\varphi}\psi + t|\psi|^2}{|\varphi + t\psi| + |\varphi|}$$

and this quotient approaches $\operatorname{Re}(\bar{\varphi}\psi/|\varphi|$ whenever $\varphi(z) \neq 0$. The lemma follows.

We need another preliminary result concerning quasiconformal mappings with harmonic Beltrami differentials. A harmonic Beltrami differential is a Beltrami differential which can be written as the complex conjugate of a holomorphic quadratic differential divided by the square of the noneuclidean metric.

Lemma 2. *Let μ and $\Delta\mu$ be harmonic Beltrami differentials defined in the upper half plane $\mathbb{H}$. Assume that $\|\mu\|_\infty < 1$ and $\|\mu + \Delta\mu\|_\infty < 1$ and w^μ is a normalized quasiconformal homeomorphism of the sphere with Beltrami coefficient μ in the upper half plane and zero in the lower half plane. Then $w_z^{\mu + \Delta\mu}$ converges uniformly on compact subsets of $\mathbb{H}$ to w_z^μ as $\|\Delta\mu\|_\infty \to 0$.*

Proof. We assume w^μ and $w^{\mu + \Delta\mu}$ have the same normalization. The result of the lemma is independent of which normalization is chosen provided the normalization for w^μ and $w^{\mu + \Delta\mu}$ is the same. From the Ahlfors–Weill extension lemma (Lemma 7 of Chapter 5), w^μ can be expressed for z in the upper half plane as

$$w^\mu(z) = \frac{\eta_1(z) + (z - \bar{z})\eta_1'(\bar{z})}{\eta_2(\bar{z}) + (z - \bar{z})\eta_2'(\bar{z})},$$

where η_1 and η_2 are two solutions to the ordinary differential equation $\eta'' = -\frac{1}{2}\varphi\eta$ normalized so that $\eta_1'\eta_2 - \eta_2'\eta_1 = 1$ and where $\mu(z) = -2y^2\varphi(\bar{z})$. A simple calculation yields

$$w_z^\mu(z) = (\eta_2(\bar{z}) + (z - \bar{z})\eta_2'(\bar{z}))^{-2}.$$

The conclusion now follows from the fact that the solutions to the ordinary differential equation $\eta'' = -\frac{1}{2}\varphi\eta$ depend continuously on the coefficient φ.

It is now possible to calculate the derivatives indicated in formulas (5) and (5′). We need notations for derivatives with respect to t at $t = 0$ of each of the expressions $L_{\mu+t\Delta\mu}$, $w^{\mu+t\Delta\mu}$, and $(w_z^{\mu+t\Delta\mu})^2$. Each of these functions depends holomorphically on t. We write

$$L_{\mu+t\Delta\mu} = L_\mu + tL' + O(t^2), \tag{6}$$

$$w^{\mu+t\Delta\mu} = w^\mu + tF + O(t^2), \tag{7}$$

$$(w_z^{\mu+t\Delta\mu}/w_z^\mu)^2 = 1 + tH + O(t^2). \tag{8}$$

In (6) the estimate is with respect to any norm for linear mappings between finite dimensional vector spaces. In (7) the estimate is uniform on compact subsets of the plane (if w^μ is normalized at 0, 1, and ∞) and F is a continuous function, which, after composition with (w^μ), is the same as the function F in Theorem 5 of Chapter 1. In (8), H is a continuous function determined by the Taylor expansion of $(w_z^{\mu+t\Delta\mu})^2$, which is a holomorphic function of t. It is important to note that the operator L' and the functions F and H depend holomorphically on μ.

The expression

$$(L_{\mu+t\Delta\mu}\varphi)(w^{\mu+t\Delta\mu})(w_z^{\mu+t\Delta\mu}/w_z^\mu)^2$$

can be written as

$$L_\mu\varphi + t\psi + O(t^2),$$

where

$$\psi = L'\varphi + (L_\mu\varphi)'F + (L_\mu(\varphi))H$$

and after a calculation one can see that the partial derivative

$$\frac{d}{dt}\tilde{G}([\mu + t\Delta\mu], \varphi)|_{t=0}$$

is equal to

$$\operatorname{Re} \iint\limits_{H/\Gamma} \varphi(w^\mu(z)) \frac{|L_\mu(\varphi)(w^\mu(z))|}{L_\mu(\varphi)(w^\mu(z))}\, dx\, dy. \tag{9}$$

In this calculation, we need the basic formula from Lemma 1. We see that the derivative is continuous in its dependence on μ and φ provided that φ is not identically zero and $[\mu]$ stays in a sufficiently small neighborhood of the origin in Teichmüller space where it can be represented by a harmonic Beltrami differential.

The calculation of the partial derivative in the variable φ of $\tilde{G}([\mu], \varphi)$ is similar and we omit it. We proceed to the next theorem.

Theorem 3. *Let Γ be a finitely generated Fuchsian group of the first kind. Then the metric F on the tangent bundle to $T(\Gamma)$ and the cometric G on the cotangent bundle to $T(\Gamma)$ are both of class C^1 except at points of the zero section.*

Proof. Since we have shown that the cometric is of class C^1, this theorem is a consequence of the following lemma from the calculus of variations.

Lemma 3. *Let U be an open set in $\mathbb{R}^n$ and $G(x, \eta)$ a continuous function on $U \times \mathbb{R}^n$, which for each x in U is a positive convex homogeneous function of η in $\mathbb{R}^n$. Define $F(x, \xi)$ on $U \times (\mathbb{R}^n)^*$, where $(\mathbb{R}^n)^*$ is the dual vector space to $\mathbb{R}^n$, by*

$$F(x, \xi) = \sup\{|\xi(\eta)|: G(x, \eta) = 1\}. \tag{10}$$

That is, F is the metric which is dual to G. Then F is continuous and, for each x, F is a positive convex homogeneous function of ξ. If G is of class C^1, in η for $\eta \neq 0$, then F is strictly convex in ξ. If G is strictly convex and has continuous first derivatives with respect to x, then $F(x, \xi)$ is C^1 except at $\xi = 0$. Moreover, G is the metric which is dual to F.

Proof. The last statement is the most trivial, because a Banach space embeds isometrically into its double dual and, for finite dimensional spaces, this isometry is surjective. The assertion that it is an isometry is simply the statement that the dual norm to F is G.

It is obvious that F is a positive convex homogeneous function, and our first step is to show that F is continuous. Let $S^{n-1} = \{\eta \text{ in } \mathbb{R}^n: \sum_{i=1}^n \eta_i^2 = 1\}$ and let

$$f_\eta(x, \xi) = \xi(\eta)/G(x, \eta) \tag{11}$$

so that $F(x, \xi) = \sup\{f_\eta(x, \xi): \eta \in S^{n-1}\}$. Given x in U choose η_0 in S^{n-1} so that $F(x, \xi) = f_{\eta_0}(x, \xi)$. Then for x' in U and ξ' in S^{n-1}, we have

$$F(x', \xi') \geqslant f_{\eta_0}(x', \xi')$$

and therefore

$$F(x, \xi) = f_{\eta_0}(x, \xi) = f_{\eta_0}(x', \xi') + (f_{\eta_0}(x, \xi) - f_{\eta_0}(x', \xi'))$$

$$\leqslant F(x', \xi') + \sup_{\eta \in S^{n-1}} |f_\eta(x, \xi) - f_\eta(x', \xi')|. \tag{12}$$

On reversing the roles of (x, ξ) and (x', ξ') and combining the two resulting inequalities, we find that

$$|F(x, \xi) - F(x', \xi')| \leqslant \sup_{\eta \in S^{n-1}} |f_\eta(x, \xi) - f_\eta(x', \xi')|. \tag{13}$$

The joint continuity of f in η, x, and ξ and the fact that η varies over a compact set imply that the right side of (13) can be made arbitrarily small if (x', ξ') is near enough to (x, ξ). Thus, F is continuous.

Next suppose G is of class C^1 in η for $\eta \neq 0$ but $F(x, \xi)$ is not strictly convex. In this part of the lemma, the dependence on x is unimportant and so we simply write $F(x, \xi) = \|\xi\|$ and $G(x, \eta) = \|\eta\|$. If $\|\xi\|$ is not strictly convex, we can find two linearly independent vectors ξ_1 and ξ_2 with $\|\xi_1\| = \|\xi_2\| = 1$ and $\|\xi_1 + \xi_2\| = 2$. This means there exists η_0 with $\|\eta_0\| = 1$ and $(\xi_1 + \xi_2)(\eta_0) = 2$. Since $\xi_1(\eta_0) \leqslant 1$ and $\xi_2(\eta_0) \leqslant 1$, we have $\xi_1(\eta_0) = \xi_2(\eta_0) = 1$. Because we assume $\|\eta\|$ is C^1 when $\eta \neq 0$, the set $\{\eta\colon \|\eta\| = 1\}$ is a manifold and has a tangent plane of dimension $n - 1$ at η_0. Now $\xi_1(\eta) \leqslant 1$ at every point on this manifold and $\xi_1(\eta_0) = 1$, and so $\xi_1(\eta) = 0$ for every η in the tangent plane. The same statement applies to ξ_2 and, since ξ_1 and ξ_2 are independent, this forces the tangent plane to have codimension at least two. This contradiction shows that F is strictly convex.

Suppose G is strictly convex, that is, $\|\eta\|$ is strictly convex. Let $\|\eta_1\| = \|\eta_2\| = 1$. The fact that these vectors are in a finite dimensional space implies uniform convexity, by which we mean for every $\varepsilon > 0$ there is a $\delta > 0$ such that $\|\eta_1 + \eta_2\| \geqslant 2 - \delta$ implies $\|\eta_1 - \eta_2\| < \varepsilon$. By strict convexity we mean that, for $0 < t < 1$, $\|t\eta_1 + (1 - t)\eta_2\| < 1$. We leave as an exercise the proof that in a finite dimensional space strict convexity implies uniform convexity.

Our next step is to show that the dual norm $\|\xi\|$ is smooth in the following sense: For every $\varepsilon > 0$, there exists a $\delta > 0$ such that if $\|\xi_1 - \xi_2\| < \delta$ then

$$\|\xi_1 + \xi_2\| \geqslant \|\xi_1\| + \|\xi_2\| - \varepsilon\|\xi_1 - \xi_2\|, \tag{14}$$

so long as $\|\xi_1\| \geqslant 1$ and $\|\xi_2\| \geqslant 1$. To prove (14) let η_1 and η_2 be elements of the unit ball such that $\xi_1(\eta_1) = \|\xi_1\|$ and $\xi_2(\eta_2) = \|\xi_2\|$. It is then elementary to verify that

$$\|\xi_1 + \xi_2\| \geqslant \|\xi_1\| + \|\xi_2\| - \|\eta_1 - \eta_2\|\, \|\xi_1 - \xi_2\|. \tag{15}$$

To prove (14) let $\|\xi_1 - \xi_2\| < \delta$. Then clearly $|(\xi_1 - \xi_2)(\eta_1)| < \delta$ and $|(\xi_1 - \xi_2)(\eta_2)| < \delta$. Since $\xi_1(\eta_1) = 1$ and $\xi_2(\eta_2) = 1$, this implies $\xi_1(\eta_2) > 1 - \delta$ and $\xi_2(\eta_1) > 1 - \delta$. Consequently,

$$(\xi_1 + \xi_2)\left(\frac{\eta_1 + \eta_2}{2}\right) = 1 + \tfrac{1}{2}\xi_1(\eta_2) + \tfrac{1}{2}\xi_2(\eta_1) \geqslant 2 - \delta.$$

Therefore, $\|\eta_1 + \eta_2\| \geqslant 2 - \delta$, which from uniform convexity implies $\|\eta_1 - \eta_2\| < \varepsilon$. Plugging this into (15), we obtain (14).

Our next objective is to show the function

$$f(t) = \|\xi + th\|, \tag{16}$$

where $\|\xi\| = 1$, is differentiable at the value $t = 0$. Since it is a real valued convex function of t, it has a derivative from the right and from the left. Let

$$H(t) = (\|\xi + th\| - \|\xi\|)/t.$$

We claim that $\lim_{t\downarrow 0}(H(t) - H(-t)) = 0$ and hence the right and left derivatives agree. But

$$H(t) - H(-t) = (\|\xi + th\| + \|\xi - th\| - 2)/t. \tag{17}$$

By substituting $\xi_1 - \xi_2 = h$ and $\xi_1 + \xi_2 = \xi$, inequality (14) can be recast in the following form: For every $\varepsilon > 0$, there exists $\alpha > 0$ such that $\|\xi\| = 1$ and $\|h\| < \alpha$ imply

$$\|\xi + h\| + \|\xi - h\| \leqslant 2 + \varepsilon\|h\|. \tag{18}$$

Since the numerator on the right side of (17) is nonnegative, from (18) we obtain

$$0 \leqslant H(t) - H(-t) \leqslant \varepsilon\|h\|.$$

Since $\varepsilon > 0$ is arbitrary, the differentiability is proved.

The next observation is that the function $f(t)$ in (16) satisfies

$$f'(0) = h(\eta), \tag{19}$$

where η is the unique element of the unit ball for which

$$\xi(\eta) = \|\xi\|. \tag{20}$$

Moreover, the mapping which assigns to any ξ with $\|\xi\| = 1$, the unique η for which (20) holds is a continuous mapping. By strict convexity, this mapping is well defined and continuity follows from the same argument used to prove (14). To prove (19), let η_t be the unique element of norm 1 such that

$$(\xi + th)(\eta_t) = \|\xi + th\|. \tag{21}$$

Let $\|\xi\| = 1$. Now

$$H(t) = \frac{(\xi + th)(\eta_t) - 1}{t} = \frac{\xi(\eta_t) - 1}{t} + h(\eta_t). \tag{22}$$

We know $\lim_{t\to 0} H(t)$ and $\lim_{t\to 0} h(\eta_t)$ both exist and hence from (22)

$$\lim_{t\to 0} \frac{\xi(\eta_t) - 1}{t} \tag{23}$$

exists. Since the numerator is nonpositive in (23), the fraction is nonpositive for $t > 0$ and nonnegative for $t < 0$. Thus the limit must be zero.

We now know that $\|\xi + h\|$ is differentiable in the variable h and has derivative $h(\eta)$, where η is the unique element in the unit ball for which $\xi(\eta) = \|\xi\|$. Moreover, η depends continuously on ξ, so this function is C^1.

To conclude the proof, we return to the function $F(x, \xi)$ and show that it is C^1 in its dependence on both variables. Let $\|\xi\| = 1$ and $\|\eta\| = 1$. Since $G(x, \eta)$ is assumed to be C^1 in both variables, using the notation of formula (11) we have

$$f_\eta(x + \Delta x, \xi) = f_\eta(x, \xi) + L_{x,\eta}(\Delta x) + o(\Delta x) \tag{24}$$

where $L_{x,\eta}$ is a linear map acting on the vector Δx and $L_{x,\eta}$ depends continuously on x and η.

A word of explanation concerning the meaning of the "little oh" notation is in order. The quantity $o(\Delta x)$ satisfies

$$|o(\Delta x)| \leqslant \varepsilon \|\Delta x\|$$

and ε approaches zero as $\Delta x \to 0$. In (24), the number ε approaches zero uniformly for all ξ and η in the unit sphere and all x in a compact neighborhood. This follows from the assumption that $G(x, \eta)$ is a C^1 function.

Let ξ with $F(x, \xi) \neq 0$ be given. Let $\eta(\Delta x)$ be the unique vector η of norm 1 for which $f_\eta(x + \Delta x, \xi) = F(x + \Delta x, \xi)$ and let η_0 be the unique vector of norm 1 for which $f_{\eta_0}(x, \xi) = F(x, \xi)$. Clearly, $\eta(\Delta x) \to \eta_0$ as $\Delta x \to 0$. The reason for this is that we know that $\eta(\Delta x)$ has a limit point (since it varies on a compact sphere) and if a limit point differed from η_0 we would get a contradiction of the strict convexity of $G(x, \eta)$.

Hence, the derivatives $L_{x,\eta}(\Delta x)$ in (24) converge uniformly and we have, for $\|\xi\| = 1$,

$$F(x + \Delta x, \xi) = F(x, \xi) + L_{x,\eta_0}(\Delta x) + o(\Delta x), \tag{25}$$

where $o(\Delta x)$ is uniform for $\|\xi\| = 1$ and x in some compact neighborhood and L_{x,η_0} depends continuously on x and η_0 depends continuously on ξ. Therefore, the first derivatives of $F(x, \xi)$ in the variable x are continuous jointly in x and ξ.

For a vector η, let $\hat{\eta}$ be the image of η under the natural mapping of a vector space into its double dual. So, by definition, $\hat{\eta}(\xi) = \xi(\eta)$. Observe that

$$F(x + \Delta x, \xi + \Delta\xi) - F(x, \xi + \Delta\xi) + F(x, \xi + \Delta\xi) - F(x, \xi)$$
$$= L_{x,\eta_0}(\Delta x) + (L_{x,\eta+\Delta x} - L_{x,\eta_0})(\Delta x) + o(\Delta x) + \hat{\eta}_0(\Delta\xi) + o(\Delta\xi).$$

Since the linear map $L_{x,\eta}$ depends continuously on (x, η) and η_0 depends continuously on ξ, we finally end up with

$$F(x + \Delta x, \xi + \Delta\xi) = F(x, \xi) + L_{x,\eta_0}(\Delta x) + \hat{\eta}_0(\Delta\xi) + o(\Delta x) + o(\Delta\xi)$$

and we see that F is C^1 in both variables as long as $\xi \neq 0$.

9.4. THE NONSMOOTHNESS OF TEICHMÜLLER'S METRIC

We have seen that the smoothness properties of Teichmüller's cometric depend on the smoothness of the function

$$f(t) = \iint_\omega |\varphi + t\psi| \, dx \, dy.$$

From this point on we assume both differentials φ and ψ are holomorphic on the surface R which is of finite analytic type. At the punctures, which are points of the completion of R but not in R, φ and ψ may have poles. We shall see that $f(t)$ has a continuous second derivative if φ has only simple zeros; however, higher-order zeros of φ cause the second derivative of $f(t)$ not to exist and the higher the order of the zero the less smooth is the first derivative of $f(t)$.

Select a conformal disk $\Delta \subset \mathbb{H}/\Gamma$ and a local parameter z such that $\Delta = \{z: |z| < 1\}$. Assume φ and ψ are holomorphic in $\bar{\Delta} - \{0\}$, and let

$$g(t) = \iint_\Delta |\varphi + t\psi| \, dx \, dy.$$

Clearly one can write $f(t)$ as a finite sum of functions of the form $g(t)$ defined with respect to appropriately defined conformal disks. Thus smoothness properties for $g(t)$ imply smoothness properties for $f(t)$. We adopt the convention that pole of order k is a zero of order $-k$.

Theorem 4. *Let $g(t)$ be the function defined above where φ and ψ are holomorphic and nonzero in $\{z: 0 < |z| \leqslant 1\}$. Let $m =$ the order of the zero of φ at $z = 0$ and $k =$ the order of the zero of ψ at $z = 0$ and assume k and $m \geqslant -1$. Then the function $g(t)$ is differentiable and*

$$g'(0) = \operatorname{Re} \iint_\Delta \psi(z) \frac{|\varphi(z)|}{\varphi(z)} \, dx \, dy.$$

If $m - k < 2 + k$, then g has a continuous second derivative and

$$g''(0) = \iint_{\Delta} \frac{1}{|\varphi(z)|^3} (\operatorname{Im} \psi(z)\overline{\varphi(z)})^2 \, dx \, dy. \tag{27}$$

If $m - k \geqslant 2 + k$, then

$$g(t) = g(0) + tg'(0) + c\varepsilon(t) + o(\varepsilon(t)), \qquad c > 0, \tag{28}$$

where

$$\varepsilon(t) = |t|^{1 + (2 + k)/(m - k)} \qquad \text{if } m - k > 2 + k,$$

$$\varepsilon(t) = t^2 \log \frac{1}{|t|} \qquad \text{if } m - k = 2 - k. \tag{29}$$

Proof. We need the inequality

$$\frac{1}{2} \frac{\left(\operatorname{Im} \dfrac{|\alpha|}{\alpha} \beta\right)^2}{|\alpha| + |\beta|} \leqslant |\alpha + \beta| - |\alpha| - \operatorname{Re} \frac{|\alpha|}{\alpha} \beta \leqslant \frac{1}{2} \frac{\left(\operatorname{Im} \dfrac{|\alpha|}{\alpha} \beta\right)^2}{|\alpha| - |\beta|}. \tag{30}$$

The left part of inequality (30) is true for complex numbers α and β with $|\beta| < |\alpha|$. The right hand part of inequality (30) is true for all α and β. Inequality (30) is a consequence of Taylor's formula with integral remainder term applied to the function $h(t) = |1 + t\tilde{\beta}|$. One finds that

$$|1 + \tilde{\beta}| = 1 + \operatorname{Re} \tilde{\beta} + \int_0^1 \frac{(\operatorname{Im} \tilde{\beta})^2}{|1 + t\tilde{\beta}|^3} (1 - t) \, dt.$$

The inequality follows by letting $\tilde{\beta} = \beta/\alpha$ and making obvious reductions which follow from the assumption that $|\beta| < |\alpha|$ and the fact that t is between zero and one.

We also need the more elementary inequality

$$0 \leqslant |\alpha + \beta| - |\alpha| - \operatorname{Re} \frac{|\alpha|}{\alpha} \beta \leqslant 2|\beta|. \tag{31}$$

In Lemma 1, we calculated the first derivative $g'(0)$. A formal calculation of $g''(0)$ yields formula (27). The integrand in (27) is clearly bounded by $|\psi|^2/|\varphi|$. Obviously, if $m - k \leqslant k + 1$, then $|\psi|^2/|\varphi|$ has at worst a simple pole at $z = 0$ and has no singularities anywhere else in the unit disk Δ. In that case, the integral (27) converges absolutely. It is then easy to see that $g''(0)$ exists.

Next we treat the case where $m - k > k + 2$. By looking at a smaller disk and changing the local parameter at the origin, we may assume $\varphi(z) = z^m$ and $\psi(z) = z^k h(z)$, where $h(z)$ is nonzero everywhere. On letting $I(t) = g(t) - g(0) - tg'(0)$ we obtain

$$I(t) = \iint_{\Delta} |z|^k \left[|z^{m-k} + th| - |z^{m-k}| - t \operatorname{Re} \frac{|z|^{m-k}}{z^{m-k}} h \right] dx\, dy \tag{32}$$

The hypothesis that $k \geqslant -1$ and the assumption that $m - k > k + 2$ imply that $m - k > 1$.

Let $j = m - k$ and $z = t^{1/j}\zeta$ and $a = t^{-1/j}$. Since $|z| < 1$, $|\zeta| < t^{-1/j} = a$. Letting $\Delta_a = \{\zeta\colon |\zeta| < a\}$, we have

$$I(t) = \iint_{\Delta_a} t^{1+[(k+2)/j]} |\zeta|^k \left[|\zeta^j + h| - |\zeta^j| - \operatorname{Re} \left(\frac{|\zeta|}{\zeta} \right)^j h \right] d\xi\, d\eta. \tag{33}$$

Notice that $a \to \infty$ as $t \to 0$. For $|\zeta|^j$ larger than one and larger than an upper bound for h, we see from (30) that $I(t)/t^{1+[(k+2)/j]}$ has an integrand bounded by a constant multiplied by $|\zeta|^{k-j} = |\zeta|^{2k-m} \geqslant |\zeta|^{-3}$. Hence, the integral $I(t)/t^{1+[(k+2)/j]}$ converges absolutely for ζ near ∞. From (31) the integrand is bounded for ζ in any bounded region. Inequality (30) and the fact that h is not zero show that the integrand is nonnegative everywhere. Thus

$$\lim_{t \to 0} I(t)/t^{1+[(k+2)/(m-k)]} = C > 0.$$

Finally, we consider the case $m - k = k + 2$. Then $j = m - k = k + 2 \geqslant 1$ and we still have $a \to \infty$ as $t \to 0$ since $a = t^{-1/j}$. We obtain

$$t^{-2} I(t) = \iint_{\Delta_a} |\zeta|^k \left[|\zeta^{k+2} + h| - |\zeta^{k+2}| - \operatorname{Re} \left(\frac{|\zeta|}{\zeta} \right)^{k+2} h \right] d\xi\, d\eta. \tag{34}$$

We break down this integral into two parts, the first part over the domain $|\zeta| < (M/\delta)^{1/j}$, where M is an upper bound for h and $0 < \delta < 1$, and the second part over the domain $(M/\delta)^{1/j} < |\zeta| < a$. Then from (31) the first integral is bounded by

$$\iint_{|\zeta| < (M/\delta)^{1/j}} 2M|\zeta|^k\, d\xi\, d\eta = \frac{2}{(k+2)} \frac{M^2}{\delta}.$$

From (30) the second integral is bounded by

$$\frac{1}{2} \iint \frac{M^2}{|\zeta|^2 (1 - \delta)}\, d\xi\, d\eta,$$

where the integral is over the domain $(M/\delta)^{1/j} < |\zeta| < t^{-1/j}$. This is equal to

$$\frac{M^2}{2(1-\delta)} \cdot \frac{1}{j}\left(\log\frac{1}{t} - \log\frac{M}{\delta}\right).$$

The estimate $t^{-2}I(t) \leqslant (\text{const})\log(1/t)$ follows. To get the other estimate needed for (28), one uses the other half of inequality (30).

9.5. WEIERSTRASS POINTS

In order to construct the conformal mapping induced by an isometry of the linear space of holomorphic quadratic differentials, we need some elementary facts about Weierstrass points.

Let A be a finite dimensional vector space of holomorphic functions on a plane domain D. Assume $\dim A = m$ and z is a point of D. Let $\operatorname{ord}_z \varphi$ be the value of $k \geqslant 0$ for which the Taylor expansion begins

$$\varphi(w) = a_k(w - z)^k + \cdots \qquad \text{with } a_k \neq 0.$$

A basis of A adapted to z is a basis $\varphi_1, \ldots, \varphi_m$ such that $\operatorname{ord}_z \varphi_1 < \operatorname{ord}_z \varphi_2 < \cdots < \operatorname{ord}_z \varphi_m$.

To construct such a basis, let φ_1 be an element of A with smallest possible order μ_1. Then let A_1 be the subspace of A whose elements have order larger than μ_1. Thus, $\dim A_1 = \dim A - 1$. Now let φ_2 be a nonzero element of A_1 with smallest possible order μ_2 and A_2 be the subspace of A_1 whose elements have order larger than μ_2. Evidently $\dim A_2 = \dim A_1 - 1$ and we can continue this process until we get elements $\varphi_1, \ldots, \varphi_m$ which are a basis for A.

Notice that $0 \leqslant \mu_1 < \mu_2 < \cdots < \mu_m$ and therefore $\mu_j \geqslant j - 1$. To the point z, we associate an integer $\tau(z)$ which we call the weight of z with respect to A,

$$\tau(z) = \sum_{j=1}^{m} (\mu_j - j + 1). \tag{35}$$

Definition. *z is called a Weierstrass point for A if $\tau(z) > 0$.*

Let $\{\varphi_1, \ldots, \varphi_m\}$ be a set of elements of A. The Wronskian of this set is the following determinant of derivatives:

$$W(z) = \det\begin{pmatrix} \varphi_1(z) & \cdots & \varphi_m(z) \\ \varphi_1'(z) & \cdots & \varphi_m'(z) \\ \varphi_1^{[m-1]}(z) & \cdots & \varphi_m^{[m-1]}(z) \end{pmatrix}.$$

It is obvious that if C is an $m \times m$ matrix of constants transforming $\{\varphi_1, \ldots, \varphi_m\}$ into $\{\tilde{\varphi}_1, \ldots, \tilde{\varphi}_m\}$, then the Wronskian for the transformed set is

$$\tilde{W}(z) = (\det C) W(z). \tag{36}$$

Thus, the order of $W(z)$ at z is invariant under change of basis.

Lemma 4. *The weight of z with respect to A is the order at z of the Wronskian of any basis of A, that is,*

$$\tau(z) = \operatorname{ord}_z W(z). \tag{37}$$

Proof. Let $\varphi_1, \ldots, \varphi_m$ be a basis for A and we adopt the notation

$$W(z) = \det[\varphi_1(z), \ldots, \varphi_m(z)].$$

By the preceding remarks, we know it is sufficient to calculate the order of $W(z)$ using any basis for A. We will use a basis adapted to z.

We leave as an exercise the proof that

$$\det[f\varphi_1, \ldots, f\varphi_m] = f^m \det[\varphi_1, \ldots, \varphi_m] \tag{38}$$

for any holomorphic function f.

Lemma 4 is proved by induction on the dimension of A. The lemma is obviously true if $\dim A = 1$. From (38), $\det[\varphi_1, \ldots, \varphi_{k+1}] = \varphi_1^{k+1} \det[1, \varphi_2/\varphi_1, \ldots, \varphi_{k+1}/\varphi_1]$. Since the first column of this last matrix has a one on top and zeros below, the expression becomes

$$\varphi_1^{k+1} \det[(\varphi_2/\varphi_1)', \ldots, (\varphi_{k+1}/\varphi_1)']$$

and, since order is additive, we see that

$$\operatorname{ord}_z W(z) = (k+1)\mu_1 + \operatorname{ord}_z(\det[(\varphi_2/\varphi_1)', \ldots, (\varphi_{k+1}/\varphi_1)']).$$

We can apply the inductive hypothesis to this last expression and we obtain

$$\operatorname{ord}_z W(z) = (k+1)\mu_1 + \sum_{j=2}^{k+1} ((\mu_j - \mu_1 - 1) - (k-2)) = \sum_{j=1}^{k+1} (\mu_j - j + 1).$$

Lemma 5. *Suppose A is a finite dimensional vector space of holomorphic functions of dimension m on a domain D. Then a set of elements $\varphi_1, \ldots, \varphi_m$ form a basis for A if, and only if, their Wronskian is not identically zero. Moreover, the set of Weierstrass points in D for the space A is a discrete set.*

Proof. If there were a linear dependence, then one of the columns of the Wronskian determinant would be a linear combination of the others, and hence the Wronskian would be zero.

Conversely, let W_k be the Wronskian of the set $\varphi_1, \ldots, \varphi_k$. If $W_m(z) \equiv 0$, we wish to show that $\varphi_1, \ldots, \varphi_m$ are linearly dependent. We may assume that $W_1(z) \neq 0$ because $W_1(z) = \varphi_1(z)$. Hence, we may take a largest value of $k < m$ for which $W_k(z) \not\equiv 0$ and $W_{k+1}(z) \equiv 0$. Form the ordinary differential equation in the unknown y given by

$$W_{k+1}(\varphi_1, \ldots, \varphi_k, y) \equiv 0. \tag{39}$$

It is a homogeneous equation of order k in y and $W_k(z)$ is the coefficient of $y^{[k]}$. Since $W_k(z) \not\equiv 0$, the equation has a k dimensional vector space of solutions. The functions φ_j, $1 \leqslant j \leqslant k$, are solutions because one takes the determinant of a matrix with two identical columns. Now φ_{k+1} is a solution because we know that $W_{k+1}(z) \equiv 0$. We conclude that $\varphi_1, \ldots, \varphi_{k+1}$ are linearly dependent.

To prove the last part of the lemma,, note that the Wronskian of a basis for A cannot be identically zero. Since the zero set of a nonconstant analytic function is discrete, we see from Lemma 4 that the set of Weierstrass points is a discrete set.

Lemma 6. *Let A be a finite dimensional space of holomorphic functions of dimension m defined on a domain D. Assume that no point of D is a Weierstrass point. To each point z of D, let φ_z be the unique differential with order $m-1$ at z and with leading Taylor coefficient* 1. *(That is, $\varphi_z(w) = (w - z)^{m-1}$ + higher terms). Then φ_z depends holomorphically on z. Moreover, the curve φ_z is not contained in any proper linear subspace of A.*

Proof. Let $\varphi_1, \ldots, \varphi_m$ be a basis for A. We know there are unique constants $c_1(z), \ldots, c_m(z)$ such that

$$c_1(z)\varphi_1(w) + \cdots + c_m(z)\varphi_m(w) = \varphi_z(w).$$

From the definition of φ_z, the function $c_j(z)$ are the unique solutions of the following system:

$$\begin{aligned}
c_1(z)\varphi_1(z) \quad &+ \cdots + c_m(z)\varphi_m(z) = 0\\
c_1(z)\varphi_1'(z) \quad &+ \cdots + c_m(z)\varphi_m'(z) = 0\\
&\vdots\\
c_1(z)\varphi_1^{[m-1]}(z) &+ \cdots + c_m(z)\varphi_m^{[m-1]}(z) = 1.
\end{aligned}$$

Since the Wronskian is nonzero and the functions $\varphi_j^{[k]}(z)$ are analytic, clearly the $c_j(z)$ are also analytic.

If the family $\varphi_z(w)$ is contained in a lower dimensional subspace, then one could choose the basis so that $c_m(z) \equiv 0$. Then the first $m - 1$ equations in this system would force $c_j(z) \equiv 0$ to be identically zero for each z, which would be a contradiction.

Our application of Lemmas 4, 5, and 6 will be to the space of quadratic differentials on a surface R. If a surface has genus $g \geqslant 2$ and no punctures, then the space of quadratic differentials has dimension $3g - 3$ (see Section 1.11). It is an easy consequence of the Riemann—Roch theorem that the space of holomorphic q-differentials (for $q \geqslant 2$) has dimension $(2q - 1)(g - 1)$ (see [FarK]). By a q-differential we mean a differential φ for which the expression $\varphi(z)\, dz^q$ is invariant. An example of a q-differential can be obtained by raising a 1-differential (abelian differential) to the qth power. Thus, the number of zeros of a q-differential counted with multiplicity is $2q(g - 1)$.

Lemma 7. *Suppose A is the space of holomorphic quadratic differentials on a surface of genus g with no punctures and $g \geqslant 2$. Let $W(z)$ be the Wronskian of a basis of A. Then $W(z)$ is a q-differential where $q = 9g(g - 1)/2$. Thus the total weight of all Weierstrass points is given by the formula*

$$\sum_{P \in R} \tau(P) = 9(g - 1)^2 g.$$

Proof. To prove the Wronskian is a q-differential we begin with the identity $\tilde{\varphi}_j(f(z))f'(z)^2 = \varphi_j(z)$, where f is the transition function from a local coordinate z to a local coordinate $\tilde{z}$ and then repeatedly apply identity (38) to the formula for the Wronskian of $\tilde{\varphi}_1, \ldots, \tilde{\varphi}_{3g-3}$.

The value of q is merely the sum of the arithmetic progression $2 + 3 + \cdots + 3g - 2$. The formula for the total weight is obtained from the formula for the total number of zeros of a q-differential which is $2q(g - 1)$.

Lemma 8. *Let A be the space of holomorphic integrable quadratic differentials on a hyperbolic Riemann surface R of type (g, n) and assume (g, n) is not one of the exceptional types $(0, 3)$, $(0, 4)$, $(1, 1)$, $(1, 2)$, or $(2, 0)$. Let $\dot{R}$ be the surface R with all of the quadratic Weierstrass points removed. Let $P(A)$ be the projective space of A and let $\dim A = m$. For each z in $\dot{R}$, let $\alpha(z)$ be the line spanned by the unique element φ_z which has order $m - 1$ at z and Taylor expansion*

$$\varphi_z(w) = (w - z)^{m-1} + \text{higher terms}$$

for some particular local parameter z. Let S be the image of α and define $\beta\colon S \to \dot{R}$ by letting $\beta(\varphi)$ be the point where φ has its highest order. Then α is injective, α and β are analytic, and $\beta \circ \alpha(z) = z$.

Proof. Since R is hyperbolic (i.e., has a Fuchsian universal covering group) and since we are excluding the exceptional cases, we know that the dimension of A is $m = 3g - 3 + n$ and

$$\begin{aligned} &n \geqslant 5 && \text{for } g = 0, \\ &n \geqslant 3 && \text{for } g = 1, \\ &n \geqslant 1 && \text{for } g = 2, \\ &n \text{ arbitrary} && \text{for } g > 2. \end{aligned} \tag{40}$$

That α is analytic easily follows from Lemma 6. Once we know α is injective, it is obvious that the image of α is a complex curve and its inverse mapping β is analytic.

To show that α is injective on $\dot{R}$, we first note that in all cases $m \geqslant 2$, and hence the projective space of A has at least dimension 1. The hypothesis that a point z of $\dot{R}$ is not a Weierstrass point and not a puncture ensures that there is an element φ in A with order exactly $m - 1 = 3g - 4 + n$. Since the total number of zeros counted with multiplicity of any element φ in A is $4g - 4 + n$, we see that the maximum order of any other zero of φ is g. But $g < 3g - 4 + n$ in all of the cases listed in (40). This shows that the mapping β which associates to an element φ in S the point where φ has its highest order zero is well defined. This, of course, is the same as showing that α is injective.

9.6. ISOMETRIES OF $T(\Gamma)$

By virtue of Royden's theorem on the equality of Teichmüller's and Kobayashi's metrics, a biholomorphic self-mapping θ of $T(\Gamma)$ must be an isometry in Teichmüller's metric and at each point p in $T(\Gamma)$ the derivative θ'_p must be a linear isometry of the fiber of the tangent bundle at p onto the fiber of the tangent bundle at $\theta(p)$. Since the dual of the metric F is the metric G, this means that there is an induced complex linear isometry θ'^*_p of the fiber of the cotangent bundle at $\theta(p)$ onto the fiber of the cotangent bundle at p.

In the next theorem we will show that the linear isometry θ'^*_p must be induced by a conformal mapping from the surface represented by p to the surface represented by $\theta(p)$ and multiplication by a constant of modulus 1. In particular, for each p in $T(\Gamma)$, there must be an element h_p of Mod Γ, possibly depending on p, such that $\theta(p) = h_p(p)$.

Assume $\mathbb{H}/\Gamma$ is not of exceptional type so that the action of the modular group on Teichmüller space is faithful (Theorem 3 of Section 8.2). We claim that for the given isometry θ, the element of the modular group h_p for which $\theta(p) = h_p(p)$ must be constant with respect to p. First we show there is a positive number δ such that $h_q(q) = h_p(q)$ if $d(p, q) < \delta$. We know the modular group acts discontinuously on Teichmüller space, and thus there is a $\delta > 0$ such that

$d(p,q) < 2\delta$ implies q is not in the orbit of p unless $q = p$. Now assume $d(p, q) < \delta$. Then $d(\theta(p), \theta(q)) = d(h_p(p), h_q(q)) < \delta$ because θ is an isometry. From the triangle inequality,

$$d(h_p(q), h_q(q)) \leqslant d(h_p(q), h_p(p)) + d(h_p(p), h_q(q)).$$

The first term on the right is less than δ because h_p is an isometry. The second term on the right is less than δ because θ is an isometry. Thus $h_p^{-1} \circ h_q(q)$ has distance less than 2δ from q and must be equal to q. We conclude that $\theta(q) = h_p(q)$ for every q with $d(q, p) < \delta$. It is an exercise to show that the points in Teichmüller space with trivial isotropy groups form a connected, open, dense subset. Thus, $\theta(q) = h_p(q)$ for all q in $T(\Gamma)$.

Except for the exceptional cases, we see that Theorem 2 is a consequence of the following theorem.

Theorem 5. *Assume L is a complex linear isometry from $A(\Gamma)$ onto $A(\Gamma')$ and that Γ and Γ' are hyperbolic covering groups of compact surfaces with a finite number of punctures and assume $\mathbb{H}/\Gamma$ and $\mathbb{H}/\Gamma'$ are not of exceptional type. Then the type of Γ and the type of Γ' are the same and there is a conformal mapping $c: \mathbb{H}/\Gamma' \to \mathbb{H}/\Gamma$ and a complex constant λ with $|\lambda| = 1$ such that*

$$L(\varphi) = \lambda c^* \varphi,$$

where $c^*\varphi(z) = \varphi(c(z))c'(z)^2$.

Remark. A complete list of isomorphisms between Teichmüller spaces of different types was given by D. B. Patterson [P]. The list is

$$T(2, 0) \cong T(0, 6),$$
$$T(1, 2) \cong T(0, 5),$$
$$T(1, 0) \cong T(1, 1) \cong T(0, 4).$$

Note that for each isomorphism pair at least one of the two is of exceptional type. The fact that there are no other isomorphisms follows from Theorem 5 and the fact that a biholomorphic mapping of two Teichmüller spaces induces isometries on the fibers of the tangent bundles.

In order to begin the proof of Theorem 5 we give the following lemma. Let (g, n) be the type of $\mathbb{H}/\Gamma$ and let $d = 3g - 3 + n$.

Lemma 9. (Due to Earle and Kra [EK1]. *Suppose $d > 3$. Let $f(t) = \|\varphi + t\psi\|$, where φ and ψ are in $A(\Gamma)$. Then ψ is holomorphic at every puncture of $\mathbb{H}/\Gamma$ if and only if*

$$f(t) = f(0) + tf'(0) + O(|t|^{1+(3/2d)}) \tag{41}$$

for all φ in $A(\Gamma)$ with $\varphi \not\equiv 0$.

Proof. If ψ is holomorphic at every puncture, then from (29) in Theorem 4 we can put $k = 0$ and m equal to the highest possible order of a zero, which, from the Riemann–Roch theorem, is $4g - 4 + n = \frac{4}{3}d - \frac{1}{3}n \leqslant \frac{4}{3}d$. Therefore, the slowest possible convergence to zero as $t \to 0$ of the expression $f(t) - f(0) - tf'(0)$ is $O(|t|^{1+(3/2d)})$.

Conversely, suppose ψ has a pole at some puncture. Then choose φ with the highest possible order zero at that puncture, which is at least of order $d - 2$. Plugging in $k = -1$ and replacing m by $d - 2$ in Theorem 4 we get

$$f(t) - f(0) - tf'(0) \geqslant c|t|^{1+(1/d-1)}.$$

But since $1/(d - 1) \leqslant 3/2d$ for $d > 3$, the lemma follows.

Corollary. *If* $\mathbb{H}/\Gamma$ *and* $\mathbb{H}/\Gamma'$ *are surfaces of nonexceptional type and if* $A(\Gamma)$ *is isometric to* $A(\Gamma')$, *then* $\mathbb{H}/\Gamma$ *and* $\mathbb{H}/\Gamma'$ *are of the same type.*

Proof. First assume $d = \dim A(\Gamma) > 3$. Note that if L is the isometry, then

$$f(t) = \|\varphi + t\psi\| = \|L(\varphi) + tL(\psi)\|.$$

Hence if $f(t)$ has a certain order of growth for φ, ψ it must have exactly the same order of growth for $L(\varphi)$ and $L(\psi)$. We conclude from Lemma 9 that L must preserve differentials which are holomorphic at the punctures. Since the holomorphic quadratic differentials have dimension

$$\begin{array}{rl} 0 & \text{if } g = 0, \\ 1 & \text{if } g = 1, \\ 3g-3 & \text{if } g > 1, \end{array}$$

we see that $\mathbb{H}/\Gamma$ and $\mathbb{H}/\Gamma'$ have the same genus. And since $\dim A(\Gamma) = \dim A(\Gamma')$, they must also have the same number of punctures.

Now suppose $d \leqslant 3$ and both $\mathbb{H}/\Gamma$ and $\mathbb{H}/\Gamma'$ are nonexceptional. This means d cannot be zero or one. If $d = 2$, the only nonexceptional type is (0, 5). If $d = 3$, there are two possible types, (0, 6) or (1, 3). To complete the proof of the corollary, we must show these two types cannot have isometric spaces of quadratic differentials. A quadratic differential on a surface of type (0, 6) is expressible in the form

$$\varphi(z)dz^2 = \frac{c_0 + c_1 z + c_2 z^2}{(z - p_1)(z - p_2)\cdots(z - p_6)} dz^2,$$

where c_0, c_1, and c_2 are arbitrary and p_1 through p_6 are the punctures. Such a differential can have at most a double zero and for $f(t) = \|\varphi + t\psi\|$, with $\varphi \not\equiv 0$, $f''(0)$ exists unless either ψ is nonzero at a double zero of φ or ψ has a pole at some puncture where φ has a zero. In these cases, Theorem 4 implies

$$\varepsilon(t) = t^2 \log \frac{1}{|t|} \quad \text{or} \quad |t|^{3/2}. \tag{42}$$

These are the possible rates of growth of $f(t) - f(0) - tf'(0)$ for pairs φ and ψ for which $f''(0)$ does not exist.

On the other hand, on a surface of type (1, 3), we may select φ with a triple zero at a point where ψ is regular and nonzero. For such a choice

$$\varepsilon(t) = |t|^{5/3}.$$

Thus, the two spaces of quadratic differentials cannot be isometric.

We now proceed with the proof of Theorem 5. Let A and A' be the spaces of holomorphic quadratic differentials on the surfaces R and R' of nonexceptional type. Assume L is a complex linear isometry of A onto A'. From the corollary to Lemma 9, we know the types of both surfaces are the same. Let $\dot{R}$ be the surface R with all of the quadratic Weierstrass points removed.

Let S be image of $\dot{R}$ under the mapping α described in Lemma 8. Repeat all of the same definitions for the surface R' and the space A'. We obtain the diagram shown in Figure 9.2. We must show that the mapping L takes S into S'. Corresponding to any point z on R, we let φ_z be the nonzero differential with highest possible order zero at z.

$$\begin{array}{ccc} P(A) & & P(A') \\ \cup & & \cup \\ S & \xrightarrow{\quad L \quad} & S \\ \alpha \uparrow\downarrow \beta & & \beta' \downarrow\uparrow \alpha' \\ \dot{R} & & \dot{R}' \end{array}$$

Figure 9.2

STEP 1. If z is a puncture of R (and R is a surface of nonexceptional type), then $L(\varphi_z)$ is a differential with its highest zero at a puncture of R'.

To show this, as usual we look at the function

$$f(t) = \|\varphi_z + t\psi\|.$$

From Theorem 4 we know that $f(t) = f(0) + tf'(0) + c\varepsilon(t) + o(\varepsilon(t))$, where $c > 0$, where $\varepsilon(t) = t^\alpha$ for certain α with $1 < \alpha < 2$, and $\varepsilon(t) = t^2 \log(1/t)$ for $\alpha = 2$.

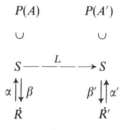

We will focus attention on α. Smaller values of α correspond to less smoothness in the function f and we will think of the smaller values as being worse. When z is a puncture, the worst value of α is achieved by letting φ_z have a zero of order at least $3g - 5 + n$ and we can make ψ have a pole at z. In Theorem 4, this means $m \geqslant 3g - 5 + n$ and $k = -1$, and

$$\alpha \leqslant 1 + \frac{1}{3g - 4 + n}.$$

Now consider the differential $L(\varphi_z)$. Let the worst value of α correspond to some point z'. If z' is not a puncture, the worst possible behavior of $\|L(\varphi_z) + tL(\psi)\|$ has

$$\alpha \geqslant 1 + \frac{2}{4g - 4 + n}$$

because $4g - 4 + n$ is the largest possible order of a zero and the differential $L(\psi)$ cannot have a pole at z'. These two conditions on α obviously show that $2g - 2 + n \leqslant 2$, and this inequality is satisfied precisely for the surfaces of exceptional type (0, 3), (0, 4), (1, 1), (1, 2), or (2, 0).

When $g = 0$ and $n = 5$ the application of Theorem 4 proceeds a little bit differently. In that case the function $\|L(\varphi_z) + tL(\psi)\|$ would be of class C^2 while $\|\varphi_z + t\psi\|$ would have $\varepsilon(t) = t^2 \log(1/t)$.

This step obviously implies that if φ_z has its highest possible zero at a nonpuncture, then so does $L(\varphi_z)$, since we can just as well apply the argument to L^{-1}.

Step 2: If z is a nonpuncture of R and φ_z has highest possible order zero at z, then $L(\varphi_z)$ has a zero of exactly the same order at some unique point of R'.

To prove this, we look again at the function

$$f(t) = \|\varphi_z + t\psi\| = \|L(\varphi_z) + tL(\psi)\|.$$

In the application of Theorem 4 we know that $k \geqslant 0$ at the nonpuncture z, and by selecting ψ to be nonzero at z, we can make $k = 0$. Now the order of the zero of φ_z at z is $m \geqslant 3g - 4 + n$ and $m > 3g - 4 + n$ precisely when z is a Weierstrass point. Since the order of any other zero is at most g and since $g < 3g - 4 + n$ for nonexceptional surfaces, we see that $\alpha = 1 + 2/m$. This forces $L(\varphi_z)$ to have a zero of the same order m at some unique point z' where $L(\psi)$ is nonzero. Thus z will be a Weierstrass point if an only if z' is. We see that L preserves the differentials φ_z corresponding to nonpunctures and non-Weierstrass points.

Step 3: Since L is complex linear, from Lemma 8, the mapping $\beta \circ L^{-1} \circ \alpha'$ from $\dot{R}'$ to $\dot{R}$ is analytic with analytic inverse $\beta' \circ L \circ \alpha$. Moreover, this mapping extends to a conformal map $c\colon R' \to R$ because punctures are removable for bounded analytic functions.

STEP 4: Let $\varphi_{c(z)}$ be a differential with highest possible order zero at $c(z)$ on $\dot{R}$. Evidently $\varphi_{c(z)}(c(w))c'(w)^2$ is a quadratic differential on $\dot{R}'$ which has highest possible order at z. Hence,

$$L(\varphi_{c(z)}) = \lambda_z \varphi_{c(z)}(c(w))c'(w)^2.$$

Since L is an isometry $|\lambda_z| = 1$. Since the dependence of $\varphi_{c(z)}$ on z is holomorphic and L is complex linear, the mapping $z \to \lambda_z$ is holomorphic and therefore must be constant.

STEP 5: By Lemma 6, the differentials φ_z, for z in $\dot{R}$, are not contained in any proper linear subspace and we must have $L(\varphi)(w) = \lambda\varphi(c(w))c'(w)^2$ for all quadratic differentials φ.

The proof is now complete.

Notes. The theorem of Bers and Greenberg appears in [BerGr]. Marden [Mar] proved the result independently and by a different method. Kra gave a proof based on Teichmüller's theorem for finite dimensional Teichmüller spaces [EK2]. The proof given in Section 9.1 differs from any that appear in the literature in that it relies on the sufficiency of Hamilton's condition for extremality. I would like to thank Irwin Kra for pointing out an essential error in the proof I originally gave. The Bers–Greenberg theorem has recently been proved in a different way by Earle and McMullen [EM].

The contents of Sections 9.2 through 9.6 are contained in Royden's paper [Ro] and the papers of Earle and Kra [EK1, EK2]. Dual relationships between convexity of a norm and smoothness of the dual norm in Banach spaces are studied exhaustively in the book by Mahlon M. Day, *Normed Linear Spaces* [D]. Theorem 4 is due to Royden [Ro] for compact surfaces and is extended to surfaces with punctures by Earle and Kra [EK1]. The section on Weierstrass points is classical. More exhaustive treatment of Weierstrass points can be found in [FarK].

Theorem 5 on isometries is due to Royden. Earle and Kra extended it to surfaces with punctures. Patterson's theorem first appears in [P].

Royden's technique of using the fact that Kobayashi's metric equals Teichmüller's metric and the nonsmoothness of this metric has been used to study holomorphic fiber spaces over Teichmüller space. See Earle and Kra [EK2], Hubbard [Hu], Nag [N1], Kra [Kr7], and Krushkal and Kuhnau [KruKu].

EXERCISES

1. Let f be a function from a domain U in $\mathbb{R}^n$ to $\mathbb{R}$ whose derivative exists at every point of U. Suppose further that x and $x + \Delta x$ are in U and that

$$|f(x + \Delta x) - \{f(x) + f'(x)(\Delta x)\}| \leqslant K \|\Delta x\|^{1+\epsilon}$$

for some positive number ε and the same value of K for every x in the domain. Show that $f'(x)$ is Hölder continuous, in the sense that

$$\|f'(x) - f'(y)\| \leqslant M\|x - y\|^{\epsilon}$$

for some constant M and every x and y in an arbitrary compact subset of U.

2. By consideration of the example $f(x) = x^2 \sin(1/x)$ and $f(0) = 0$, show that the result of Exercise 1 is false if one permits the constant K to depend on the point x in the domain.
3. (Open Problem) Determine whether or not the Teichmüller cometric $G([\mu], \varphi)$ is of class $C^{1+\epsilon}$.
4. (Open Problem) Determine whether, for infinite dimensional Teichmüller spaces, the Teichmüller cometric is of class C^1.
5. Show that if a norm on $\mathbb{R}^n$ is strictly convex, then it is uniformly convex. A norm is uniformly convex if for every $\varepsilon > 0$ there exists a $\delta > 0$, such that if $\|x\| = \|y\| = 1$ and $\|x + y\| \geqslant 2 - \delta$, then $\|x - y\| < \varepsilon$. A norm is strictly convex if for every t with $0 < t < 1$ and every x and y with $\|x\| = \|y\| = 1$, one has $\|tx + (1 - t)y\| < 1$.
6. (Open Problem) For an infinite dimensional Teichmüller space $T(\Gamma)$ determine whether it is possible to find a biholomorphic bijective mapping of $T(\Gamma)$ onto $T(\Gamma)$ which is not an element of the modular group.
7. Show the following biholomorphic isomorphisms between Teichmüller spaces:

$$T(1, 0) \cong T(1, 1) \cong T(0, 4),$$

$$T(2, 0) \cong T(0, 6),$$

$$T(1, 2) \cong T(0, 5).$$

10

QUADRATIC DIFFERENTIALS WITH CLOSED TRAJECTORIES

Let R be a Riemann surface of finite analytic type (g, n). Assume the dimension $d = 3g - 3 + n$ of the space $A(R)$ of integrable, holomorphic, quadratic differentials on R is positive. In the first section we show that d also has a topological meaning. It is the maximal number of simple closed curves on R whose homotopy classes can be represented by nonintersecting curves which are not homotopic to a puncture or to each other. A family of d or fewer such curves on R will be called an admissible family.

Our objective is to show that corresponding to any such admissible family of curves, $\alpha_1, \ldots, \alpha_k$, $(k \leqslant d)$ on the Riemann surface R and to any set of positive constants $b_1, \ldots, b_k$ there exists a unique quadratic differential φ in $A(R)$ with the following properties:

(a) The noncritical horizontal trajectories of φ are closed.
(b) The classes of homotopic horizontal closed trajectories of φ partition R-{critical trajectories} into k annuli $A_1, \ldots, A_k$.
(c) Any closed trajectory lying in the annulus A_i is homotopic to α_i.
(d) The height of each annulus A_i measured along any vertical trajectory in the metric $|\varphi(z)dz^2|^{1/2}$ is b_i.

We call differentials with these properties Jenkins–Strebel differentials. We show that the existence of such differentials can be viewed either as a consequence of Weyl's lemma or of the theorem on the existence of trivial curves, Theorem 6 of Chapter 5. The uniqueness comes from the length area method or from the second minimal norm principle of Chapter 2.

10.1. ADMISSIBLE SYSTEMS

A system of closed curves $\{\alpha_i: 1 \leqslant i \leqslant k\}$ on R is called admissible if

(a) each α_i is simple and no α_i intersects any α_j for $i \neq j$;
(b) no α_i is homotopic to any α_j for $j \neq i$; and
(c) no α_i is homotopically trivial or homotopic to a puncture.

Lemma 1. *Suppose $d = 3g - 3 + n$ is the dimension of the space of integrable holomorphic quadratic differentials, $A(R)$. Assume that when the genus of R is zero, the number of punctures is four or more, when the genus is one, the number of punctures is one or more, and, when the genus is two or more, the number of punctures is arbitrary. Then the maximal number of elements of an admissible curve system is d (i.e., $k \leqslant d$). Moreover, given any admissible curve system $\{\alpha_i: 1 \leqslant i \leqslant k\}$ with $k < d$, one can find $d - k$ additional curves, $\alpha_{k+1}, \ldots, \alpha_d$, such that $\{\alpha_i: 1 \leqslant i \leqslant d\}$ is an admissible curve system.*

Proof. First check that the hypothesis $d > 0$ ensures that there is a simple curve not homotopically trivial and not homotopic to a puncture. Given the system $\{\alpha_i: 1 \leqslant i \leqslant k\}$, cut the surface R along each one of the curves in this system. After R is cut along these curves, it breaks up into components. If any of these components is not homeomorphic to a sphere with three punctures, we may cut them up along additional curves $\alpha_{k+1}, \ldots, \alpha_m$ none of which is homotopically trivial or homotopic to a puncture until every component is homeomorphic to a sphere with three punctures.

Let j_0 be the number of such components for which all three boundary contours are holes. Let j_1 be the number of such components which have one boundary contour which is a puncture and two which are holes. Let j_2 be the number which have two boundary punctures and one boundary hole. If R is reconstructed by sewing together boundary curves, the boundary curves are paired together and, therefore,

$$3j_0 + 2j_1 + j_2 = 2m.$$

Moreover, the total number of punctures on R is $n = 2j_2 + j_1$ and it is therefore obvious that

$$2m + n = 3j, \tag{1}$$

where $j = j_0 + j_1 + j_2$ is the total number of genus zero components.

On selecting a triangulation of R whose vertices include the n punctures and whose sides include the curves $\alpha_1, \ldots, \alpha_m$ and on using the Euler characteristic formula $2 - 2g =$ faces $-$ edges $+$ vertices, one easily finds that

$$2 - 2g = -j + n. \tag{2}$$

Figure 10.1

Therefore, the number of components is $j = 2g - 2 + n$ and, plugging this result into $2m + n = 3j$, we see that $m = 3g - 3 + n$. Q.E.D.

Two admissible systems $\{\alpha_i\colon 1 \leqslant i \leqslant k\}$ and $\{\tilde{\alpha}_j\colon 1 \leqslant j \leqslant k\}$ on R are called compatible if each α_i is homotopic to one of the $\tilde{\alpha}_j$ and if each $\tilde{\alpha}_j$ is homotopic to one of the α_i. Compatibility of admissible systems is obviously an equivalence relation. Notice that there are many incompatible systems on a given surface. Even in the case when the type is $(g, n) = (1, 0)$ or $(g, n) = (0, 4)$ which lead to a one complex dimensional space $A(R)$ and for which the curve family can have only one element, there are many incompatible systems. Figure 10.1 shows an example of two incompatible systems when $(g, n) = (0, 4)$.

10.2. AN EXTREMAL PROBLEM FOR ADMISSIBLE SYSTEMS

Let R be a Riemann surface of type (g, n) with $d = 3g - 3 + n > 0$ and let $\{\alpha_i\colon 1 \leqslant i \leqslant k\}$ be an admissible system on R. We will call a Jordan curve contained in an annulus a core curve for the annulus if its homotopy class generates the fundamental group of the annulus. Let A_i be an open annulus embedded in R in such a way that a core curve of A_i is homotopic in R to α_i. Given k such annuli which do not intersect each other, we call the system $\{A_i; 1 \leqslant i \leqslant k\}$ compatible with the system of curves $\{\alpha_i\colon 1 \leqslant i \leqslant k\}$.

Let the admissible system $\{\alpha_i\colon 1 \leqslant i \leqslant k\}$ and a set of k positive constants b_i, $1 \leqslant i \leqslant k$, be given. For any compatible system $\{A_i; 1 \leqslant i \leqslant k\}$ form the number

$$P\{A_i\} = \sum_{=1}^{k} \frac{b_i^2}{m_i}, \tag{3}$$

where m_i is the modulus of the annulus A_i. Recall from Section 1.9 that if an annulus is uniformized to be a domain of the form $\{z\colon 1 \leqslant z \leqslant R\}$, then its modulus is $(\log R)/2\pi$.

We pose the following extremal problem. For a given admissible curve system $\{\alpha_i; 1 \leqslant i \leqslant k\}$ *and a given system of constants* $b_i > 0$, $1 \leqslant i \leqslant k$, *find a compatible system of annuli* $\{A_i; 1 \leqslant i \leqslant k\}$ *for which* (3) *is minimum.*

First, note that there always is a compatible system. One simply has to take tubular neighborhoods of a given system of nonintersecting curves α_i in such a way that the tubular neighborhoods are nonintersecting. Second, observe that for a given simple curve α_i which is not homotopic to a puncture or to a point, the modulus m_i of any annulus A_i compatible with α_i is bounded above.

To prove this fact, let $\rho(z)|dz|$ be the noneuclidean metric for the surface R and let $\rho_i(z)|dz|$ be the noneuclidean metric for the annulus A_i. Then, by the comparison principle for noneuclidean metrics (see Ahlfors's lemma in the appendix to Chapter 9), $\rho_i(z) \geqslant \rho(z)$. Therefore, the noneuclidean length of the core curve of A_i in the metric ρ_i is larger than the length of the geodesic α_i on R in the same homotopy class. Let $\ell(\alpha_i)$ be the length of this geodesic on R. A simple calculation shows that for an annulus, π/m_i is the noneuclidean length of the geodesic core curve on A_i in the metric ρ_i. Thus $\pi/m_i \geqslant \ell(\alpha_i)$ and $m_i \leqslant \pi/\ell(\alpha_i)$. Since $\ell(\alpha_i)$ is assumed to be positive, m_i is bounded.

The expression $P\{A_i\}$ in (3) cannot have zero as a minimum. In fact, each of the terms b_i^2/m_i is bounded below by $b_i^2\ell(\alpha_i)/\pi$ and so each of the summands in (1) is bounded away from zero. Let

$$c = \inf P\{A_i\}, \tag{4}$$

where the infinum is taken over all compatible systems of annuli.

Theorem 1. *Let b_i be a set of positive constants for $1 \leqslant i \leqslant k$ and let $\{\alpha_i$: $1 \leqslant i \leqslant k\}$ be an admissible system of curves on a Riemann surface R of type (g, n) with $0 < k \leqslant d = 3g - 3 + n$. Then there exists a family of compatible annuli $\{A_i$; $1 \leqslant i \leqslant k\}$ for which the infimum in* (4) *is achieved.*

Proof. Select a sequence of compatible annuli $\{A_{in}$: $1 \leqslant i \leqslant k\}$ such that $P\{A_{in}\}$ decreases to the number c in (2). For each annulus A_{in} select a conformal homeomorphism z_{in} which maps A_{in} onto an annulus in the plane whose inner boundary is the circle of radius 1 and whose outer boundary is the circle of radius r_{in} where $(1/2\pi)\log r_{in} =$ modulus of $A_{in} = m_{in}$. Since there is an upper bound M for $P\{A_{in}\}$, we see that $b_i^2/m_{in} \leqslant M$ and thus $m_{in} \geqslant b_i^2/M$. On the other hand, we have already shown that $m_{in} \leqslant \pi/\ell(\alpha_i)$. Thus any subsequence of m_{in} has a subsequence which converges to a positive number. Let f_{in} be the inverse mapping to z_{in}. The region $\{z$: $1 < |z| < b_i^2/M\}$ is contained in the domains of all the functions f_{in}. Thus the functions f_{in} form a normal family of univalent functions and have a subsequence which converges to a univalent function f_i. (The limit f_i cannot be constant without having m_{in} converge to zero.) Let z_i be the inverse function to f_i. By taking a subsequence, we may assume the numbers r_{in} converge to a number $r_i > 1$ and the domain of the function f_i is $\{z$: $1 < |z| < r_i\}$. By taking further subsequences, we obtain the function f_i for each curve α_i. Let $A_i =$ the image of f_i. It is a simple matter to show that the A_i are disjoint from each other, because, if not, one could find n such that the intersection of A_{in} and A_{jn} is nonempty for some $i \neq j$. By construction, the annuli A_i achieve the minimum in (4) and are a system of annuli compatible with the system of curves α_i.

10.3. WEYL'S LEMMA

Ultimately, we wish to show that the extremal property (2) satisfied by the family of annuli in Theorem 1 yields a holomorphic quadratic differential with closed trajectories. It turns out that the extremal property forces an infinitesimal condition which, by Weyl's lemma [Ah1], causes a particular quadratic differential to be holomorphic.

We first state Weyl's lemma in local form and then give a corollary which is adapted to L_1-quadratic differentials on a Riemann surface. The proof of Weyl's lemma is relatively elementary and appears in many places. The reader is referred to Bers [Ber1], Farkas and Kra [FarK], Kra [Kr5], and Springer [Sp].

We need several definitions. Let D be a plane domain and suppose f is an L_1-function on D. If f is of class C^1, then the complex derivative $f_{\bar{z}}$ is defined by

$$f_{\bar{z}}(z) = \frac{1}{2}\left(\frac{\partial f}{\partial x} + i\,\frac{\partial f}{\partial y}\right) \tag{5}$$

and if φ is any C^1 function with compact support in D, then integration by parts yields

$$\iint_D f_{\bar{z}}(z)\varphi(z)\,dx\,dy = -\iint_D f(z)\varphi_{\bar{z}}(z)\,dx\,dy. \tag{6}$$

In order to avoid indication of the domain of integration, we assume the functions φ are extended to the whole plane by letting them be identically zero outside D. We adopt the convention that when the domain of integration is not indicated it is assumed to be the whole complex plane.

If f is an L_1-function (but not necessarily C^1), we say $f_{\bar{z}} = h$ in D in the sense of distributions if

$$\iint h\varphi\,dx\,dy = -\iint f\varphi_{\bar{z}}\,dx\,dy \tag{7}$$

for all C^1-functions with compact support in D.

Theorem 2. (*Weyl's Lemma*). *Suppose f is an L_1-function in a domain D and $f_{\bar{z}} \equiv 0$ in the sense of distributions. Then there exists a function $\tilde{f}$ holomorphic in D such that $\tilde{f} = f$ almost everywhere.*

Corollary. *Let φ be an L_1-quadratic differential on a Riemann surface R. Suppose that every point p has a neighborhood N and a local coordinate $z\colon N \to \mathbb{C}$ such that*

$$\iint \varphi(z)g_{\bar{z}}(z)\,dx\,dy = 0 \tag{8}$$

for all C^1-functions g with compact support in $z(N)$. Then $\varphi(z)$ is equal almost everywhere to a holomorphic quadratic differential $\tilde{\varphi}$ on R.

Proof. It suffices to show that the expression for φ in terms of any local coordinate z is holomorphic in the coordinate neighborhood where z is defined. In that coordinate neighborhood the hypothesis of the corollary assures that $\varphi_{\bar{z}} \equiv 0$ in the sense of distributions. The conclusion follows.

10.4. EXISTENCE OF JENKINS–STREBEL DIFFERENTIALS WITH PRESCRIBED HEIGHTS

Our objective is to show that the annuli A_i which achieve the minimum in Theorem 1 determine a holomorphic quadratic differential φ in $A(R)$ satisfying (a), (b), (c), and (d) listed in the introduction to this chapter.

Theorem 3. *Let b_j be a set of positive constants for $1 \leqslant j \leqslant k$ and let $\{\alpha_j: 1 \leqslant j \leqslant k\}$ be an admissible system of curves on a Riemann surface R of type (g, n) with $3g - 3 + n > 0$. Then there exists a holomorphic quadratic differential φ with properties* (a), (b), (c), *and* (d) *listed in the introduction to this chapter. The maximal annuli swept out by homotopic noncritical horizontal closed trajectories of φ are the same annuli that realize the minimum in* (2).

Proof. We select a set of annuli $A_1, \ldots, A_k$ with moduli $m_1, \ldots, m_k$ which are given by Theorem 1. By the uniformization theorem, there are conformal mappings z_j from A_j minus an arc β_j connecting the two boundary contours of A_j onto a rectangle R_j which has width a_j and height b_j, where b_j is given and a_j is chosen so the modulus m_j of A_j is equal b_j/a_j.

In the domain A_j we define $\varphi(z)dz^2 = dz_j^2$. Since the domains A_j do not overlap, we obtain an L_1-quadratic differential φ on R by letting $\varphi = 0$ on $R - \bigcup_{j=1}^k A_j$. Thus, $\|\varphi\| = \sum_{j=1}^k a_j b_j$. We must show that φ is equal almost everywhere to a holomorphic quadratic differential whose regular horizontal trajectories sweep out the annuli A_j and such that the set $R - \bigcup_{j=1}^k A_j$ consists of a finite number of critical trajectories of finite length and a finite number of critical points of φ.

Select any point p on R and a neighborhood N of p and a local parameter $z: N \to \mathbb{C}$. Let h be any C^1 complex valued function with compact support in N. The function $z \mapsto w = z + \varepsilon h(z)$ in $z(N)$ and extended to be the identity outside $z(N)$ is a quasiconformal homeomorphism for sufficiently small ε since its dilatation is

$$\nu(z) = \frac{w_{\bar{z}}}{w_z} = \frac{\varepsilon h_{\bar{z}}}{1 + \varepsilon h_{\bar{z}}}. \tag{9}$$

If we let N be simply connected, then the mapping $z \mapsto w$ is homotopic to the identity. This mapping transforms the ring domains A_j into ring domains A_j'

and, since these ring domains form an admissible and competing system, we obtain an inequality from the fact that the A_j achieve the minimum in (4).

Let $A'_j = w(A_j)$ and $z'_j\colon A'_j \to R'_j$ be a holomorphic mapping from the annulus A'_j cut along an analytic arc joining its two boundary contours onto the rectangle R'_j with base a'_j, height b_j, and modulus $m'_j = b_j/a'_j$.

Now $z'_j \circ w \circ (z_j)^{-1}$ is a mapping from R_j to R'_j. If we view v as a function of the local parameter z_j, we see that

$$a'_j \leqslant \int_{w(\alpha)} |dz'_j| = \int_\alpha |w_z\, dz + w_{\bar z} d\bar z| = \int_\alpha |w_z||1 + v|\, dx_j, \tag{10}$$

where α is a horizontal line segment going across the rectangle R_j. Integration of (10) with respect to dy_j from 0 to b_j leads to

$$a'_j b_j \leqslant \iint_{R_j} |w_z||1 + v|\, dx_j\, dy_j.$$

Summing over j yields

$$\sum a'_j b_j \leqslant \iint_{\bigcup R_j} |w_z||1 + v|\, dx\, dy. \tag{11}$$

In the right-hand side of (11), introduce a factor of $(1 - |v|^2)^{1/2}$ in the numerator and denominator and apply Schwarz's inequality. The result is

$$\left(\sum a'_j b_j\right)^2 \leqslant \left(\iint_{\bigcup R_j} |w_z|^2(1 - |v|^2)\, dx\, dy\right)\left(\iint_{\bigcup R_j} \frac{|1 + v|^2}{1 - |v|^2}\, dx\, dy\right). \tag{12}$$

Each of the integrals $\iint_{R_j} |w_z|^2(1 - |v|^2)\, dx\, dy$ is equal to $\iint_{R_j} dx'_j\, dy'_j = a'_j b_j$ and so we obtain

$$\sum a'_j b_j \leqslant \iint_{\bigcup R_j} \frac{|1 + v|^2}{1 - |v|^2}\, dx\, dy. \tag{13}$$

Over each part R_j of the domains of integration in the right-hand sides of (11), (12), and (13) the variable of integration is $z_j = x_j + iy_j$ and $dx\, dy = dx_j\, dy_j$.

The fact that $\sum b_j^2/m_j$ is minimum among all admissible systems implies $\sum a_j b_j \leqslant \sum a'_j b_j$ and, therefore,

$$\sum a_j b_j \leqslant \iint_{\bigcup R_j} \frac{|1 + v|^2}{1 - |v|^2}\, dx\, dy. \tag{14}$$

Recall that the quadratic differential $\varphi(z)dz^2$ is defined by

$$\varphi(z)dz^2 = \begin{cases} dz_j^2 & \text{in } A_j, \quad 1 \leqslant j \leqslant k \\ 0 & \text{in } R - \bigcup_{j=1}^{k} A_j. \end{cases} \tag{15}$$

Therefore, in R_j, $dz_j = \pm\sqrt{\varphi(z)}\,dz$ and $dx_j\,dy_j = (1/2i)\,dz_j\,\overline{dz_j} = (1/2i)|\varphi(z)|\,dz\,\overline{dz} = |\varphi(z)|\,dx\,dy$. Moreover, in R_j

$$\begin{aligned} \nu(z_j) &= \nu(z)\left(\frac{\overline{dz}}{\overline{dz_j}}\right)\Big/\left(\frac{dz}{dz_j}\right) = \nu(z)\varphi^{1/2}(z)/(\overline{\varphi(z)})^{1/2} \\ &= \nu(z)\varphi(z)/|\varphi(z)|. \end{aligned}$$

Since φ is zero outside all of the annuli, inequality (14) can be rewritten in the invariant form

$$\iint_R |\varphi(z)|\,dx\,dy \leqslant \iint_R \frac{|1 + \nu(z)\varphi(z)/|\varphi(z)||^2}{1 - |\nu(z)|^2}\,|\varphi(z)|\,dx\,dy. \tag{16}$$

Rewriting the numerator in the right-hand side as

$$\{1 - |\nu|^2 + 2\,\mathrm{Re}\,\nu\varphi/|\varphi| + 2|\nu|^2\}|\varphi(z)|$$

and simplifying yields

$$0 \leqslant \iint_R \frac{\mathrm{Re}\,\nu\varphi + |\nu|^2|\varphi|}{1 - |\nu|^2}\,dx\,dy. \tag{17}$$

Substituting (9) into (17), dividing by ε, and taking the limit as $\varepsilon \to 0$ we find that

$$0 = \iint_N \varphi(z)h_{\bar{z}}(z)\,dx\,dy$$

for every C^1 complex valued function $h(z)$ defined in the arbitrary simply connected coordinate chart $z(N)$. By the corollary to Weyl's lemma, this shows there is a holomorphic quadratic differential $\tilde{\varphi}$ which is equal almost everywhere to $\varphi(z)$ in (15). Thus $\tilde{\varphi}(z) \equiv dz_j^2$ in A_j and the set $R - \bigcup_{j=1}^{k} A_j$ has measure zero. Moreover, A_j is swept out by closed trajectories of $\tilde{\varphi}$. If any boundary point of A_j lies on a horizontal ray of $|\varphi|$-length more than a_j, then that ray must be closed and the ring domain A_j would lie in a larger ring domain $\tilde{A}_j$ which does not intersect any of the other A_k, for $k \neq j$. This would contradict the fact that the sum for (2) is minimal. Thus all critical horizontal trajectories have finite length and $R - \bigcup_{j=1}^{k} A_j$ consists of these critical trajectories and singular points of the quadratic differential. It is clear that the annuli A_j and the quadratic differential φ have properties (a), (b), (c), and (d) listed in the introduction to this chapter.

10.5. UNIQUENESS OF JENKINS–STREBEL DIFFERENTIALS

Let the system of admissible curves α_j, $1 \leqslant j \leqslant k$, on R and the positive constants b_j, $1 \leqslant j \leqslant k$, be given. Then every simple closed curve γ on R determines a height by the formula

$$h(\gamma) = \sum_{j=1}^{k} [b_j \times (\text{the number of points in } \gamma \cap \alpha_j)]. \tag{18}$$

Let

$$h[\gamma] = \inf h(\gamma'), \tag{19}$$

where the infinum is taken over all curves γ' freely homotopic to γ. It is clear that for a curve γ transversal to the horizontal trajectories of the quadratic differential φ constructed in Theorem 3 one has

$$h[\gamma] = \int_{\gamma} |\mathrm{Im}\sqrt{\varphi}\, dz|. \tag{20}$$

In this way, the following theorem can be seen as a consequence of Theorem 8 of Chapter 2. We prefer to prove it directly.

Theorem 4. *The annuli obtained as solutions to the extremal problem* (2) *for the admissible curve system $\alpha_1, \ldots, \alpha_k$ and for the positive constants $b_1, \ldots, b_k$ are uniquely determined. Moreover, the associated holomorphic quadratic differential is uniquely determined by the curve system $\alpha_1, \ldots, \alpha_k$ and the heights b_i of the annuli A_i.*

Proof. From Theorem 3, we know that if we have another system of annuli $A'_1, \ldots, A'_k$ with core curves homotopic to $\alpha_1, \ldots, \alpha_k$ and with moduli $m'_1, \ldots, m'_k$ which realize the minimum in (2), then there is an associated holomorphic quadratic differential φ'. The height of each annulus A'_j measured in the metric $|\sqrt{\varphi'}\, dz|$ is b_j. The same is true for the height of each annulus A_j measured in the metric $|\sqrt{\varphi}\, dz|$.

Let the modulus of A'_j be $m'_j = b_j/a'_j$ and the modulus of A_j be $m_j = b_j/a_j$. Let α_j and α'_j be horizontal closed trajectories for the quadratic differentials φ_j and φ'_j going around the annuli A_j and A'_j. Since α has the shortest width of any curve freely homotopic to it measured in the metric $|\mathrm{Re}\sqrt{\varphi}\, dz|$, we see that

$$a_j = \int_{\alpha_j} |\mathrm{Re}\sqrt{\varphi}\, dz| \leqslant \int_{\alpha_j'} |\mathrm{Re}\sqrt{\varphi}\, dz|. \tag{21}$$

Using a natural parameter $z' = x' + iy'$ for the quadratic differential φ' and integrating (21) over the region A_j', we obtain

$$a_j b_j \leqslant \iint_{A_j'} |\mathrm{Re}\sqrt{\varphi(z')}|\, dx'\, dy'. \tag{22}$$

[The integral (22) is not invariant under changes of local parameter.] Since a natural parameter z' is given by the formula

$$z' = \int_{z_0'}^{z'} \sqrt{\varphi'(\zeta)}\, d\zeta, \tag{23}$$

one sees that $\varphi'(z')\, dz'^2 = dz'^2$. Thus (22) yields the following invariant form:

$$a_j b_j \leqslant \iint_{A_j'} |\sqrt{\varphi(z)}\sqrt{\varphi'(z)}|\, dx\, dy. \tag{24}$$

From (22) and (24), one deduces

$$\begin{aligned}
\sum_{j=1}^{k} a_j b_j &\leqslant \sum_{j=1}^{k} \iint_{A_j'} |\mathrm{Re}\sqrt{\varphi(z')}|\, dx'\, dy' \leqslant \iint_R |\sqrt{\varphi}\sqrt{\varphi'}|\, dx\, dy \\
&\leqslant \left(\iint_R |\varphi|\right)^{1/2} \left(\iint |\varphi'|\right)^{1/2} \\
&= \left(\sum_{j=1}^{k} a_j b_j\right)^{1/2} \left(\sum_{j=1}^{k} a_j' b_j\right)^{1/2}.
\end{aligned} \tag{25}$$

By hypothesis, we know that the minimum in (2) is achieved by both $\sum_{j=1}^{k} b_j^2/m_j = \sum_{j=1}^{k} a_j b_j$ and by $\sum_{j=1}^{k} b_j^2/m_j' = \sum_{j=1}^{k} a_j' b_j$. This implies $\sum a_j' b_j = \sum a_j b_j$ and we have equality everywhere in (25). By the uniqueness part of Schwarz's inequality, we see that $\varphi' = c\varphi$ for some $c \neq 0$. Since $\|\varphi'\| = \sum a_j' b_j = \sum a_j b_j = \|\varphi\|$, this implies $|c| = 1$. Moreover, $|\mathrm{Re}\sqrt{\varphi(z')}| \equiv 1$ for any natural parameter z' for the quadratic differential φ'. It is also obvious from these inequalities that $\mathrm{Im}\sqrt{\varphi(z')} \equiv 0$. Since $\varphi'(z') \equiv 1$, we conclude that $\varphi' \equiv \varphi$.

Notes. Most of the theorems of this chapter are proved in greater generality in Strebel's book [St4]. Strebel finds differentials all of whose noncritical trajectories are closed as solutions to several different extremal problems. We refer to his book for further references on the topic of differentials with closed trajectories. Jenkins also proved the existence of holomorphic quadratic differentials with closed trajectories [Je1]. The fact that the heights of the cylinders for an admissible system can be arbitrarily prescribed is proved by Hubbard and Masur [HuM] and by Renelt [Ren] as well as in Strebel's book [St4]. The fact that these heights uniquely determine the holomorphic differential follows from the "heights principle" of Marden and Strebel [MarS] and is also a consequence of Strebel's technique [St4]. The result of exercise 4 in this chapter appears in [Ga2], Also see [ScSp].

EXERCISES

1. Consider the quadratic differential $\varphi(z)\,dz^2 = dz^2/z(z-1)(z-\alpha)$ on the sphere with four punctures at 0, 1, ∞, and α. Assume $\alpha > 1$. Show that this is a Jenkins–Strebel differential and sketch its regular closed trajectories.
2. Find the values of θ for which $e^{i\theta}\varphi(z)\,dz^2$ is a Jenkins–Strebel differential where $\varphi(z)\,dz^2$ is the differential in Exercise 1.
3. Let R be a compact Riemann surface and μ an infinitesimally trivial Beltrami coefficient on R with support in a closed subset S of R. Show that there is a curve ν_t of trivial Beltrami coefficients with support in S and such that $\nu_t(z) = t\mu(z) + o(t)$, where the estimate is uniform in the variable z.
4. Let f be a quasiconformal homomorphism from an annulus $R = \{z\colon 1 \leqslant |z| \leqslant \rho\}$ onto an annulus $\tilde{R} = \{z\colon 1 \leqslant |z| \leqslant \tilde{\rho}\}$ and let $\mu(z) = f_{\bar{z}}/f_z$. For $|t| \leqslant 1$ let $f^{t\mu}$ be a quasiconformal mapping with Beltrami coefficient $t\mu$ which maps R onto the annulus $R_{t\mu} = \{z\colon 1 \leqslant |z| \leqslant \rho_t\}$. Thus the modulus of the annulus $R_{t\mu}$ is $M(t\mu) = (\log \rho_t)/2\pi$. Show that

$$\log M(t\mu) = \log M(0) + 2\,\mathrm{Re} \iint_R \mu\varphi \, dx\, dy + o(t)$$

where $\varphi(z) = z^{-2}(2\pi \log \rho)^{-1}$.

5. Let γ be a simple closed curve, not homotopically trivial, on a compact surface R. Let A be an annulus on R with core curve homotopic to γ and suppose the modulus of A is at least as large as the modulus of any other annulus embedded in R with core curve homotopic to γ. Let z be a holomorphic homeomorphism from A onto $\{z\colon 1 < |z| < \rho\}$. Show that dz^2/z^2 is the restriction to A of a holomorphic quadratic differential on R and that the boundary of A consists of critical trajectories.

11

MEASURED FOLIATIONS

The main objective of this chapter is to solve a type of Dirichlet problem associated to measured foliations on a Riemann surface of finite analytic type. The data for the Dirichlet problem consists of the complex structure for the surface and the heights with respect to a given measured foliation of the homotopy classes of simple closed curves on the surface.

The associated Dirichlet problem is an extremal problem, namely, to find the minimum possible L_1-norm of a continuous quadratic differential ψ on the Riemann surface subject to certain side conditions. The side conditions are that for every simple closed curve γ on R, not homotopically trivial and not homotopic to a puncture, the height of the homotopy class of γ with respect to ψ is larger than or equal to the height of the homotopy class of γ with respect to the given measured foliation.

The Dirichlet principle states that there exists a unique extremal quadratic differential φ which realizes the minimum. Moreover, that quadratic differential is holomorphic except for possibly simple poles at the punctures of R and its heights are exactly equal to the heights of the given measured foliation. The uniqueness of the solution φ depends on the length–area method and the second minimum norm principle as expressed in Theorem 8 of Chapter 2.

The existence depends on two important facts. The first is the existence of Jenkins–Strebel differentials with prescribed heights, as expressed in the results of Chapter 10. The second is a topological result due to Thurston which indicates how the measure classes of measured foliations appear as a kind of completion of the set of homotopy classes of simple closed curves on the surface. By multiplying such a homotopy class by an arbitrary positive number there is a natural way in which it becomes a ray of measured foliations. Thurston shows that the set of such rays is dense in the space of all measure classes of measured foliations. We do not prove the result in this book. It follows most naturally after the introduction of local coordinates for the space of measured foliations, the so-called *MST*-coordinates. This approach is described in complete detail in

Travaux de Thurston sur les Surfaces by A. Fathi, F. Laudenbach, and V. Poénaru [FatLP].

We point out here why the L_1-norm of a continuous quadratic differential ψ is a generalization of the Dirichlet integral. On the surface R the L_1-norm of ψ is

$$\|\psi\| = \iint_R |\psi(z)|\, dx\, dy.$$

In a neighborhood of any point where ψ is not zero select a local square root $\sqrt{\psi}$. Then $\sqrt{\psi}$ is, up to plus or minus sign, a differential 1-form at points of R where ψ is not zero. Locally we can solve the equation

$$g_z = \sqrt{\psi},$$

where $g_z = \frac{1}{2}(g_x - ig_y)$ and $g = u + iv$. We see that

$$\iint_R |\psi|\, dx\, dy = \iint_R |g_z|^2\, dx\, dy. \tag{1}$$

This is the quantity to be minimized. The side conditions are inequalities for line integrals over certain curves γ. The inequalities take the form

$$\int_\gamma |\mathrm{Im}(\sqrt{\psi(z)}\, dz)| = \frac{1}{2} \int_\gamma |(v_x - u_y)\, dx + (u_x + v_y)\, dy| \geqslant c(\gamma), \tag{2}$$

where $c(\gamma)$ depends only on the homotopy class of γ. Notice that when the locally defined functions u and v satisfy the Cauchy–Riemann equations $u_x = v_y$ and $u_y = -v_x$, the integral (1) becomes

$$\iint_\omega (v_y^2 + v_x^2)\, dx\, dy$$

and the line integral (2) becomes

$$\int_\gamma |v_x\, dx + v_y\, dy| = \int_\gamma |dv|.$$

After establishing the Dirichlet principle we investigate how the quantity to be minimized in the Dirichlet problem varies over Teichmüller space. In particular, we find a formula for the first variation which contains as a special case certain formulas for the variation of extremal length as a function on Teichmüller space.

11.1. DEFINITION OF A MEASURED FOLIATION

Let R be a compact Riemann surface of genus g. Fix once and for all a set of n distinct points $\{q_1, \ldots, q_n\}$ on R and let $\dot{R} = R - \{q_1, \ldots, q_n\}$. We allow, as a special case, the set $\{q_1, \ldots, q_n\}$ to be empty. The points $q_1, \ldots, q_n$ are called the punctures of $\dot{R}$.

A measured foliation $|dv|$ on $\dot{R}$ with singularities $p_1, \ldots, p_m$ of order $k_1, \ldots, k_m$ at points of R is given by an open covering $\{U_i\}$ of $\dot{R} - \{p_1, \ldots, p_m\}$ and C^1 real valued functions v_i on U_i such that

(a) $|dv_i| = |dv_j|$ on each $U_i \cap U_j$;

(b) at each point p of R there is a neighborhood V of p in R and a local C^1-chart (u, v): $V \to \mathbb{R}^2$ such that, for $z = u + iv$,

$$|dv_i| = |\mathrm{Im}(z^{k/2}\, dz)| \text{ on } U_i \cap V \qquad \text{for some integer } k,$$

(c) the integer k in (b) depends on the location of the point p in the following way:
If p is not a singularity of the measured foliation, then $k = 0$.
If p is one of the singular points p_j but not one of the punctures $\{q_1, \ldots, q_n\}$, then $k = k_j > 0$.
If p is one of the singular points p_j and p_j coincides with some q_j, then $k = k_j \geqslant -1$ and $k \neq 0$.

The leaves of the foliation are curves along which v is constant. The height of an arc γ with respect to the foliation is defined by

$$h_v(\gamma) = \int_\gamma |dv|.$$

The height of a homotopy class $[\gamma]$ of simple closed curves is

$$h_v[\gamma] = \inf h_v(\gamma'),$$

where the infimum is taken over all simple closed curves γ' homotopic to γ. Two measured foliations $|dv_1|$ and $|dv_2|$ are called measure equivalent if $h_{v_1}[\gamma] = h_{v_2}[\gamma]$ for all simple closed curves γ on $\dot{R}$ which are not homotopically trivial and not homotopic to a puncture of $\dot{R}$.

We denote the space of measure equivalence classes of measured foliations on $\dot{R}$ by $\mathfrak{M}\mathscr{F}(\dot{R})$, or simply by $\mathfrak{M}\mathscr{F}$, if there is no possible ambiguity. Let $\mathscr{S}$ be the set of homotopy classes of simple closed curves on $\dot{R}$ which are not homotopically trivial and not homotopic to punctures. Any measure class of measured foliations in $\mathfrak{M}\mathscr{F}$ represented by a measure $|dv|$ determines a function from $\mathscr{S}$

into $\mathbb{R}$ by

$$[\gamma] \mapsto h_v[\gamma].$$

Thus h_v is an element of the product space $\mathbb{R}^{\mathscr{S}}$ and the mapping

$$h^*: \mathfrak{M}\mathscr{F} \to \mathbb{R}$$

given by $|dv| \mapsto (\gamma \mapsto h_v[\gamma])$ is injective because of the equivalence relation which defines $\mathfrak{M}\mathscr{F}$. Therefore, the product topology on $\mathbb{R}^{\mathscr{S}}$ induces a topology on $\mathfrak{M}\mathscr{F}$. We shall call the mapping h^* the heights mapping.

A nonzero holomorphic quadratic differential $\varphi(z)\,dz^2$ on $\dot{R}$ which has at most simple poles at the distinguished points $\{q_1, \ldots, q_n\}$ yields a measured foliation in a natural way. In a sufficiently small simply connected neighborhood V of any point p of R we define the natural parameter

$$\zeta(p) = \int_{p_0}^{p} \sqrt{\varphi(z)}\,dz.$$

Of course, this can be a multivalued function when φ has a zero in the neighborhood V, but in any case ζ is well defined up to plus or minus sign. On letting $\zeta = u + iv$, the absolute value $|dv| = |\mathrm{Im}(\sqrt{\varphi(z)}\,dz)|$ is well defined and gives a measure for a measured foliation. The leaves of the foliation are the horizontal trajectories of the quadratic differential φ. We denote by $h_\varphi[\gamma]$ the infimum of the integrals

$$\int_{\tilde{\gamma}} |\mathrm{Im}(\sqrt{\varphi(z)}\,dz)|,$$

where $\tilde{\gamma}$ can be any curve homotopic to γ.

Let $A(\dot{R})$ be the vector space of all holomorphic quadratic differential on $\dot{R}$ which have at most simple poles at the distinguished points $q_1, \ldots, q_n$. The above construction defines the mapping

$$\Phi: (A(\dot{R}) - \{0\}) \to \mathfrak{M}\mathscr{F}(\dot{R}), \tag{3}$$

which associates to any nonzero holomorphic quadratic differential φ the measured foliation $|\mathrm{Im}(\sqrt{\varphi}\,dz)|$. One of the objectives of this chapter is to show that Φ is a homeomorphism and to prove, in particular, the following theorem first proved by Hubbard and Masur for compact Riemann surfaces ([HuM]).

Theorem 1. *Given a measured foliation $|dv|$ on $\dot{R}$ and a complex structure on $\dot{R}$, there exists a unique holomorphic quadratic differential in $A(\dot{R}) - \{0\}$ such that the foliation given by the horizontal trajectories of φ and the measure $|\mathrm{Im}(\sqrt{\varphi}\,dz)|$ is measure equivalent to $|dv|$. Moreover, the mapping Φ in (3) is a homeomorphism.*

The next theorem is the Dirichlet principle described in the chapter introduction. Let the complex structure on $\dot{R}$ be given and let $|dv|$ be a measured foliation representing some equivalence class in $\mathfrak{M}\mathscr{F}(\dot{R})$. For each continuous quadratic differential ψ on $\dot{R}$ and each γ in $\mathscr{S}$ let $h_\psi[\gamma]$ be the infimum of the numbers

$$\int_{\tilde{\gamma}} |\operatorname{Im} \sqrt{\psi(z)}\, dz|,$$

where $\tilde{\gamma}$ is any simple closed curve homotopic to γ. Let

$$M(|dv|) = \inf\left\{\iint_R |\psi(z)|\, dx\, dy\colon h_\psi[\gamma] \geqslant h_v[\gamma] \text{ for all } \gamma \text{ in } \mathscr{S}\right\}. \tag{4}$$

Theorem 2. *Given any measured foliation $|dv|$ on a Riemann surface $\dot{R}$ of finite analytic type, the infimum in* (4) *is realized by a unique quadratic differential $\varphi(z)\, dz^2$. That quadratic differential is holomorphic on $\dot{R}$ and has at most simple poles at the punctures of $\dot{R}$. For any homotopy class of simple closed curves γ in $\mathscr{S}$, one has $h_\varphi[\gamma] = h_v[\gamma]$.*

The proofs of Theorems 1 and 2 will be given in Sections 11.2 through 11.6.

11.2. INJECTIVITY OF THE HEIGHTS MAPPING

Theorem 3. (*The Heights Theorem*). *The mapping*

$$\Phi\colon A(\dot{R}) \to \mathbb{R}^{\mathscr{S}}$$

which assigns to a holomorphic quadratic differential φ in $A(\dot{R})$ the element

$$(h_\varphi[\gamma])_{\gamma \in \mathscr{S}}$$

of the product space $\mathbb{R}^{\mathscr{S}}$ is injective.

Proof. Apply Theorem 8 of Chapter 2. Suppose ψ is another element of $A(\dot{R})$ and $h_\psi[\gamma] = h_\varphi[\gamma]$ for all γ in $\mathscr{S}$. Then, on the one hand, $\|\varphi\| \leqslant \|\psi\|$. On applying the same theorem to the holomorphic quadratic differential ψ, we obtain $\|\psi\| \leqslant \|\varphi\|$. But the theorem also says one can have equality only if $\varphi = \psi$. Q.E.D.

11.3. CONTINUITY OF THE HEIGHTS MAPPING

Our objective in this section is to show the mapping Φ in Theorem 3 is continuous. A much stronger result is true as has been pointed out by Strebel [St4]. For the benefit of the reader we state Strebel's result without proof and then go on to give a direct method for proving the continuity of the mapping Φ defined in Theorem 3.

A Theorem of Strebel. *Let φ_n be a sequence of holomorphic quadratic differentials on an arbitrary Riemann surface R which converge locally uniformly to a quadratic differential φ. Then, for every homotopy class of closed loops γ on R,*

$$\lim_{n\to\infty} h_{\varphi_n}[\gamma] = h_\varphi[\gamma].$$

We point out that one part of this result is elementary, namely, to show that $\overline{\lim}\, h_{\varphi_m}[\gamma] \leqslant h_\varphi[\gamma]$. To see this, choose $\varepsilon > 0$ and select γ_0 in the homotopy class of γ such that

$$\int_{\gamma_0} |\mathrm{Im}\sqrt{\varphi}\, dz| < h_\varphi[\gamma] + \varepsilon.$$

and such that γ_0 does not pass through any singular points of φ. Since φ_n converges uniformly on γ_0 to φ, we see that

$$\int_{\gamma_0} |\mathrm{Im}\sqrt{\varphi_n}\, dz| \to \int_{\gamma_0} |\mathrm{Im}\sqrt{\varphi}\, dz|.$$

Therefore $\overline{\lim}\, h_{\varphi_n}[\gamma] \leqslant h_\varphi[\gamma]$. For the other half of Strebel's theorem we refer to Strebel [St4, pp. 162–165]. The chief difficulty in proving Strebel's result is that the quantity $h_\varphi[\gamma]$ need not be realized by any curve in the homotopy class of γ. To see this, consider the most elementary example: R is the annulus $\{z: 1 < |z| < 2\}$, $\varphi(z) \equiv 1$ (so the associated φ-metric is the Euclidean metric) and γ is a curve which winds once around the annulus. As curves in the homotopy class of γ are selected so as to more and more nearly realize the infimum $h_\varphi[\gamma]$ they must approach the inner boundary contour of the annulus R in such a way that they lie between the horizontal lines $y = 1 + \varepsilon$ and $y = -1 - \varepsilon$.

For the proof of Theorem 4 this is not a difficulty because in any homotopy class of $\mathscr{S}$ for the surface there always exists a geodesic (which, however, may not be unique).

Theorem 4. *Let $\dot{R}$ be a Riemann surface which is compact except for a finite number of punctures. Let $\gamma \in \mathscr{S}$ and $\varphi_n \in A(\dot{R})$ and suppose $\varphi_n \to \varphi$. Then*

$$\lim_{n\to\infty} h_{\varphi_n}[\gamma] = h_\varphi[\gamma].$$

Proof. The proof is divided into several steps which we only outline.

STEP 1. In any homotopy class γ in $\mathscr{S}$ and for any nonzero holomorphic quadratic differential φ in $A(\dot{R})$, there exists a geodesic γ_0 with respect to the metric $|\varphi|^{1/2}|dz|$, where γ_0 is homotopic to γ. The curve γ_0 is simple and is made up of straight line segments in the φ-metric. If γ_0 has any vertices, they can occur only at the singularities of φ.

STEP 2. For the φ-geodesic γ_0 constructed in (1) and any curve γ_1 homotopic to γ,

$$\int_{\gamma_1} |\mathrm{Im}\sqrt{\varphi}\,dz| \geqslant \int_{\gamma_0} |\mathrm{Im}\sqrt{\varphi}\,dz|.$$

That is, a φ-geodesic γ_0 realizes the height of γ.

STEP 3. For φ_n converging to φ, there exists a sequence of φ_n-polygons γ_n which are homotopic to γ, which realize the φ_n-height of γ, and which converge to some φ-geodesic γ_0 in the homotopy class of γ.

STEP 4. The uniform convergence of φ_n to φ in a tubular neighborhood of γ_0 implies

$$\int_{\gamma_n} |\mathrm{Im}\sqrt{\varphi_n}\,dz| \to \int_{\gamma_0} |\mathrm{Im}\sqrt{\varphi}\,dz|,$$

which from step 2 shows that $\lim h_{\varphi_n}[\gamma] = h_{\varphi}[\gamma]$.

11.4. CONVERGENCE OF HEIGHTS IMPLIES CONVERGENCE OF QUADRATIC DIFFERENTIALS

From the previous two sections we know that the mapping

$$\Phi\colon A(\dot{R}) \to \mathbb{R}^{\mathscr{S}}$$

is injective and continuous, where $\mathbb{R}^{\mathscr{S}}$ is assumed to have the product topology. We now show that Φ^{-1}, defined on the image of Φ, is also continuous. Although the continuity of Φ does not depend on the finite dimensionality of $A(\dot{R})$, for the continuity of Φ^{-1}, compactness of the unit ball in $A(\dot{R})$ is essential.

Lemma 1. *Suppose φ_n is a sequence of quadratic differentials in $A(\dot{R})$ and suppose for every γ in $\mathscr{S}$ the sequence $h_{\varphi_n}[\gamma]$ converges. Then φ_n converges.*

Proof. We first show that if for every γ in $\mathscr{S}$ there exists a constant $M(\gamma)$ for which

$$h_{\varphi_n}[\gamma] \leqslant M(\gamma) \qquad \text{for all } n,$$

then φ_n is a bounded sequence. If not, one could find a subsequence (which we may also denote by φ_n) such that $\|\varphi_n\| \to \infty$. Then $\tilde{\varphi}_n = \varphi_n/\|\varphi_n\|$ is a sequence of vectors of norm one. By taking a subsequence (for which we again use the same subscript) we obtain $\tilde{\varphi}_n \to \tilde{\varphi}$, where $\|\tilde{\varphi}\| = 1$. Since $\|\tilde{\varphi}\| = 1$, there must be at least one curve γ for which $h_{\tilde{\varphi}}[\gamma] > 0$. For that curve, we have $h_{\tilde{\varphi}_n}[\gamma] \to h_{\tilde{\varphi}}[\gamma]$ and

$$h_{\tilde{\varphi}_n}[\gamma] = \frac{1}{\|\varphi_n\|} h_{\varphi_n}[\gamma] \leqslant \frac{M(\gamma)}{\|\varphi_n\|}.$$

Thus $\overline{\lim} \|\varphi_n\| \leqslant M(\gamma)/h_{\tilde{\varphi}}[\gamma]$, and therefore it is not possible for $\|\varphi_n\|$ to approach ∞.

The hypotheses of the lemma imply that $\{\|\varphi_n\|\}$ is a bounded set. Thus, any subsequence of φ_n has a subsequence which converges to some holomorphic quadratic differential. If two such subsequences have limits φ' and φ'', then since $h_{\varphi_n}[\gamma]$ converges for all γ in $\mathscr{S}$, both φ' and φ'' must have the same heights. By Theorem 3, the "heights theorem," this implies $\varphi' = \varphi''$ and thus the original sequence φ_n converges.

11.5. INTERSECTION NUMBERS

Definition. *For any two elements α and β in $\mathscr{S}$ the intersection number $i(\alpha, \beta)$ is the minimum number of points that any two simple loops α' and β' homotopic, respectively, to α and β have in common.*

As usual, we assume $\dot{R}$ is realized as the upper half-plane $\mathbb{H}$ factored by the universal covering group Γ. We know that every element of $\mathscr{S}$ has a unique representative which is a Poincaré geodesic in $\mathbb{H}/\Gamma$. This geodesic is the image under the covering group Γ of the axis of a hyperbolic transformation in Γ.

Lemma 2. *Suppose α_0 and β_0 are closed Poincaré geodesics on $\dot{R}$ representing elements of $\mathscr{S}$. Then $i(\alpha, \beta)$ is the number of points in $\alpha_0 \cap \beta_0$.*

Proof. By a choice of normalization we may select the covering group Γ for $\dot{R}$ so that $\dot{R} = \mathbb{H}/\Gamma$ and α_0 is the image of the axis of the hyperbolic transformation $A(z) = \lambda z$ with $\lambda > 1$. In particular, the segment $[i, i\lambda)$ on the imaginary axis projects to the simple closed geodesic α_0 on $\dot{R}$ and no two points in $[i, i\lambda)$ are identified by an element of Γ.

Let n be the number of points in $\alpha_0 \cap \beta_0$. Then β_0 lifts to n hyperbolic lines $\tilde{\beta}_{01}, \ldots, \tilde{\beta}_{0n}$ which intersect the interval $[i, i\lambda)$ on the imaginary axis. Moreover, since β_0 is simple, no two of the hyperbolic lines $\tilde{\beta}_{0j}$ and $\tilde{\beta}_{0k}$ with j not equal to k can intersect. Now suppose α is a simple curve on $\mathbb{H}/\Gamma$ homotopic to α_0. Then α has a lifting $\tilde{\alpha}$ which is a simple arc in $\mathbb{H}$ approaching 0 at one end and approaching ∞ at the other end. Obviously, the curve $\tilde{\alpha}$ must intersect each of the geodesics $\tilde{\beta}_{0j}$ at least once.

Moreover, if β is simple and homotopic to β_0, then β has liftings to simple arcs $\tilde{\beta}_1, \ldots, \tilde{\beta}_n$ in $\mathbb{H}$ such that each $\tilde{\beta}_j$ has the same endpoints as $\tilde{\beta}_{0j}$ on the boundary of $\mathbb{H}$. Clearly, $\tilde{\alpha}$ must intersect each $\tilde{\beta}_j$, $1 \leqslant j \leqslant n$, at least once. Let $\tilde{P}_j$ be a point in $\tilde{\alpha} \cap \tilde{\beta}_j$ for $1 \leqslant j \leqslant n$. Suppose there is a transformation in Γ which identifies two of the points $\tilde{P}_j$ and $\tilde{P}_k$. Then that transformation identifies two points on $\tilde{\alpha}$ and must therefore be a power of the transformation $A(z) = \lambda z$. Thus, a power of A would identify a point on the curve $\tilde{\beta}_j$ and with a point on the curve $\tilde{\beta}_k$.

Now, any transformation C in Γ which takes a point of $\tilde{\beta}_j$ to a point of $\tilde{\beta}_k$ must transform the whole arc $\tilde{\beta}_j$ into the arc $\tilde{\beta}_k$. The reason for this is that we know the group Γ contains an element D such that $D(\tilde{\beta}_j) = \tilde{\beta}_k$. Therefore, $D \circ C^{-1}$ takes a point of $\tilde{\beta}_k$ to a point of $\tilde{\beta}_k$. Now, β_k is a simple closed curve on $\mathbb{H}/\Gamma$ and every transformation in Γ which identifies two points on $\tilde{\beta}_k$ must be some power of a primitive hyperbolic transformation B_k which has the two endpoints of $\tilde{\beta}_k$ as fixed points and which preserves $\tilde{\beta}_k$. We see that $D \circ C^{-1} = B_k^n$. Thus, $C = B_k^{-n} \circ D$ and so C also takes the arc $\tilde{\beta}_k$ into $\tilde{\beta}_k$.

The transformation C must also take the hyperbolic line $\tilde{\beta}_{0j}$ into the hyperbolic line $\tilde{\beta}_{0k}$ because the endpoints of $\tilde{\beta}_j$ coincide with the endpoints of $\tilde{\beta}_{0j}$ and the endpoints of $\tilde{\beta}_k$ coincide with the endpoints of $\tilde{\beta}_{0k}$.

From the previous two paragraphs we conclude that if the points $\tilde{P}_j$ and $\tilde{P}_k$ are identified by an element of Γ then a power of the transformation $A(z) = \lambda z$ will identify the two nonintersecting hyperbolic lines $\tilde{\beta}_{0j}$ and $\tilde{\beta}_{0k}$ which pass through the segment $[i, i\lambda)$. This is clearly impossible so all of the points $\tilde{P}_1, \ldots, \tilde{P}_n$ determine distinct points on the quotient surface $\mathbb{H}/\Gamma$. We conclude that

$$i(\alpha, \beta) \geqslant \operatorname{card}\{\alpha_0, \beta_0\},$$

and the lemma is proved.

Recall that a set of Jordan curves $\alpha = \{\alpha_1, \ldots, \alpha_n\}$is admissible if no two α_j's intersect, no two α_j's are homotopic to each other, and none of them is homotopically trivial or homotopic to a puncture.

Lemma 3. *Let R be a Riemann surface of finite analytic type. Let $\alpha = \{\alpha_1, \ldots, \alpha_k\}$ be an admissible system of k simple Jordan curves on R ($k \leqslant 3g - 3 + n$). Let $b_1, \ldots, b_k$ be a set of positive numbers and let $\varphi[\alpha]$ be the Jenkins–Strebel differential associated to α whose cylinders corresponding to α_j have heights b_j. Then for every curve σ in $\mathscr{S}$,*

$$h_{\varphi[\alpha]}[\alpha] = \sum_{j=1}^{k} b_j i(\alpha_j, \sigma).$$

Proof. Let $\varphi = \varphi[\alpha]$. For an arbitrary closed loop σ on $\dot{R}$, construct a homotopic loop $\tilde{\sigma}$ which is a φ-polygon (composed of horizontal and vertical segments for φ) and for which

$$\int_{\tilde{\sigma}} |\mathrm{Im}\sqrt{\varphi}\, dz| \leqslant \int_{\sigma} |\mathrm{Im}\sqrt{\varphi}\, dz|.$$

Let $\tilde{\alpha}_j$ be a regular closed trajectory of φ in the cylinder A_j for φ with $\tilde{\alpha}_j$ homotopic to α_j. Any subarc $\tilde{\sigma}_{jm}$ of $\tilde{\sigma}$ which is a crosscut of the cylinder A_j must enter and leave A_j from the same side of A_j or from opposite sides. In the former case $\tilde{\sigma}_j$ can be replaced by a homotopic subarc which does not meet α_j. In the latter case $\tilde{\sigma}_{jm}$ can be replaced by a subarc which intersects α_j only once. The integral of $|\mathrm{Im}\sqrt{\varphi}\, dz|$ over a subarc of this type is $\geqslant b_j$. Therefore,

$$\int_{\sigma} |\mathrm{Im}\sqrt{\varphi}\, dz| \geqslant \int_{\tilde{\sigma}} |\mathrm{Im}\sqrt{\varphi}\, dz| \geqslant \sum_j \sum_m \int_{\tilde{\sigma}_{jm}} |\mathrm{Im}\sqrt{\varphi}\, dz|$$

$$\geqslant \sum_j b_j i(\alpha_j, \sigma).$$

Since σ is arbitrary, $h_\varphi[\sigma] \geqslant \sum_j b_j i(\alpha_j, \sigma)$.

To prove the opposite inequality we choose a φ-polygon $\tilde{\sigma}$ in the same homotopy class as σ in $\dot{R}$ for which $i(\sigma, \alpha_j) = \mathrm{card}\, \tilde{\sigma} \cap \alpha_j$ for each of a previously selected set of horizontal closed trajectories $\tilde{\alpha}_j$ of φ in the associated cylinders A_j. This can be done by using Lemma 2 and constructing a homeomorphism from $\dot{R}$ onto $\dot{R}$ which is homotopic to the identity and takes the curves $\tilde{\alpha}_j$ onto Poincaré geodesics in the same homotopy class. Every subarc $\tilde{\sigma}_{jm}$ of $\tilde{\sigma}$ which joins two sides of A_j can be replaced by a homotopic subarc which is monotone and intersects α_j only once. Those subarcs which join the same sides may be pushed arbitrarily near to one of the sides in such a way that $\tilde{\sigma}$ remains a simple curve and the total height of all of these subarcs is arbitrarily small. For the modified curve $\tilde{\sigma}$, we obtain

$$h_\varphi[\sigma] \leqslant \int_{\tilde{\sigma}} |\mathrm{Im}\sqrt{\varphi}\, dz| \leqslant \sum_j b_j(\mathrm{card}(\tilde{\sigma} \cap \alpha_j)) + \varepsilon.$$

We conclude that

$$h_{\varphi[\alpha]}[\alpha] \leqslant \sum_j b_j i(\sigma, \alpha).$$

It is useful to enlarge the space $\mathscr{S}$ to allow finite unions of simple closed curves which do not intersect each other. Accordingly, we make the following definition.

Definition. *$\mathscr{S}'$ is the set of homotopy classes of finite unions of simple closed curves of $\dot{R}$ with the following properties. If $\gamma_1, \ldots, \gamma_m$ are the components of γ representing an element of $\mathscr{S}'$ then*

(a) *each γ_j is a simple closed curve;*
(b) *no two of the curves γ_j intersect one another; and*
(c) *none of the curves γ_j on $\dot{R}$ is homotopically trivial or homotopic to a puncture.*

A representative γ of an element of $\mathscr{S}'$ does not need to be an admissible system since we permit different components of γ to be in the same homotopy class. Obviously an element of $\mathscr{S}'$ is determined by an admissible system $\alpha_1, \ldots, \alpha_k$ representing k-homotopy classes ($k \leqslant 3g - 3 + n$) and a set of positive integers $m_1, \ldots, m_k$. Each m_j counts the number of components of γ in the homotopy class of α_j.

It is useful to allow the integers $m_1, \ldots, m_k$ to be positive real numbers. Accordingly, we make a further definition.

Definition. *$J\mathscr{S}$ is the set of all formal sums $\sum_{j=1}^{k} b_j \alpha_j$, where b_j are positive numbers and $\alpha_1, \ldots, \alpha_k$ are homotopy classes for an admissible system of curves on $\dot{R}$.*

In an obvious way, $\mathscr{S} \subset \mathscr{S}' \subset J\mathscr{S}$.

By the theorem on the existence of Jenkins–Strebel differentials (Theorem 3 of Section 10.4), $J\mathscr{S}$ is isomorphic to the space of all Jenkins–Strebel differentials on $\dot{R}$. The set $J\mathscr{S}$ determines an element of $\mathbb{R}^{\mathscr{S}}$ by the intersection mapping i^*:

$$i^*\left(\sum_{j=1}^{k} b_j \alpha_j\right) = \left(\sigma \mapsto \sum_{j=1}^{k} b_j i(\alpha_j, \sigma)\right).$$

To show that i^* is injective, consider each of the curves α_j. If α_j divides R into two components, it is possible to find a curve σ with $i(\alpha_j, \sigma) = 2$ and $i(\alpha_m, \sigma) = 0$ for $m \neq j$. If α_j is not dividing, it is possible to find a curve σ with $i(\alpha_j, \sigma) = 1$ and $i(\alpha_m, \sigma) = 0$ for $m \neq j$. In either case, the constants b_j are determined.

Lemma 3 asserts that under the identification of $J\mathscr{S}$ with the space of Jenkins–Strebel, the inclusion i^* of $J\mathscr{S}$ into $\mathbb{R}^{\mathscr{S}}$ is identified with the heights mapping from $A(\dot{R}) - \{0\}$ into $\mathbb{R}^{\mathscr{S}}$. We have the commutative diagram shown in Figure 11.1. Here, the heights mappings h^* and h_1^* are defined by

$$h^*(\varphi) = (\sigma \mapsto h_\varphi[\sigma])$$

and

$$h_1^*(|dv|) = (\sigma \mapsto h_v[\sigma])$$

and all mappings in the diagram are injective.

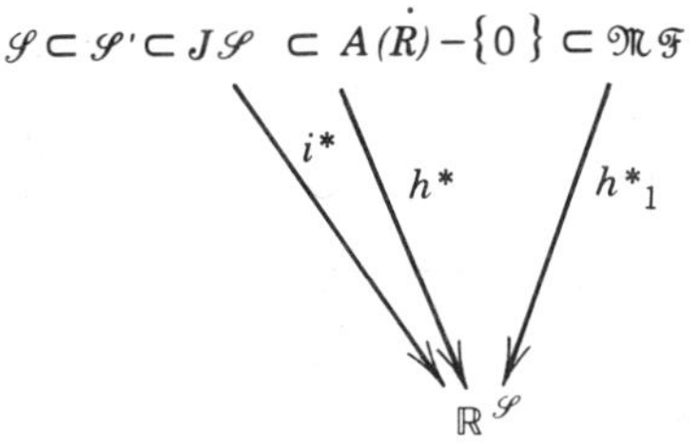

Figure 11.1

11.6. PROJECTIVIZATIONS

The spaces $J\mathcal{S}$, $A(\dot{R}) - \{0\}$, and $\mathfrak{M}\mathcal{F}$ are preserved under multiplication by positive real numbers. We define the projective spaces by specifying that two vectors are equivalent if one can be obtained from the other by multiplication by a positive real number. The diagram in the preceding section yields the diagram in Figure 11.2 of injective mappings of the corresponding projective spaces.

At this point we invoke a result of Thurston concerning the relationships between $\mathcal{S}$, $\mathcal{S}'$, and $\mathfrak{M}\mathcal{F}$. By introducing what he calls the *MST*-coordinates for $\mathfrak{M}\mathcal{F}$ [FatLP], it becomes transparent that the image of $\mathcal{S}'$ in $P\mathfrak{M}\mathcal{F}$ is dense. This is because the elements of $\mathcal{S}'$ are realized as a set of points on an integer lattice in the *MST*-codinates and the set of possible rays determined by this integer lattice form a dense set in the space of corresponding coordinates for elements of $\mathfrak{M}\mathcal{F}$. The rays are dense even when some of the components in the pants decomposition of the surface $\dot{R}$ have boundary components which are isolated points.

Although it is not essential for the proofs of Theorems 1 and 2 of this chapter, Thurston also shows that the image of $\mathcal{S}$ is dense in $P\mathfrak{M}\mathcal{F}$. This is done by creating a very long curve in $\mathcal{S}$ corresponding to a system of curves representing an element of $\mathcal{S}'$ which has approximately the same intersection properties up to projective equivalence.

A corollary to the fact that the image of $\mathcal{S}'$ is dense in $P\mathfrak{M}\mathcal{F}$ is that any projective class of measured foliations can be realized by a projective class of

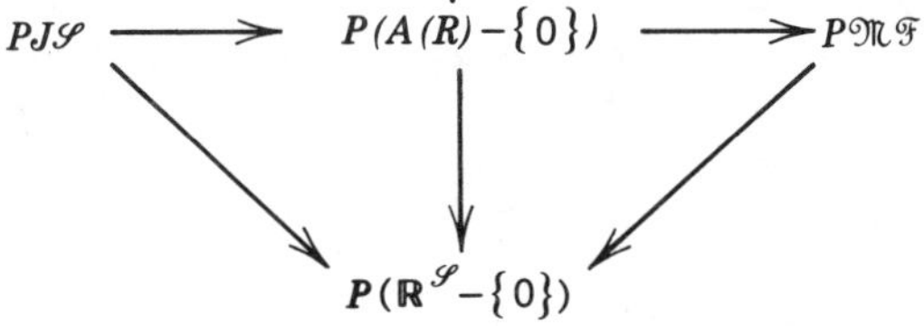

Figure 11.2

holomorphic quadratic differentials. For, given the projective class of a measured foliation $|dv|$, we approximate it in the product topology on $P(\mathbb{R}^{\mathcal{S}} - \{0\})$ by a sequence of projective classes of elements of $\mathcal{S}'$. From Lemma 3, we know that each of these projective classes is represented by a Jenkins–Strebel (holomorphic quadratic) differential φ_n on $\dot{R}$. Since the projective classes of the heights of φ_n converge in the topology for $P(\mathbb{R}^{\mathcal{S}} - \{0\})$, by multiplying φ_n by the positive number $c_n = \|\varphi_n\|^{-1}$, we obtain a sequence $\tilde{\varphi}_n = c_n\varphi_n$ whose heights converge. Then Lemma 1 tells us that $\tilde{\varphi}_n$ converges to a holomorphic quadratic differential φ and this differential must give the same projective class of heights as the given measured foliation $|dv|$. Putting this result together with the heights theorem, Theorem 1 is now proved.

We remark that since the image of $\mathcal{S}'$ is dense in $P\mathfrak{M}\mathcal{F}$, it follows that space of Jenkins–Strebel differentials is dense in $A(\dot{R})$. Since Thurston shows that $\mathcal{S}$ is dense in $P\mathfrak{M}\mathcal{F}$, it also follows that the space of simple Jenkins–Strebel differentials (those which have only one characteristic cylinder) is dense in $A(\dot{R})$.

To prove Theorem 2, we start with a measured foliation $|dv|$ on $\dot{R}$ and use Theorem 1 to obtain a unique holomorphic quadratic differential φ which realizes the same heights as $|dv|$. Then Theorem 8 of Chapter 2 shows that the infimum $M(|dv|)$ is realized uniquely by the holomorphic quadratic differential φ.

11.7. THE HEIGHTS MAPPING BETWEEN QUADRATIC DIFFERENTIALS ON DIFFERENT RIEMANN SURFACES IN THE SAME TEICHMÜLLER SPACE

As usual, let $\dot{R}$ be a Riemann surface of finite analytic type. Consider the Teichmüller space $T(\dot{R})$ of $\dot{R}$. Corresponding to any point of Teichmüller space $T(\dot{R})$ represented by a Beltrami coefficient $[\mu]$, there is a mapping $w = u + iv = f^{\mu}(z)$ where

$$f^{\mu}\colon \dot{R} \to \dot{R}_{\mu} \tag{5}$$

and $f = f^{\mu}$ is a homeomorphism which satisfies the Beltrami equation

$$f_{\bar{z}} = \mu f_z. \tag{6}$$

An arbitrary equivalence class $[\mu]$ representing a point in Teichmüller space has a representative μ which is C^1 on the surface $\dot{R}$. This is easily seen by using the Ahlfors–Weill section (Lemma 7 of Chapter 5) for μ with $\|\mu\|_{\infty} < \frac{1}{3}$. Then f^{μ} can be expressed as a composition $f^{\mu} = f^{\mu_1} \circ f^{\mu_2} \circ \cdots \circ f^{\mu_k}$ with each $\|\mu_j\|_{\infty} < \frac{1}{3}$. It follows from the theory of the Beltrami equation (6) that if μ is C^1, then f^{μ} is also C^1.

Therefore, a measured foliation $|dv|$ on $\dot{R}$ induces a measured foliation $|d\tilde{v}|$ on $\dot{R}_{\mu}$. Given an open covering U_j of $\dot{R}$, we get an open covering $\tilde{U}_j = f(U_j)$ of $\dot{R}_{\mu}$.

The functions v_j which determine the measure $|dv|$ on $\dot{R}$ are taken into functions $\tilde{v}_j$ defined on $\tilde{U}_j$ by $\tilde{v}_j = v_j \circ f^{-1}$.

We leave as an exercise the task of showing that $|d\tilde{v}|$ is a measured foliation on $\dot{R}_\mu$. Note that $h_{\tilde{v}}(f(\gamma)) = h_v(\gamma)$ for every rectifiable closed curve on $\dot{R}$, and therefore $h_{\tilde{v}}[f(\gamma)] = h_v[\gamma]$ for every homotopy class in $\mathscr{S}$.

If f_1 is any C^1 mapping from R to $\dot{R}_\mu$ homotopic to f, then $f(\gamma)$ is homotopic to $f_1(\gamma)$ on $\dot{R}_\mu$ for every closed curve γ on $\dot{R}$. Therefore, the mapping induced by f which takes $|dv|$ on $\dot{R}$ into $|d\tilde{v}|$ on R_μ induces a mapping between measure classes of measured foliations which depends only on the Teichmüller class of the mapping f.

Since the space $\mathfrak{M}\mathscr{F}(\dot{R})$ of measure equivalence classes of measured foliations on $\dot{R}$ is isomorphic to $A(\dot{R})$, the homeomorphism

$$f: \dot{R} \to \dot{R}_\mu$$

induces a correspondence between quadratic differentials,

$$f^*: A(\dot{R}) \to A(\dot{R}_\mu).$$

If we let $f^*(\varphi) = \psi$, then ψ is determined by the conditions

$$h_\varphi[\gamma] = h_\psi[f(\gamma)] \qquad \text{for all } \gamma \text{ in } \mathscr{S}.$$

Here f^* is a positive homogeneous homeomorphism.

The Dirichlet infimum problem of Theorem 2 yields a real positive value $M(|dv|)$ associated to the Riemann surface $\dot{R}$ and the foliation $|dv|$. Associated to the induced foliation $|d\tilde{v}|$ on $\dot{R}_\mu$, there is the possibly different number

$$M_\mu(|dv|).$$

It is given by the extremal problem

$$M_\mu(|dv|) = \inf \left\{ \iint_{\dot{R}_\mu} |\psi(w)| \, du \, dv \right\}, \tag{7}$$

where $w = u + iv$ is a complex analytic coordinate on $\dot{R}_\mu$ and the infimum in (7) is taken over all continuous quadratic differentials ψ on the Riemann surface $\dot{R}_\mu$ satisfying the side conditions

$$h_\psi[f(\gamma)] \geqslant h_v[\gamma]$$

for all γ in $\mathscr{S}$. Clearly, the infimum in (7) depends only on the Teichmüller class of μ.

11.8. VARIATION IN THE DIRICHLET NORM

Theorem 5. *Let $\dot{R}$ be a Riemann surface of finite analytic type. Let $|dv|$ be the measure for a measured foliation on $\dot{R}$. Then $M_\mu(|dv|)$ is a differentiable function on Teichmüller space and*

$$\log M_\mu(|dv|) = \log M(|dv|) + 2 \operatorname{Re} \frac{1}{\|\varphi\|} \iint_R \mu\varphi \, dx \, dy + o(\|\mu\|_\infty),$$

where φ is the unique holomorphic quadratic differential on $\dot{R}$ for which $|\operatorname{Im}\sqrt{\varphi}\, dz|$ is in the same measure class of measured foliations as $|dv|$.

In order to prove this result we need yet another form of the minimum norm principle given in Chapter 2. Given a holomorphic quadratic differential φ, we single out certain polygonal simple closed curves, associated with φ. Define a simple closed curve γ representing an element of $\mathscr{S}$ to be *allowable for* φ if it is one of the following types:

(a) γ is a regular closed vertical trajectory of φ.
(b) γ consists of two regular segments α and β, where α is horizontal and β is vertical and β departs from one endpoint of α and returns to the other endpoint α from the opposite side,
(c) γ consists of four segments α_1, α_2, β_1, β_2, where the α's and β's are horizontal and vertical and β_1 and β_2 emanate from the endpoints of each α_j on opposite sides of α_j. (We permit either of the segments α_1 or α_2 to degenerate to a point.)

Lemma 4. *Set $\varphi \in A(\dot{R})$, where $\dot{R}$ is compact except for a finite number of punctures. Let ψ be another quadratic differential which is locally an L_1-function on $\dot{R}$. Suppose that for almost every curve γ which is allowable with respect to φ we have the inequality*

$$\int_\gamma |\operatorname{Im}(\sqrt{\varphi}\, dz)| \leqslant \int_\gamma |(\sqrt{\psi}\, dz)|. \tag{8}$$

Then $\|\varphi\| \leqslant \iint_{\dot{R}} |\sqrt{\varphi}\sqrt{\psi}|\, dx\, dy \leqslant \iint_{\dot{R}} |\psi|\, dx\, dy$.

Proof. The proof depends on looking at the decomposition of $\dot{R}$ induced by φ into spiral domains and ring domains. If R_j is a ring domain, we may take the curve γ to be a closed vertical trajectory and it is straightforward to see that (8) implies

$$\iint_{R_j} |\varphi(\zeta)|\, d\xi\, d\eta \leqslant \iint_{R_j} |\sqrt{\psi(\zeta)}|\, d\xi\, d\eta,$$

where ζ is a natural parameter for φ. Passing to an arbitrary local parameter z, we obtain

$$\iint_{R_j} |\varphi(z)|\, dx\, dy \leqslant \iint_{R_j} |\sqrt{\psi(z)}\sqrt{\varphi(z)}|\, dx\, dy.$$

The case of a spiral domain R_k is more difficult. We pick a short regular horizontal segment lying in the interior of R_k with the property that the line integral

$$\int_\alpha |\sqrt{\psi}\, dz|$$

exists and is finite.

The existence of such an interval α is assured by Fubini's theorem and the hypothesis that ψ is a locally L_1-function. Then, just as in Theorem 8 of Chapter 2, we construct a decomposition of R_k induced by φ and the horiontal segment α. The inequality (8) leads to the following inequality:

$$\iint_{R_k} |\varphi|\, dx\, dy \leqslant \iint_{R_k} |\sqrt{\psi(\zeta)}|\, d\xi\, d\eta + 2\ell_\varphi(\alpha) \int_\alpha |\sqrt{\psi(z)}\, dz|,$$

where $\ell_\varphi(\alpha)$ is the length of α with respect to φ. Since we can let α shrink to a shorter and shorter segment, we obtain

$$\iint_{R_k} |\varphi|\, dx\, dy \leqslant \iint_{R_k} |\sqrt{\psi(\zeta)}|\, d\xi\, d\eta$$

for a natural parameter ζ. The conclusion of the lemma follows by adding up over all ring domains and spiral domains in the decomposition of $\dot{R}$ and applying Schwarz's inequality in the same way we did in the proof of Theorem 9 of Section 2.6.

Lemma 5. *Let* $K = (1 + \|\mu\|_\infty)/(1 - \|\mu\|_\infty)$. *Then*

$$K^{-1} M_\mu(|dv|) \leqslant M(|dv|) \leqslant K M_\mu(|dv|).$$

Proof. Let φ_μ be the unique holomorphic quadratic differential on R_μ for which $h_v[\gamma] = h\varphi_\mu[f(\gamma)]$. We know that $M_\mu(|dv|) = \|\varphi_\mu\|$. Let γ be an allowable curve for φ, let $k = \|\mu\|_\infty$, and let

$$\psi(z) = (1 + k)^2 \varphi_\mu(f(z)) f_z^2.$$

Then

$$\int_\gamma |\mathrm{Im}\sqrt{\varphi}\, dz| \leqslant \int_{f(\gamma)} |\mathrm{Im}\sqrt{\varphi_\mu(f)}\, df|$$

$$\leqslant \int_\gamma \left|\sqrt{\varphi_\mu(f(z))} f_z \left(1 + \mu \frac{d\bar{z}}{dz}\right) dz\right|$$

$$\leqslant \int_\gamma |\sqrt{\psi(z)}\, dz|.$$

Thus, from Lemma 4, $\|\varphi\| \leqslant \|\psi\|$, and so

$$\|\varphi\| \leqslant (1+k)^2 \iint_R |\varphi_\mu(f(z))|\, |f_z|^2\, dx\, dy$$

$$\leqslant \frac{(1+k)^2}{1-k^2} \iint_{R_\mu} |\varphi_\mu(w)|\, du\, dv = K\|\varphi_\mu\|.$$

This yields $M(|dv|) \leqslant KM_\mu(|dv|)$. The opposite inequality follows by applying the same reasoning to f^{-1}.

Remark. Lemma 5 shows that $M_\mu(|dv|)$ is a continuous function on $T(\dot{R})$ since clearly

$$K^{-1}M_\sigma(|dv|) \leqslant M_\mu(|dv|) \leqslant KM_\sigma(|dv|)$$

where K is the dilatation of the mapping $f^\sigma \circ (f^\mu)^{-1}$.

We are now ready to proceed to the proof of Theorem 5. Form the differential

$$\psi(z) = \varphi_\mu(f(z)) f_z^2 \left(1 - \mu(z)\frac{\varphi(z)}{|\varphi(z)|}\right)^2.$$

Here ψ is a quadratic differential on R and note that for any segment vertical with respect to φ, we have the identity

$$\int_{f(\beta)} |\mathrm{Im}\sqrt{\varphi_\mu(f)}\, df| = \int_\beta |\mathrm{Im}\sqrt{\psi(z)}\, dz|.$$

Let γ be an allowable simple closed curve for φ. Since $h_\varphi[\gamma] = h_{\varphi_\mu}[f(\gamma)]$ for all γ

in $\mathscr{S}$, we obtain

$$\int_{\gamma} |\mathrm{Im}\sqrt{\varphi}\, dz| \leqslant \int_{f(\gamma)} |\mathrm{Im}\sqrt{\varphi_\mu(\zeta)}\, d\zeta| = \int_{\beta_1 \cup \beta_2} |\mathrm{Im}\sqrt{\psi(z)}\, dz| + (1+k)^2 \int_{\alpha_1 \cup \alpha_2} |\sqrt{\varphi_\mu(f(z))} f_z|. \tag{9}$$

By making the α-curves arbitrarily small and summing over ring domains and spiral domains corresponding to φ, we obtain

$$\iint |\varphi|\, dx\, dy \leqslant \iint |\sqrt{\psi}\sqrt{\varphi}|\, dx\, dy. \tag{10}$$

The last term in (9) can be made arbitrarily small because from Lemma 5 there is a uniform bound on $\|\varphi_\mu\|$. On multiplying the integrand on the right side of (10) in the numerator and denominator by $|f_z|(1-|\mu|^2)^{1/2}$ and applying Schwarz's inequality [with the term $|\sqrt{\varphi_\mu f(z))} f_z(1-|\mu|^2)^{1/2}|$ lumped together], we find that

$$\|\varphi\| \leqslant \|\varphi_\mu\|^{1/2} \left(\iint_R |\varphi(z)| \frac{(1-\mu\varphi/|\varphi|)^2}{1-|\mu|^2}\, dx\, dy \right)^{1/2}.$$

Squaring both sides and dividing by $\|\varphi\|\ \|\varphi_\mu\|$, we get

$$\frac{\|\varphi\|}{\|\varphi_\mu\|} \leqslant \frac{1}{\|\varphi\|} \iint |\varphi| \frac{|1-\mu(\varphi/|\varphi|)|^2}{1-|\mu|^2}\, dx\, dy \leqslant 1 - \frac{2}{\|\varphi\|} \mathrm{Re} \iint \mu\varphi\, dx\, dy + O(\|\mu\|_\infty^2)$$

and so

$$\log\|\varphi_\mu\| \geqslant \log\|\varphi\| + 2\,\mathrm{Re} \frac{1}{\|\varphi\|} \iint_R \mu\varphi + O(\|\mu\|_\infty^2).$$

To get a reverse inequality we apply a similar argument to the inverse mapping f_1 of f. Here $f_1 \circ f(z) = z$ and μ_1 is the Beltrami coefficient of f_1 related to μ by $\mu_1(f(z)) = -\mu(z)/\theta$, where $\theta = \overline{f_z}/f_z$. Note that $\|\mu_1\|_\infty = \|\mu\|_\infty$. The analogous argument yields

$$\frac{\|\varphi_\mu\|}{\|\varphi\|} \leqslant \frac{1}{\|\varphi_\mu\|} \iint_{R_\mu} |\varphi_\mu| \frac{|1-\mu_1(\varphi_\mu/|\varphi_\mu|)|^2}{1-|\mu_1|^2}\, dx\, dy$$

and so

$$\log\|\varphi\| \geqslant \log\|\varphi_\mu\| + 2\,\mathrm{Re}\,\frac{1}{\|\varphi_\mu\|}\iint_{R_\mu} \mu_1\varphi_\mu + O(\|\mu\|_\infty^2). \tag{11}$$

The integral on the right-hand side of this inequality transforms into

$$-2\,\mathrm{Re}\,\frac{1}{\|\varphi_\mu\|}\iint_R \mu\varphi_\mu(f(z))f_z^2(z)\,dx\,dy. \tag{12}$$

Let $\hat{\varphi} = \varphi_\mu(f^\mu(z))(f_z^\mu)^2(z)$. Inequality (11) with the substitution from (12) implies

$$\log\|\varphi_\mu\| \leqslant \log\|\varphi\| + 2\,\mathrm{Re}\,\frac{1}{\|\varphi_\mu\|}\iint_R \mu\hat{\varphi}\,dx\,dy + O(\|\mu\|_\infty^2).$$

We wish to replace the first order term in this inequality by

$$2\,\mathrm{Re}\,\frac{1}{\|\varphi\|}\iint_R \mu\varphi$$

and, in so doing, we are permitted to weaken the second order term to a term which is $o(\|\mu\|_\infty)$ as $\|\mu\|_\infty \to 0$. From Lemma 5, if we replace $\|\varphi_\mu\|$ by $\|\varphi\|$, we introduce a multiplicative error which has order $\|\mu\|_\infty$. We are left with the term $\iint_R \mu(\hat{\varphi} - \varphi)\,dx\,dy$ which is $o(\|\mu\|_\infty)$ if we can show that $\iint |\hat{\varphi} - \varphi|\,dx\,dy$ approaches zero as $\|\mu\|_\infty \to 0$. From the theory of quasiconformal mappings, $f^\mu(z)$ converges locally uniformly to z and $f_z^\mu(z)$ converges to 1 locally in $L^p\,(p \geqslant 1)$. Moreover, φ_μ may be viewed as a holomorphic differential form defined on the unit disk and automorphic with respect to a Fuchsian covering group Γ_μ. Since from Lemma 5 the norms of φ_μ over any compact subset of the unit disk are bounded, they form a normal family (for $\|\mu\|_\infty \leqslant k_0 < 1$). Thus φ_μ converges to the differential φ as $\|\mu\|_\infty \to 0$ (since any limit of φ_μ must be the unique differential with the same heights as the heights of φ).

Let $\omega_r = \omega \cap \{z\colon |z| \leqslant r\}$. The following inequality is elementary:

$$\iint_\omega |\hat{\varphi} - \varphi| \leqslant (\|\hat{\varphi}\| - \|\varphi\|) + 2\iint_{\omega - \omega_r} |\varphi| + 2\iint_{\omega_r} |\hat{\varphi} - \varphi|. \tag{13}$$

It is easy to see that $\|\hat{\varphi}\|$ converges to $\|\varphi\|$ as $\|\mu\|_\infty \to 0$ since from, Lemma 5, $\|\varphi_\mu\| \to \|\varphi\|$. The second term on the right-hand side of (13) can be made arbitrarily small by choosing r sufficiently near to 1 since $\iint_{\omega} |\varphi| < \infty$. The third term on the right can be made arbitrarily small since $f^{\mu}(z)$ converges locally uniformly to z, f_z^{μ} converges locally in L^2, and $\varphi_{\mu}(z)$ converges locally uniformly to $\varphi(z)$.

Notes. Theorem 1 is due to Hubbard and Masur [HuM]. It was also proved by Kerckhoff [Ker]. Both Kerckhoff and Hubbard and Masur proved the theorem for surfaces without punctures. The Dirichlet principle for measured foliations as stated in Theorem 2 is due to Gardiner [Ga6]. The proof of the injectivity of the heights mapping given in Theorem 3 appears in the paper of Marden and Strebel [MarS]. The continuity of the heights mapping and its inverse (Theorem 4 and Lemma 1) is proved by Hubbard and Masur [HuM] and by Strebel [St3]. The fact that integrable holomorphic quadratic differentials with closed trajectories are dense in $A(\dot{R})$ was proved by Douady and Hubbard [DH] for compact surfaces without punctures. It was also realized by Kerckhoff [Ker] as being an elementary consequence of the analogous theorem for measured foliations. Masur [Masu3] improved the result of Douady and Hubbard to show that differentials with only one characteristic ring domain are dense. This result is also a consequence of the analogous theorem of Thurston for measured foliations ([FatLP], [Ker]).

The open problem suggested in Exercise 5 in this chapter would generalize the uniqueness of axis theorem for pseudo-Anosov diffeomorphisms, see [FatLP] and [Ber9].

EXERCISES

1. Let $R = \mathbb{C}/\langle z \mapsto z + m + in \rangle$, where m and n are arbitrary integers. R is called the square torus because a fundamental domain for the lattice $L = \{m + in\colon m \text{ and } n \text{ integers}\}$ is the unit square. Let $|dv| = |\mathrm{Im}(e^{i\theta}\, dz)|$. Then $|dv|$ is the measured foliation associated to the quadratic differential $\varphi(z)\, dz^2 = e^{2i\theta}\, dz^2$. Show that the parallel lines at angle $-\theta$ to the real axis are the leaves of the foliation and show that $M(|dv|) = 1$.
2. Continuing with the notation of Exercise 1, let

$$\tilde{w}(z) = \frac{(\tau + i)z - (\tau - i)z}{2i},$$

where $\mathrm{Im}\,\tau > 0$. Then $\tilde{w}$ induces a mapping w from the square torus R to the torus R_τ, where $R_\tau = \mathbb{C}/\langle z \mapsto z + m + n\tau \rangle$. The mapping w and the measured foliation $|dv|$ induce a measured foliation $|dv_\tau|$ on R_τ. The measured foliation $|dv_\tau|$ is determined by the condition that it has the same corresponding heights on R_τ that $|dv|$ has on R. Show that the measure class of $|dv_\tau|$ is representable in the form

$$|dv_\tau| = |\mathrm{Im}(\rho e^{i\theta_\tau}\, dz)| = \rho|\sin\theta_\tau\, dx + \cos\theta_\tau\, dy|.$$

Let $\tau = \tau_1 + i\tau_2$ where τ_1 and τ_2 are real. Show that

$$\tan \theta_\tau = \frac{\tau_2 \tan \theta}{1 + \tau_1 \tan \theta}$$

$$\rho \sin \theta_\tau = \sin \theta,$$

$$M_\tau(|dv_\tau|) = \tau_2 \sin^2\theta + \frac{1}{\tau_2}(\cos \theta + \tau_1 \sin \theta)^2.$$

3. Continuing with the notation of Exercises 1 and 2, show that the locus in the τ-plane of solutions to the equation $M_\tau(|dv_\tau|) = C$, a constant, is a circle in the upper half plane tangent to the real axis with center at $(-\cot \theta, \frac{1}{2}C\, csc^2\theta)$ and radius $\frac{1}{2}C \csc^2\theta$.
4. Continuing with the notations of Exercises 1, 2, and 3, let $|du|$ and $|dv|$ be two foliations on the square torus associated to the horizontal trajectories of the quadratic differentials $e^{2i\theta}\, dz^2$ and $e^{2i\theta_2}\, dz^2$ with $\theta_1 \neq \theta_2$ and $0 \leqslant \theta_1 < \pi$ and $0 \leqslant \theta_2 < \pi$. Assume that on the torus R_τ, $|du|$ goes over into a foliation with angle $\theta_{1\tau}$ and $|dv|$ goes over into a foliation with angle $\theta_{2\tau}$. From Exercise 2 we have the formula

$$\tan \theta_{1\tau} = \frac{\tau_2 \tan \theta_1}{1 + \tau_1 \tan \theta_1}$$

and the analogous formula relating $\theta_{2\tau}$ and θ_2. Find the minimum value of $M_\tau(|dv|)$ subject to the constraint that $M_\tau(|du|) = 1$. Show there is a unique τ solving this minimum problem and that τ lies on the hyperbolic line with endpoints at $-\cot \theta_1$ and $-\cot \theta_2$ on the real axis. Moreover, the point lies on the unique point of tangency of two horocycles determined by the equations $M_\tau(|du_\tau|) = 1$ and $M_\tau(|dv_\tau|) =$ a minimum.
5. (Open Problem) Carry out the analogy to Exercises 1 through 4 for pairs of transversal measured foliations on surfaces of finite analytic type.

BIBLIOGRAPHY

[Ab] Abikoff, W., *The real analytic theory of Teichmüller space. Springer Lect. Notes Math.* **820** (1980).

[Ah1] Ahlfors, L. V., On quasiconformal mappings. *J. Anal. Math.* **4**, 1–58 (1954).

[Ah2] Ahlfors, L. V., Quasiconformal reflections. *Acta Math.* **109**, 291–301 (1964).

[Ah3] Ahlfors, L. V., Finitely generated Kleinian groups. *Am. J. Math.* **86**, 413–429 (1964): **87**, 759 (1965).

[Ah4] Ahlfors, L. V., *Lectures on Quasiconformal Mappings.* Van Nostrand-Reinhold, Princeton, New Jersey, 1966.

[Ah5] Ahlfors, L. V., The structure of finitely generated Kleinian groups. *Acta Math.* **122**, 1–17 (1969).

[Ah6] Ahlfors, L. V., *Conformal Invariants: Topics in Geometric Function Theory.* McGraw-Hill, New York, 1973.

[AhBer] Ahlfors, L. V., and Bers, L., Riemann's mapping theorem for variable metrics. *Ann. Math.* **72**, 385–404 (1960).

[AhBeu] Ahlfors, L. V., and Beurling, A., The boundary correspondence under quasiconformal mappings, Acta Math., **96**, 125–142 (1956).

[AhS] Ahlfors, L. V., and Sario, L., *Riemann Surfaces.* Princeton Univ. Press, Princeton, New Jersey, 1960.

[AhW] Ahlfors, L., and Weill, G., A uniqueness theorem for Beltrami equations. *Proc. Am. Math. Soc.* **13**, 975–978 (1962).

[Bea] Beardon, A. F., *The Geometry of Discrete Groups.* Springer-Verlag, Berlin and New York, 1983.

[Ber1] Bers, L., *Riemann Surfaces, Lecture Notes.* Inst. Math. Sci. Lect. Notes. New York University, New York, 1957–1958.

[Ber2] Bers, L., Quasiconformal mappings and Teichmüller's theorem. In Analytic Functions (R. Nevanlinna *et al.*, eds.), pp. 89–119. Princeton Univ. Press, Princeton, New Jersey, 1960.

[Ber3] Bers, L., On Moduli of Riemann Surfaces, Summer Lect. Forschunginst. Math., Eidgenössische Technische Hochschule, Zurich, 1964.

[Ber4] Bers, L., An approximation theorem. *J. Anal. Math.* **14**, 1–4 (1965).

[Ber5] Bers, L., Automorphic forms and Poincaré series for infinitely generated Fuchsian groups. *Am. J. Math.* **87**, 196–214 (1966).

[Ber6] Bers, L., A non-standard integral equation with applications to quasiconformal mappings. *Acta Math.* **116**, 113–134 (1966).

[Ber7] Bers, L., Extremal quasiconformal mappings. *Ann. Math. Stud.* **66**, 27–52 (1971).

[Ber8] Bers. L., Fibre spaces over Teichmüller spaces. *Acta Math.* **130**, 89–126 (1973).

[Ber9] Bers, L., An extremal problem for quasiconformal mappings and a theorem by Thurston. *Acta Math.* **141**, 73–98 (1978).

[Ber10] Bers, L., A new proof of a fundamental inequality for quasiconformal mappings. *J. Anal. Math.* **36**, 15–30 (1979).

[BerGa] Bers, L., and Gardiner, F. P., Fricke spaces. *Adv. Math.* **62**, 1–36 (1987).

[BerGr] Bers, L., and Greenberg, L., Isomorphisms between Teichmüller spaces. *Ann. Math. Stud.* **66**, 53–79 (1971).

[BerR] Bers, L., and Royden, H. L., Holomorphic families of injections. *Acta Math.* **157**, 259–286 (1986).

[BM] Brooks, R., and Matelski, P., Dynamics of 2-generator subgroups of PSL(2, C). *Ann. Math Stud.* **97**, 65–69 (1981).

[Bu] Bujalance, E., NEC groups and Klein surfaces. *Springer Lect. Notes*, **971**, 6–14 (1981).

[D] Day, M. M., *Normed Linear Spaces.* Academic Press, New York, 1962.

[DE] Douady, A., and Earle, C. J., Conformally natural extension of homeomorphisms of the circle. *Acta Math.* (1987) (to be published).

[DH] Douady, A., and Hubbard, J., On the density of Strebel forms. *Invent. Math.* **30**, 175–179 (1975).

[E1] Earle, C. J., Some remarks on Poincaré series. *Compos. Math.* **21**, 167–176 (1969).

[E2] Earle, C. J., The Teichmüller distance is differentiable. *Duke Math. J.* **44**, 389–397 (1977).

[EE1] Earle, C. J., and Eells, J., The diffeomorphism group of a compact Riemann surface. *Bull. Am. Math. Soc.* **73**, 557–559 (1967).

[EE2] Earle, C. J., and Eells, J., Foliations and fibrations. *J. Differ. Geom.***1**(1), 33–41 (1967).

[EE3] Earle, C. J., and Eells, J., On the differential geometry of Teichmüller spaces. *J. Anal. Math.* 35–52 (1967).

[EE4] Earle, C. J., and Eells, J., A fibre bundle description of Teichmüller theory. *J. Differ. Geom.* **3**(1), 19–43 (1969).

[EK1] Earle, C. J., and Kra, I., On isometries between Teichmüller spaces. *Duke Math. J.* **41**(3), 583–591 (1974).

[EK2] Earle, C. J., and Kra, I., On holomorphic mappings between Teichmüller spaces. *In "Contributions to Analysis"* (L. Ahlfors, I. Kra, B. Maskit, and L. Nirenberg, eds.), pp. 107–124. Academic Press, New York, 1974.

[EM] Earle, C. J., and McMullen, C., Quasiconformal Isotopies, prep., Math. Sciences Res. Inst. Conference on Holomorphic Functions and Moduli, Berkeley, April (1986).

[EN] Earle, C. J., and Nag, S., "Conformally Natural Reflections in Jordan curves with Applications to Teichmüller Spaces." Prepr. Math. Sciences Res. Inst., Conference on Holomorphic Functions and Moduli, Berkeley, April (1986).

[Ep] Epstein, D. B. A., Curves on 2-manifolds and iosotopies, *Acta Math.* **115**, 83–107 (1966).

[FarK] Farkas, H., and Kra, I., *Riemann Surfaces.* Springer-Verlag, Berlin and New York, 1980.

[FatLP] Fathi, A., Laudenbach, F., and Poénaru, V., *Travaux de Thurston sur les Surfaces,* Asterisque, pp. 66–67. Soc. Math. Fr., Paris, 1979.

[Feh] Fehlmann, R., On a fundamental variational lemma for extremal quasiconformal mappings. *Comment. Math. Helv.* (to be published).

[FehS] Fehlmann, R., and Sakan, K., On extremal quasiconformal mappings with varying dilation bounds, Osaka J. Math. **23**, 751–764 (1986).

[FenN] Fenchel, W., and Nielsen, J., "Discontinuous groups of non-Euclidean motions" (unpublished manuscript).

[For] Ford, L., *Automorphic Functions*, Chelsea, New York, 1929.

[Fox] Fox, R. H., On Fenchel's conjecture about F-groups. *Mat. Tiddskr. B* pp. 61–65 (1952).

[FrK] Fricke, R., and Klein, F., *Vorlesungen über die theorie der automorphen funktionen*, Vol. 1. Teubner, Leipzig, 1896 (Vol. 2, 1912).

[Ga1] Gardiner, F. P., Analysis of the group operation in universal Teichmüller space. *Trans. Am. Math. Soc.* **132**, 471–486 (1968).

[Ga2] Gardiner, F. P., Schiffer's interior variation and quasiconformal mappings. *Duke Math. J.* **42**, 371–380 (1975).

[Ga3] Gardiner, F. P., On the variation of Teichmüller's metric. *Proc. R. Soc. Edinburgh, Sect. A* **85**, 143–152 (1980).

[Ga4] Gardiner, F. P., On partially Teichmüller Beltrami differentials. *Mich. Math. J.* **29**, 237–242 (1982).

[Ga5] Gardiner, F. P., Approximation of infinite dimensional Teichmüller spaces. *Trans. Am. Math. Soc.* **282**(1), 367–383 (1984).

[Ga6] Gardiner, F. P., Measured foliations and the minimal norm property for quadratic differentials. *Acta Math.* **152**, 57–76 (1984).

[Ge1] Gehring, F. W., Univalent functions and the Schwarzian derivative. *Comment. Math. Helv.* **52**, 561–572 (1977).

[Ge2] Gehring, F. W., Injectivity of local quasi-isometries. *Comment. Math. Helv.* **57**, 202–220 (1982).

[Ge3] Gehring, F. W., Characteristic Properties of Quasidisks, Semin. Math. Super. Presses de l'Universite de Montreal, 1982.

[GeO] Gehring, F. W., and Osgood, B. G., Uniform domains and the quasihyperbolic metric. *J. Anal. Math.* **36**, 50–74 (1979).

[Gi1] Gilman, J., A geometric approach to Jorgensen's inequality, Bull. of Amer. Math. Soc. **16**(1), 91–92 (1987).

[Gi2] Gilman, J., Inequalities and discrete subgroups of PSL(2, R), Prepr. Math. Sciences Research Inst. (1986).

[Go] Godement, R., "Series de Poincaré et Spitzenformen," Semin. Henri Cartan, 10 année, 1957–1958 Expos. 10.

[Gr1] Grötzsch, H., Über einige extremalprobleme der konformen abbildung. I. *Ber. Verh. Saechs. Akad. Wiss. Leipzig, Math.-Naturwiss. Kl.* **80**, 367–376 (1928).

[Gr2] Grötzsch, H., Über möglichst konforme abbildungen von schlichten Bereichen. *Leipz. Ber.* **84**, 114–120 (1932).

[H] Hamilton, R. S., Extremal quasiconformal mappings with prescribed boundary values. *Trans. Am. Math. Soc.* **138**, 399–406 (1969).

[Ha1] Harvey, W. J., Chabauty spaces of discrete groups. *Ann. Math. Stud.* **79**, 239–246 (1974).

[Ha2] Harvey, W. J., Spaces of discrete groups. In *Discrete Groups and Automorphic Functions.* (W. J. Harvey, ed.), Academic Press, New York, 1977.

[Ha3] Harvey, W. J., Boundary Structure of the Modular Group, Proc. Stony Brook Conf. Ann. Math. Stud., pp. 245–252. Princeton Univ. Press, Princeton, New Jersey, 1981.

[HS] Hawley, N. S., and Schiffer, M., Half-order differentials on Riemann surfaces. *Acta Math.* **115**, 199–236 (1966).

[Hu] Hubbard, J. H., Sur le non-existence de sections analytique a la curve universelle de Teichmüller. *C.R. Acad. Sci. Paris*, Ser. A-B, **274**, A978–A979 (1972).

[HuM] Hubbard, J., and Masur, H., Quadratic differentials and foliations. *Acta Math.* **142**, 221–274 (1979).

[Je1] Jenkins, J. A., On the existence of certain general extremal metrics. *Ann. Math.* **66**, 440–453 (1957).

[Je2] Jenkins, J. A., *Univalent functions and conformal mapping. Ergeb. Math. Grenzgeb.* [N.S.] **18**, 1–167 (1958).

[Je3] Jenkins, J. A., On the global structure of the trajectories of a positive quadratic differential. *Ill. J. Math.* **4**, 405–412 (1960).

[Jo] Jørgensen, T., On discrete groups of Möbius transformations. Am. J. Math. **98**, 739–749 (1976).

[JoK] Jørgensen, T., and Klein, P., Algebraic convergence of finitely generated Kleinian groups. *Q. J. Math. Oxford Ser.* [2] **33**, 325–332 (1982).

[Ke1] Keen, L., Canonical polygons for finitely generated Fuchsian groups. *Acta Math.* **115**, 1–16 (1966).

[Ke2] Keen, L., Intrinsic moduli on Riemann surfaces. *Ann. Math.* **84**, 404–420 (1966).

[Ke3] Keen, L., On Fricke moduli." *Ann. Math. Stud.* **66**, 205–224 (1971).

[Ke4] Keen, L., Collars on Riemann surfaces. *Ann. Math. Stud.* **79**, 263–268 (1974).

[Ker] Kerckhoff, S., The asymptotic geometry of Teichmüller space. *Topology* **19**, 23–41 (1980).

[Ko] Kobayashi, S., *Hyperbolic Manifolds and Holomorphic Mappings.* Dekker, New York, 1970.

[Kr1] Kra, I., On cohomology of Kleinian groups. *Ann. Math.* **89**, 533–556 (1969).

[Kr2] Kra, I., On cohomology of Kleinian groups: II. *Ann. Math.* **90**, 575–589 (1969).

[Kr3] Kra, I., Eichler cohomology and the structure of finitely generated Kleinian groups. *Ann. Math. Stud.* **66**, 225–261 (1971).

[Kr4] Kra, I., On cohomology of Kleinian groups: III. Singular Eichler integrals. *Acta Math.* **127**, 23–40 (1971).

[Kr5] Kra, I., *Automorphic Forms and Kleinian Groups.* Benjamin, Reading, Massachusetts, 1972.

[Kr6] Kra, I., On new kinds of Teichmüller spaces. *Isr. J. Math.* **16**(3), 237–257 (1973).

[Kr7] Kra, I., On the Nielsen-Thurston-Bers type of some self-maps of Riemann surfaces. *Acta Math.* **146**, 231–269 (1981).

[Kr8] Kra, I., The Caratheodory metric on abelian Teichmüller disks. *J. Anal. Math.* **40**, 129–143 (1981).

[Kr9] Kra, I., On cohomology of Kleinian groups: IV. The Ahlfors-Sullivan construction of holomorphic Eichler integrals. *J. Anal. Math.* **43**, 51–87 (1983–1984).

[Kr10] Kra, I., On the vanishing of and spanning sets for Poincaré series for cusp forms. *Acta Math.* **153**, 47–116 (1984).

[Krau] Kraus, W., Uber den Zusammenhang einiger characteristiken eines einfach zusammenhängenden Bereiches mit der Kreisabbildung. *Mitt. Math. Semin. Giessen* **21**, 1–28 (1932).

[Krav] Kravetz, S., On the geometry of Teichmüller spaces and the structure of their modular groups. *Ann. Acad. Sci. Fenn., Ser. A2: Math.-Phys.* **278**, 1–35 (1959).

[Kru1] Krushkal, S., Extremal quasiconformal mappings. *Sib. Math. J.* **10**, 411–418 (1969).

[Kru2] Krushkal, S., *Quasiconformal Mappings and Riemann Surfaces*, translation (edited by I. Kra, ed.). Winston, Washington, D.C., 1979.

[KruKu] Krushkal, S., and Kuhnau, R., *Quasiconformal Mappings—New Methods and Applications.* Nauka, Novosibirsk, 1984.

[Ku] Kulkarni, R. S., Normal subgroups of Fuchsian groups. *Q. J. Math. Oxford Ser.* [2] **36**, 325–344 (1985).

[Leh] Lehner, J., *Discontinuous Groups and Automorphic Functions*, Math. Surv., No. 8. Am. Math. Soc., Providence, Rhode Island, 1964.

[LehtV] Lehto, O., and Virtanen, K. I., *Quasiconformal Mapping*, Springer-Verlag, Berlin and New York, 1965.

[Leu] Leutbecher, A., Uber Spitzen diskontinuierlichen Gruppen von lineargebrochenen Transformationen. *Math. Z.* **100**, 183–200 (1967).

[Mac1] Macbeath, A. M., "Discontinuous Groups and Birational Transformations," Proc. Summer Sch. Math. Queens College, Dundee, 1961.

[Mac2] Macbeath, A. M., The classification of non-Euclidean plane crystallographic groups. *Can. J. Math.* **19**, 1192–1205 (1967).

[MañSS] Mañé, R., Sad, P., and Sullivan, D., On the dynamics of rational maps. *Ann. Éc. Norm. Super.* **16**, 193–217 (1983).

[Mar] Marden, A., On homotopic mappings of Riemann surfaces. *Ann. Math.* **90**, 1–8 (1969).

[MarS] Marden, A., and Strebel, K., The heights theorem for quadratic differentials on Riemann surfaces. *Acta Math.* **153**, 153–211 (1984).

[Mask] Maskit, B., On Poincaré's theorem for fundamental polygon. *Adv. Math.* **7**, 219–230 (1971).

[Masu1] Masur, H., The curvature of Teichmüller space. *Springer Lect. Notes Math.* **400**, 122–123 (1974).

[Masu2] Masur, H., On a class of geodesics in Teichmüller space. *Ann. Math.* **102**, 205–221 (1975).

[Masu3] Masur, H., The Jenkins-Strebel differentials with one cylinder are dense. *Comment. Math. Helv.* **54**, 179–184 (1979).

[Masu4] Masur, H., Interval exchange transformations and measured foliations. *Ann. Math.* **115**, 169–200 (1982).

[Mat] Matelski, P., A compactness theorem for Fuchsian groups of the second kind. *Duke Math. J.* **43**, 829–840 (1976).

[Mc] McKean, H., Selberg's trace formula as applied to a compact Riemann surface. *Commun. Pure Appl. Math.* **25**, 225–246 (1972).

[Men] Mennicke, J., Eine Bemerkung über Fuchsische Gruppen. Invent. Math. **2**, 301–305 (1967); corrigendum, *ibid.* **6**, 106 (1968).

[N1] Nag, S., Non-geodesic discs embedded in Teichmüller spaces. *Am. J. Math.* **104**(2), 399–408 (1982).

[N2] Nag. S., Schiffer variation of complex structure and coordinates for Teichmüller spaces. *Proc. Indian Acad. of Sci.*, **94**(2&3), 111–122 (1985).

[Ne] Nehari, Z., The Schwarzian derivative and schlicht functions. *Bull. Am. Math. Soc.* **66**, 545–551 (1949).

[O] O'Byrne, B., On Finsler geometry and applications to Teichmüller spaces. *Ann. Math. Stud.* **66**, 317–328 (1971).

[P] Patterson, D. B., The Teichmüller spaces are distinct. *Proc. Am. Math. Soc.* **35**, 179–182 (1972); **38**, 668 (1973).

[Po1] Poincaré, H., Sur les groupes des équations linéaires. *Acta Math.* **4**, 201–312 (1884).

[Po2] Poincaré, H., Les fonctions fuchsiennes et l'équation $\Delta u = e^u$. *J. Math. Pures Appl.* [5] **4**, 137–230 (1898).

[Rei1] Reich, E., On the decomposition of a class of plane quasiconformal mappings. *Comment. Math. Helv.* **53**, 15–27 (1978).

[Rei2] Reich, E., A criterion for unique extremality of Teichmüller mappings. *Indiana Univ. Math. J.* **30**(3), 441–447 (1981).

[Rei3] Reich, E., Nonuniqueness of Teichmüller extremal mappings. *Proc. Am. Math. Soc.* **88**(3), 513–516 (1983).

[ReiS1] Reich, E., and Strebel, K., Teichmüller mappings which keep the boundary pointwise fixed. *Ann. Math. Stud.* **66**, 365–367 (1971).

[ReiS2] Reich, E., and Strebel, K., Extremal quasiconformal mappings with given boundary values. *In* "Contributions to Analysis" (L. Ahlfors, I. Kra, B. Maskit, and L. Nirenberg, eds.), pp. 375–392. Academic Press, New York, 1974.

[Ren] Renelt, H., Konstruktion gewisser quadratischer Differentiale mit von Dirichletintegralen. *Math. Nachr.* **73**, 125–142 (1976).

[Ri] Rickman, S., Characterization of quasiconformal arcs. *Ann. Acad. Sci. Fenn., Ser.* A **395**, 7–30 (1966).

[Ro] Royden, H. L., Automorphisms and isometries of Teichmüller space. *Ann. Math. Stud.* **66**, 369–384 (1971).

[Sa] Sakan, K., Necessary and sufficient conditions for extremality in certain classes of quasiconformal mappings. *J. Math. Kyoto Univ.* **26**(1), 31–37 (1986).

[ScSp] Schiffer, M., and Spencer, D. C., *Functionals of Finite Riemann Surfaces.* Princeton Univ. Press, Princeton, New Jersey, 1954.

[Se] Selberg, A., On discontinuous groups in higher-dimensional spaces. "Contributions to Function Theory." Tata Institute, Bombay, 1960.

[Sh] Shimizu, H., On discontinuous groups operating on the product of upper half-planes. *Ann. Math.* **77**, 33–71 (1959).

[Si1] Singerman, D., Finitely maximal Fuchsian groups. *J. London Math. Soc.* **6**, 29–38 (1972).

[Si2] Singerman, D., On the structure of non-Euclidean crystallographic groups. *Proc. Cambridge Philos. Soc.* **76**, 223–240 (1974).

[Sp] Springer, G., *Introduction to Riemann Surfaces.* Addison-Wesley, Reading, Massachusetts, 1957.

[St1] Strebel, K., On the trajectory structure of quadratic differentials. *Ann. Math. Stud.* **79**, (1974), 419–438.

[St2] Strebel, K., On the existence of extremal Teichmüller mappings. *J. Anal. Math.*, **30**, 441–447 (1976).

[St3] Strebel, K., On quasiconformal mappings of open Riemann surfaces. *Comment. Math. Helv.* **53**, 301–321 (1978).

[St4] Strebel, K., *Quadratic Differentials.* Springer-Verlag, Berlin and New York, 1984.

[T1] Teichmüller, O., Extremale quasikonforme Abbildungen und quadratische Differentiale. *Ahh. Preuss. Akad. Wiss., Math.-Naturwiss.* K1.**22**, 1–197 (1939).

[T2] Teichmüller, O., *Bestimmung der extremalen quasikonformen Abbildungen bei geschlossenen orientierten Riemannschen Flächen*, Collect. Pap. Springer-Verlag, Berlin and New York, 1982.

[Th1] Thurston, W. P., *The Geometry and Topology of 3-Manifolds*, Notes. Princeton Univ. Press, Princeton, New Jersey, 1981.

[Th2] Thurston, W. P., Minimal stretch maps between hyperbolic surfaces (unpublished preprint) (1985).

[Th3] Thurston, W. P., Zippers and schlicht functions (unpublished) (1985).

[Ts] Tsuji, M., *Potential Theory in Modern Function Theory.* Maruzen, Tokyo, 1959.

[Tu1] Tukia, P., On discrete groups of the unit disk and their isomorphisms. *Ann. Acad. Sci. Fenn., Ser. A1: Math.-Phys.* **504**, (1972).

[Tu2] Tukia, P., Extension of boundary homeomorphisms of discrete groups of the unit disk. *Ann. Acad. Sci. Fenn., Ser. A1: Math.-Phys.* **548**, (1973).

[Tu3] Tukia, P., On isomorphisms of geometrically finite Möbius groups. Inst. Hautes Études Sci. Publ. Math. **61**, 171–214 (1985).

[V] Velling, John A. A geometric interpretation of the Ahlfors–Weill mappings and an induced foliation of $\mathbb{H}^3$, to appear in the *Proceedings of the Conference on Riemann Surfaces and Moduli*, Math. Sci. Res. Inst., Berkeley, March (1986).

[We] Weyl, H., "Der Idee der Riemannschen Fläche." Teubner, Leipzig, 1913.

[Wo1] Wolpert, S., The length spectrum of a compact Riemann surface. I. (1977).

[Wo2] Wolpert, S., The length spectrum of a compact Riemann surface. II. (1977).

[Wo3] Wolpert. S., The Fenchel-Nielsen twist deformation. *Ann. Math.* **115**, 501–528 (1982).

[Wo4] Wolpert, S., On the symplectic geometry of deformation of a hyperbolic surface. *Ann. Math.* **117**, 207–234 (1983).

[Wo5] Wolpert, S., "Geodesic Length Functions and the Nielsen Problem," Prepr. University of Maryland, College Park, 1985.

[Wo6] Wolpert, S., "Thurston's Riemannian Metric for Teichmüller Space," Prepr. University of Maryland, College Park, 1985.

INDEX